COMPREHENSIVE CHEMICAL KINETICS

COMPREHENSIVE

CHEMICAL KINETICS

EDITED BY

C.H. BAMFORD

M.A., Ph.D., Sc.D. (Cantab.), F.R.I.C., F.R.S.
Campbell-Brown Professor of Industrial Chemistry,
University of Liverpool

AND

C.F.H. TIPPER

Ph.D. (Bristol), D.Sc. (Edinburgh)
Senior Lecturer in Physical Chemistry,
University of Liverpool

VOLUME 16
LIQUID-PHASE OXIDATION

ELSEVIER SCIENTIFIC PUBLISHING COMPANY
AMSTERDAM — OXFORD — NEW YORK
1980

ELSEVIER SCIENTIFIC PUBLISHING COMPANY
335 Jan van Galenstraat
P.O. Box 211, 1000 AE Amsterdam, The Netherlands

Distributors for the United States and Canada

ELSEVIER NORTH-HOLLAND INC.
52 Vanderbilt Avenue
New York, N.Y. 10017

QD 501
B242
vol 16

ISBN 0-444-41631-5 (Series)
ISBN 0-444-41860-1 (Vol. 16)

with 9 illustrations and 70 tables

Printed in The Netherlands

COMPREHENSIVE CHEMICAL KINETICS

Contributors to Volume 16

E.T. DENISOV   Institute of Chemical Physics,
USSR Academy of Sciences,
Noginsk, Moscow 142432, USSR

D.G. HENDRY   Physical Organic Chemistry Department,
SRI International,
Menlo Park,
CA 94025, U.S.A.

T. MILL   Physical Organic Chemistry Department,
SRI International,
Menlo Park,
CA 94025, U.S.A.

L. SAJUS   TECHNIP,
Place Henri Regnault,
92090 Paris la Defense,
France

I. SÉRÉE DE ROCH   IFP,
Avenue de Bois Préau,
92500 Rueil Malmaison,
France

D.L. TRIMM   Department of Chemical Technology,
University of New South Wales,
Sydney,
N.S.W., Australia

Preface

Section 6 deals with the autocatalytic reactions of inorganic and organic compounds with molecular oxygen in the liquid phase and the highly exo-thermic processes in the gas phase, collectively known as combustion, which may involve oxygen, other oxidants or decomposition flames and are so important technologically. Catalysis, retardation and inhibition are covered. The kinetic parameters of the elementary steps involved are given, when available, and the reliability of the data discussed.

Volume 16 covers oxidation in the liquid phase by ground state and singlet molecular oxygen, and by ozone. The free-radical chain mechanisms involved and the complex role of hydroperoxides, dihydroperoxides, peroxides and polyoxides, together with the mechanism of the action of catalysts and inhibitors (in particular metal salts, amines and phenols) are discussed in detail. The important role of hydrogen bonding is considered. Chapter 1 deals with the oxidation of alkanes, aralkanes and olefins, Chapter 2 with the oxidation of saturated and unsaturated aliphatic alde-hydes and aromatic aldehydes, Chapter 3 with the oxidation of alcohols, ketones, ethers, carboxylic acids, esters and phenols, and Chapter 4 with the oxidation of organic compounds of nitrogen (mainly amines), sulphur (sulphides, thiols and the reactions of the sulphoxide products) and chlorine. Cooxidation of various organic reactants, e.g. hydrocarbons and alcohols, is also discussed.

C.H. Bamford
C.F.H. Tipper

Liverpool
December 1979

Contents

Chapter 4 (D.L. Trimm)

Chapter 1

Kinetics and Mechanisms of Free Radical Oxidation of Alkanes and Olefins in the Liquid Phase

THEODORE MILL and DALE G. HENDRY

1. Introduction

Reactions of oxygen with organic compounds occupy a central position in the scheme of living things, producing the energy that drives all biochemical machines and most of the mechanical and heat energy used in technology. Over a wide temperature range, bounded roughly by enzyme-mediated oxygenations at low temperatures and fast combustion reactions at high temperatures, are a host of relatively slow oxidation processes, involving free radicals, which are responsible for the conversion of hydrocarbons to useful industrial intermediates as well as unwanted degradation of lipids and polymers, and the intensification of environmental pollution.

The major objective of this chapter is to provide a critical review of the kinetics and mechanisms of free radical oxidation of alkanes and alkenes and the techniques for their measurement and determination under mild conditions in the liquid phase. A brief discussion of photooxygenation (singlet oxygen) reactions is included for completeness. Literature has been reviewed carefully through 1975 and updated with references to mid-1978.

Our principal concern is to utilize both kinetics and product formation as diagnostic tools for elucidating the detailed mechanisms of oxidation reactions in terms of elementary steps, rate coefficients, thermochemistry and structure—reactivity relationships. Accordingly, our emphasis throughout the chapter will be on these relationships, exemplified by reactions of simple molecules and the way in which they may be used to interpret and predict the rates and products of oxidation reactions involving more complex molecules or extreme conditions *.

2. Historical basis of oxidation kinetics

Although autoxidation reactions have been studied for well over a hundred years, clear understanding of the processes had to await development of the concept of free radical chain reactions in the 1920s [1—3]. Credit for first recognizing the radical chain nature of an autoxidation

---

* All kinetic rate coefficients in this review are in units of molar-second.

reaction, that of benzaldehyde forming perbenzoic acid, seems to belong to Backstrom [2] who suggested first an "energy chain" but then later what we now recognize as a typical radical chain transfer process involving benzoylperoxy radicals, viz.

$$C_6H_5\overset{\overset{O}{\|}}{C}O_2\cdot + C_6H_5CHO \rightarrow C_6H_5\overset{\overset{O}{\|}}{C}O_2H + C_6H_5\dot{C}O \qquad (1)$$

A contemporary and widely held theory, occasionally advanced even today, proposed direct addition of ground-state oxygen to carbon—carbon double bonds to form endoperoxides or dioxetanes [3]

$$(2)$$

Despite a diligent search for evidence of dioxetane formation, no one has been able to verify the occurrence of this proposed reaction. Moreover, since benzaldehyde could not undergo this type of reaction, an alternate explanation which could also be applied to olefins was desirable.

That peroxy radicals were important and common radical intermediates in autoxidation became clearer from the work of Criegee et al. [4]. These workers showed that the peroxide resulting from UV-initiated autoxidation of cyclohexene was a hydroperoxide formed by removal of a reactive allylic hydrogen followed by addition of oxygen and chain transfer between cyclohexenyl peroxy radical and cyclohexene, viz.

$$(3)$$

$$(4)$$

Later studies by Farmer and Sutton [5] greatly extended the concept of allylic hydroperoxidation to a variety of cyclic olefins.

Work at the National Rubber Producers laboratory in England in the 1940s by Barnard et al. [6] provides the important bridge between the qualitative recognition of the free radical character of autoxidation chain reactions and the quantitative kinetic framework in which we shall discuss the subject. By using added free radical sources to provide more or less steady rates of radical production, they avoided the troublesome induction periods characteristic of autoxidation and in so doing developed quantitative kinetic relationships among several variables of the olefin—oxygen systems, including olefin concentration and structure, rates of initiation, and oxygen pressure. Non-steady-state systems in which radical concentrations vary with time were used by Bateman et al. [7] to evaluate the absolute rate coefficients for the propagation and disappearance of radicals in these systems.

These basic concepts and techniques were further extended in the fifties and sixties by Russell and coworkers [8] to structure reactivity relationships for aromatic compounds, by Mayo et al. [9] to copolymerization of oxygen with many vinyl monomers, and by Ingold and Howard to extensive measurements of absolute rate coefficients for peroxy and alkoxy radicals [10]. During this same period, an active group in the Soviet Union including Emanuel et al. [11] examined many complex oxidation systems.

Throughout this period, perhaps best summarized in the proceedings of the International Oxidation Symposium in 1968 [12], there had been a gradual shift of emphasis from studies of complex oxidation systems to studies of those elementary reactions that comprise most individual oxidation reactions. Benson [13] has actively applied thermochemical kinetic analysis to oxidation systems and has shown the power of this tool for providing boundary conditions on possible reaction paths and intermediates. This same period marked the emergence of a clear understanding of the role of singlet oxygen in photooxygenation reactions and the diversity of processes exhibited by this unique reagent [14].

From these and later studies has come a better understanding of how different chain transfer and radical interaction processes compete, how these competitions change with reaction conditions, structure and temperature and how rate parameters (Arrhenius parameters) for elementary oxidation reactions may be predicted from the structures of the reactants and products.

3. The elementary rate steps in oxidation

In this section, we discuss some specific autoxidation reactions to illustrate the kinds of important elementary rate steps common to a variety of simple and complex systems. The examples cited here have been studied in considerable detail and, in most cases, the detailed kinetic behavior of the system has been elucidated.

3.1 OXIDATION OF CYCLOHEXENE. THE HYDROPEROXIDE CHAIN

Early investigators found that cyclohexene oxidized readily at 30—40°C if exposed to daylight or UV and used this method to prepare the hydroperoxide in high yield. The hydroperoxide was first prepared in 1928 by Stephens [15] who followed the then current view in assigning the structure as an endoperoxide (as did Hock and Schrader [3] in 1936). The correct structure was finally assigned by Criegee et al. [4] in 1939.

Bolland and Gee [16] reported the first detailed kinetic investigation of the oxidation of cyclohexene in 1946 using photoinitiation with the hydroperoxide at several concentrations of oxygen and cyclohexene, mostly at $10°C$. They showed that the rate of oxygen consumption corresponded very closely to the rate of formation of hydroperoxide with a rate law

$$R_o = kI_a^{1/2}[RH]f[O_2] \tag{5}$$

The form of this expression is identical to that developed by Bolland [17] in his classic study of the thermal initiated oxidation of ethyl lineolate where the detailed mechanism proposed was

$$ROOH \underset{h\nu}{\overset{k_i}{\rightarrow}} RO\cdot + HO\cdot \tag{6}$$

$$RO\cdot(HO\cdot) + RH \rightarrow ROH(H_2O) + R\cdot \tag{7}$$

$$R\cdot + O_2 \overset{k_3}{\rightarrow} RO_2\cdot \tag{8}$$

$$RO_2 + RH \overset{k_p}{\rightarrow} RO_2H + R\cdot \tag{9}$$

$$2\,RO_2\cdot \overset{k_t}{\rightarrow} \text{termination} \tag{10}$$

In eqn. (5), I_a is the absorbed light leading to dissociation of ROOH and is equivalent to the rate of initiation, R_i, k is a composite rate coefficient, $k = k_p/(2k_t)^{1/2}$, and $f[O_2] = k_3[O_2]/(1 + k_3[O_2])$. Experimentally, at $p_{O_2} >$ 50 torr, the latter term reduces to unity and eqn. (5) reduces to the general form of the rate law for autoxidation of a great variety of organic compounds

$$R_o = \left(\frac{R_i}{2k_t}\right)^{1/2} k_p[RH] \tag{11}$$

Two characteristic features of the kinetics are significant in establishing the foregoing mechanism. These are quantum yields much greater than unity and half power dependence on R_i (or I_a), which demonstrate that oxidation proceeds by way of a chain reaction terminated by interaction of two chain carriers. Moreover, since the principal product in the oxidation of cyclohexene is cyclohexenyl hydroperoxide, the principal chain carrier must be the cyclohexenylperoxy radical.

Thus cyclohexene illustrates the simplest kind of oxidation system where only one hydroperoxide can be formed by H-atom transfer from an allylic position to a peroxy radical

$$(12)$$

and this peroxide is found to be the principal product (>95%) under mild conditions in neat cyclohexene.

Under less than ideal conditions, a variety of complications attend the oxidation of cyclohexene including complex termination under low oxygen pressure, autocatalysis owing to thermal dissociation of the hydroperoxide into radicals, and a competitive propagation step in which the $RO_2 \cdot$ radical adds to the double bond.

The latter two complications in the oxidation of cyclohexene were examined in detail by Van Sickle et al. [18] in the early 1960s and all three are discussed in more detail in succeeding sections of this chapter.

3.2 OXIDATION OF STYRENE. THE PEROXY RADICAL ADDITION MECHANISM

$$CH_2 = CHPh + O_2 = [-CH_2CH(Ph)O_2-]_n$$

Styrene, like many reactive olefins which do not have reactive allylic carbon—hydrogen bonds, none the less reacts readily with oxygen even at room temperature [19,20]. Because the chain reaction can proceed only by addition of a peroxy radical to the double bond, it is a useful model for demonstrating the addition mechanism of oxidation in olefins. If the styrene oxidation reaction mixture is worked up without the use of excessive heat, the major product is a polymeric material with an average molecular weight of 5000 [21]. The polymer is sensitive to heat and even at room temperature slowly decomposes to benzaldehyde and formaldehyde [21]. Apparently for this reason, the oxidation of styrene was originally reported to give these compounds as initial products [19]. These observations, together with the fact that the polymer can be reduced to phenylethylene glycol [21], indicates that it is composed predominantly of alternating units of styrene and molecular oxygen

$$
(PhCHCH_2O_2-)_x \xrightarrow{H_2} PhCHCH_2OH \atop \qquad\qquad\quad OH
$$

$$(13)$$

$$\xrightarrow{\Delta} PhCHO + CH_2O$$

$$(14)$$

At low pressures of oxygen (below 100 torr at 50°C), more styrene units than oxygen are incorporated in the polymer product so that some styrene units appear adjacent to each other. The important elementary oxidation steps, initiated by a radical $X\cdot$, are

$$X\cdot + S \rightarrow XS\cdot \; (= S\cdot) \tag{15}$$

$$S\cdot + O_2 \rightarrow SO_2\cdot \tag{16}$$

$$SO_2\cdot + S \rightarrow SO_2S\cdot \; (= S\cdot) \tag{17}$$

$$S\cdot + S \rightarrow SS\cdot \; (= S\cdot) \tag{18}$$

where S represents styrene and $S\cdot$ represents a polymer radical ending with a styrene unit.

Low molecular weight products also accompany styrene polyperoxide and account for almost 50% of the styrene at 50 torr, but decrease to about 10% at 750 torr [20,23]. These products include styrene oxide, benzaldehyde, and formaldehyde formed during propagation by the competition between addition of oxygen to a β-peroxystyryl group and cleavage to a new β-peroxystyryloxy group and styrene oxide. Further cleavage to aldehydes and alkoxy radicals then follows with resumption of propagation by addition of $RO\cdot$ to styrene, viz. [23]

$$SO_2\cdot + S \rightarrow SO_2CH_2\dot{C}HPh \tag{19}$$

$$SO_2CH_2\dot{C}HPh + O_2 \rightarrow SO_2CH_2CH(Ph)O_2\cdot \tag{20}$$

$$SO_2CH_2CH(Ph)O_2\cdot + S \rightarrow SO_2CH_2CH(Ph)O_2CH_2\dot{C}HPh \tag{21}$$

$$SO_2CH_2CHPhO_2\dot{C}HPh \rightarrow SO_2CH_2CH(Ph)O\cdot + CH_2\underset{\diagdown O\diagup}{CHPh} \tag{22}$$

$$SO_2CH_2CH(Ph)O\cdot \rightarrow SO\cdot + CH_2O + PhCHO \tag{23}$$

$$SO\cdot + S \rightarrow SOS\cdot \; (= S\cdot) \tag{24}$$

The detailed competition among propagation steps was worked out by Mayo and coworkers [20—24]; however, the oxidation process is complicated still further by self-initiation [25,26] and by first-order termination [27].

3.3 THE OXIDATION OF ISOBUTANE. COMPETITION AMONG ALKYLPEROXY AND ALKOXY RADICAL CHAINS AND OLEFIN FORMATION

$$(CH_3)_3CH + O_2 \rightarrow (CH_3)_3COOH + (CH_3)_3COH$$

The relatively high reactivity of the tertiary hydrogen in isobutane and the stability of the derivative t-butyl hydroperoxide make isobutane an ideal but unusual substrate for studies of the oxidation of alkanes. Winkler and Hearne [28] reported that the initiated oxidation of liquid isobutane at 125°C gave 75% t-BuOOH, 21% t-BuOH with small amounts of acetone and isobutyl derivatives.

A detailed study of both the liquid and gas phase oxidations at 50—155°C was reported by Allara et al. [29] in 1968. The kinetics and products of the liquid phase oxidation are largely explained by the steps

$$\text{Initiator} \rightarrow 2 \text{ X·} \tag{25}$$

$$\text{X· (or } XO_2·) + RH \rightarrow R· + H \text{ (or } O_2H) \tag{26}$$

$$R· + O_2 \rightleftharpoons RO_2· \tag{27}$$

$$RO_2· + RH \overset{k_p}{\rightarrow} RO_2H + R· \tag{28}$$

$$2 RO_2· \overset{k_x}{\rightarrow} O_2 + (2 RO·)_{cage} \overset{1-a}{\underset{a}{\diagup \diagdown}} \begin{matrix} 2 RO· & (29) \\ \\ R_2O_2 & (30) \end{matrix}$$

$$RO· + RH \rightarrow R· + ROH \tag{31}$$

A significant feature of this mechanism is the inclusion of step (29), a non-terminating interaction of t-butyl peroxy radicals to give t-butoxy radicals. Thus the oxidation of isobutane proceeds via two competing chain carriers whose relative concentrations depend on the rate of initiation.

In reactions (29) and (30), a represents the fraction of radical interactions that terminate and k_x is rate-determining in the sequence. Reactions (27), (29), and (30) have also been studied at 25°C in the gas phase by photogeneration of t-butyl radicals in the presence of oxygen [30]. The rate expression for oxygen consumption corresponding to reactions (24)—(31) is

$$\frac{\Delta[O_2]}{dt} = R_o = \frac{R_i}{2a} + \left(\frac{R_i}{2ak_x}\right)^{1/2} k_p[RH] \tag{32}$$

where R_i = rate of production of initiating radicals (X·). The last term is the oxygen uptake associated with formation of hydroperoxide; it includes XO_2H when X· radicals react with oxygen (not when X· is an alkoxy radical). The $R_i/2a$ term is the sum of two others, $(R_i/2) + \{R_i(1-a)/2a\}$, corresponding, respectively, to oxygen absorbed and appearing in R_2O_2 and ROH. When rewritten, eqn. (32) can be used to evaluate a, viz.

$$R_o = \frac{R_i}{2a} + \frac{\Delta[RO_2H]}{dt} \quad \text{or} \quad \Delta O_2 = \frac{\Delta[\text{initiator}]}{a} + \Delta[RO_2H] \tag{33}$$

A quantitative treatment of steps (29) and (30) is presented in Sect. 4.4.3. For now, we simply note that the fraction of total interactions of t-BuO$_2$· radicals ($k_{29} + k_{30}$) that terminate, a, is approximately 0.1 at 50°C and 0.05 at 100°C and, consistent with the proposed mechanism, the ratio of two primary products, t-BuO$_2$H to t-BuOH, varies inversely with the rate of initiation. At very low rates of initiation, t-BuO$_2$H is the principal product at 100°C even in the gas phase at moderate concentrations of i-BuH. At 155°C in the gas phase, another reaction to produce isobutene

$$t\text{-Bu·} + O_2 \rightarrow C_4H_8 + HO_2·$$ (34)

starts to become competitive with the formation of t-BuO$_2$H. Benson [31] has accounted for the shift from hydroperoxide-based products (including alcohol and carbonyl) at low temperature to olefin-based products (including carbonyl from secondary reactions of the olefin) at high temperatures on the basis of a reversal of reaction (3) and a slight activation energy for reaction (10). Thus at low concentrations in the gas phase at temperatures above 250°C, most alkanes give olefins as major products [32]. But as the concentration of alkane is increased, abstraction by RO$_2$· [reaction (28)] competes more favorably with steps (27) and (34) with a shift toward oxygenated products. The kinetic expression for this competition is described by [31]

$$\frac{d(RO_2H)}{d(\text{olefin})} = \frac{k_3 k_4 [RH]}{k_{10}(k_{-3} + k_4[RH]}$$ (35)

above 250°C, where [RH] is small, $k_{-3} \gg k_4[RH]$ and

$$\frac{d(RO_2H)}{d(\text{olefin})} = \frac{k_3 k_4 [RH]}{k_{10} k_{-3}}$$ (36)

This analysis is consistent with the results of Medley and Cooley [33] on the effect of pressure on product composition in the oxidation of isobutane.

3.4 OXIDATION OF CUMENE (ISOPROPYLBENZENE). THE ROLE OF ALKOXY RADICALS

The free radical oxidation of cumene can give high yields of cumene hydroperoxide by a sequence similar to that found in the oxidation of isobutane (Sect. 3.3) and the reaction is of industrial importance as a source of acetone and phenol via acid-catalyzed rearrangement of the hydroperoxide. High yields are obtained as a result of the high reactivity of the isopropyl tertiary hydrogen and the small termination rate coefficient [34]. The alkoxy chain found in the oxidation of isobutane (Sect. 3.3) is also important for this oxidation [reaction (31)], but the cleavage of the cumoxy radical to give acetophenone and a methyl radical is much more rapid (Sect. 4.3.5) than cleavage of t-butoxy radical to give acetone and a methyl radical. As a result, once cumoxy radical is formed it usually cleaves rather than abstracts, viz.

$$2 \text{ PhCMe}_2\text{O}_2\cdot \rightarrow 2 \text{ PhCMe}_2\text{O}\cdot + \text{O}_2 \tag{37}$$

$$\searrow \text{PhCMe}_2\text{O}_2\text{CMe}_2\text{Ph} \tag{38}$$

$$\text{PhCMe}_2\text{O}\cdot \rightarrow \text{PhCOMe} + \text{Me}\cdot \tag{39}$$

$$\text{PhCMe}_2\text{O}\cdot + \text{RH} \rightarrow \text{PhCMe}_2\text{OH} + \text{R}\cdot \tag{40}$$

Under some experimental conditions, including low hydrocarbon concentration and high rate of radical formation, the interaction of methylperoxy and cumylperoxy radicals rather than self-reaction of cumylperoxy radicals accounts for most of the terminating interactions. The competing reaction for methylperoxy is abstraction from cumene, viz.

$$\text{CH}_3\text{O}_2\cdot + \text{PhCMe}_2\text{O}_2\cdot \rightarrow \text{termination} \tag{41}$$

$$\text{CH}_3\text{O}_2\cdot + \text{RH} \rightarrow \text{CH}_3\text{O}_2\text{H} + \text{R}\cdot \tag{42}$$

At higher conversions of cumene, cumene hydroperoxide can trap the cumyloxy radical efficiently prior to fragmentation and in those cases where fragmentation does occur, cumyl hydroperoxide also traps the methylperoxy radical

$$\text{PhCMe}_2\text{O}\cdot + \text{PhCMe}_2\text{O}_2\text{H} \rightarrow \text{PhCMe}_2\text{OH} + \text{PhCMe}_2\text{O}_2\cdot \tag{43}$$

$$\text{MeO}_2\cdot + \text{PhCMe}_2\text{O}_2\text{H} \rightarrow \text{MeO}_2\text{H} + \text{PhCMe}_2\text{O}_2\cdot \tag{44}$$

The basic expression for the rate of oxidation is the same as for isobutane

$$R_\text{o} = \frac{R_\text{i}}{a} + k_\text{p}\left(\frac{R_\text{i}}{2k_\text{t}}\right)^{1/2}[\text{RH}] \tag{45}$$

where a and k_t depend on the reaction conditions. The value of k_t is

$$k_\text{t} = k_{38} + 2k_{37}\{[k_{39}/(k_{39} + k_{40}[\text{RH}] + k_{43}[\text{ROOH}])][k_{41}[\text{RO}_2\cdot]/$$

$$(k_{41}[\text{RO}_2\cdot] + k_{42}[\text{RH}] + k_{44}[\text{ROOH}])]\} \tag{46}$$

The term a is the fraction of terminations per self-reaction of two cumylperoxy radicals. When there are no methylperoxy radicals, a equals 0.1 at 60°C. At high rates of initiation and low [RH], the value of a increases and in principle could reach 2.1 if every cumoxy is converted to methylperoxy and terminates with another cumylperoxy radical. Thus k_t can range only from k_{38} to 2.1 k_{38}.

From inspection of eqn. (46), one can see how, at high conversions or where cumyl hydroperoxide is added, the rate expression simplifies and $k_\text{t} \rightarrow k_{38}$. Similar results occur at high [RH] and low rates of initiation, since [RO$_2\cdot$] is proportional to $R_\text{i}^{1/2}$.

3.5 n-BUTANE. OXIDATION AT SECONDARY AND PRIMARY C—H BONDS

$$n\text{-}C_4H_{10} + O_2 \rightarrow sec\text{-}C_4H_9OOH$$

n-Alkanes are generally considered to be unreactive towards oxidation at low temperatures and, compared with branched alkanes and most olefins, are so. n-Butane and n-pentane oxidize at about 1/30 of the rate of oxidation of isobutane at the same rate of initiation at 100°C and generally tend to give more fragmentation and secondary oxidation products. However, when oxidized at low rates of initiation and to low conversions, fairly long chain lengths and good yields of sec-hydroperoxide can be obtained [35]. At moderate temperatures, only small amounts of primary carbon oxidation products are found even from n-butane.

The detailed mechanism of oxidation of butane [35] is well accounted for by the same elementary steps as for isobutane but with two important differences. First, the rate coefficient for propagation at 100°C in n-butane is only 1/10 as large as for isobutane, as expected for abstraction of a sec-C—H bond stronger by 3.5 kcal mole^{-1} than the t-C—H bond. The second, more important, difference arises in the self-reaction of sec-RO$_2$· radicals, viz.

$$2\,R_2CHO_2\cdot \rightleftharpoons R_2CHO_4CHR_2 \xrightarrow{(1-a)k_x} 2\,R_2CHO\cdot + O_2 \qquad (29)$$

$$\searrow^{ak_x} R_2CHOH + R_2CHO + O_2 \qquad (47)$$

Unlike the corresponding self-reaction of t-RO$_2$· radicals [steps (29) and (30)] where only one in two to twenty interactions gives termination at 100°C, almost every self-reaction of sec-RO$_2$· leads to termination by disproportionation ($a \sim 1.0$); that is, $k_{29} \sim 0$. This shift in termination mechanism has two results: one is that few alkoxy radicals are formed as chain carriers at temperatures below 120—130°C and the second is that the rate of termination is nearly 100 times as fast as for the t-BuO$_2$· radicals at 100°C.

The great commercial utility of n-butane for producing acetic acid rests on the fact that as the temperature increases to 160—200°C, a high proportion of alkoxy radicals is formed both in the self-reaction of sec-BuO$_2$· and by homolysis of initially formed sec-BuO$_2$H; these in turn lead to two-carbon fragment precursors of acetic acid, viz.

$$\begin{array}{c} sec\text{-}BuO_2H \\ \searrow^{-OH} \\ 2\,sec\text{-}BuO_2\cdot \end{array} \xrightarrow[-O_2]{} sec\text{-}BuO\cdot \rightarrow CH_3CHO + Et\cdot \qquad (48)$$

Mayo [36] has shown that all of the products from the oxidation of

n-butane at $180°C$ can be accounted for quite readily using the set of elementary reactions used to describe the oxidation at $100°C$ with the addition of oxidation steps for intermediates such as acetaldehyde and Et· radicals.

3.6 2,4-DIMETHYLPENTANE. INTRAMOLECULAR OXIDATION

The unusual feature of the oxidation of 2,4-dimethylpentane is the formation, even at the lowest measurable conversions, of the dihydroperoxide in yields of over 90%, viz.

$$2 O_2 + (CH_3)_2 CHCH_2 CH(CH_3)_2 = (CH_3)_2 \underset{O_2H}{C} CH_2 \underset{O_2H}{C} (CH_3)_2 \qquad (49)$$

Rust [37] showed that among several branched alkanes which gave difunctional products on oxidation at $120°C$, 2,4-dimethylpentane gave the highest yield of the dihydroperoxide and on the basis of this selectivity he proposed that the key reaction involved intramolecular H-atom transfer from C-4 through a sterically favorable six-center transition state

$$\qquad (50)$$

Mill and Montorsi [38], in a more detailed kinetic study, showed that, not only was intramolecular abstraction the dominant process, but the ratio of rates of intra- and intermolecular abstraction was almost unchanged with temperature indicating little (<1 kcal mole^{-1}), if any, difference in activation energy between the two steps. Moreover, at very low oxygen concentrations, some oxetane formed by ring closure of I in competition with (the much faster) addition of oxygen, viz.

$$\qquad (51)$$

(I)

The rate law for the oxidation of 2,4-dimethylpentane (HRH) (at long chain lengths) obeys closely the relation

$$R_o = \left(\frac{R_i}{2k_t}\right)^{1/2} 2k_p [\text{HRH}] \qquad (52)$$

which, with the exception of the factor of 2 to account for the formation of the dihydroperoxide, is identical with the rate law for the oxidation of isobutane [eqn. (32)]. In addition to dihydroperoxide, small amounts of monohydroperoxide and acetone were found with the ratio of di- to monohydroperoxide increasing from 4 at 50°C to 7.5 at 125°C.

Intramolecular oxidation proceeds with unusual facility in dimethylpentane compared with most other simple normal and branched alkanes because of structural and kinetic features that are particularly favorable. In the following reaction scheme (where HRH is dimethylpentane)

$$HRO_2 \cdot + HRH \xrightarrow{k_p} HRO_2H + HR \cdot \tag{53}$$

$$HRO_2 \cdot \xrightarrow{k_r} \cdot RO_2H \text{ (intramolecular abstraction)} \tag{54}$$

$$\cdot RO_2H + O_2 \rightarrow \cdot O_2RO_2H \text{ (fast)} \tag{55}$$

$$\cdot O_2RO_2 + HRH \rightarrow HO_2RO_2H + HR \cdot \tag{56}$$

the ratio of di- (D) to monohydroperoxide (M) concentration is simply

$$\frac{[D]}{[M]} = \frac{k_r}{k_p[HRH]} \tag{57}$$

Experiments [38,39] show that k_p is only a third as large as for isobutane (per hydrogen) which suggests that there is some steric hindrance to abstraction in dimethylpentane. On the other hand, the difference in activation energy for external and internal hydrogen abstraction, eqns. (53) and (54), is only one kcal mole^{-1} (equivalent to a k_r/k_p ratio of 3.9 at 100°C), so the ratio of hydroperoxides is mostly dependent on the ratio of A-factors for the unimolecular and bimolecular processes and on the concentration of dimethylpentane (12.1 M in t-CH at 100°C). For the internal six-center process, $A_r \sim 10^{11.5}$ s^{-1} [40] and for the bimolecular process, $A_p \sim 10^{9.0}$ 1 mole^{-1} s^{-1} [39]. Therefore the ratio of rate coefficients at 100°C is

$$\frac{k_r}{k_p} = 10^{2.5} e^{-1000/RT} = 82 \text{ mole } l^{-1} \tag{58}$$

in good agreement with the experimental value of 85 mole l^{-1}.

The comparable intramolecular oxidation is not observed in 2,3-dimethylbutane ([D]/[M] < 0.1), even though it oxidizes almost three times as fast as dimethylpentane [38,39] and the A-factor for such a process is at least a half-power of ten more favorable than for the comparable six-center process. We rationalize this result most readily in terms of an activation energy of nearly 7 kcal mole^{-1} of ring strain for the formation of the five-center transition state which reduces the rate of the intramolecular process to about one hundredth the rate of the bimolecular process at 100°C.

These generalizations still fail to account for the fact that, among n-alkanes such as octane or pentane [41], not more than a 10% of the products arise from intramolecular oxidation, even at low concentrations to favor the unimolecular process. However, at very low concentrations and high temperatures, intramolecular abstraction is, for many branched and normal alkanes, a major process [42].

4. Elementary rate steps. Absolute rate coefficients

Most oxidation reactions proceed by way of elementary steps involving alkylperoxy and alkoxy radicals; therefore quantitative descriptions of oxidation processes require reliable absolute rate coefficients for all important elementary steps. This section provides a compilation of rate coefficients and rate parameters for H-atom transfer (abstraction), addition, ring closures and combinations by peroxy radicals, and for abstraction and cleavage by alkoxy radicals.

No attempt has been made to provide the intensive detail found in Howard's review of oxyradicals [10], the review of Hendry et al. [43] of H-atom transfer to several radicals or Anbar and Neta's review of HO· radical reactions [44]. Instead, we have attempted to extend the scope of those reviews in two ways: (i) rate coefficients are provided for $RO_2·$ radical addition to many olefins, for ring closures to form cyclic ethers, and for intramolecular abstraction; (ii) for each reaction, we have estimated the "best value" Arrhenius parameters (A-factor and E) and, where such values have been measured they are also listed. We believe the value of absolute rate coefficients is improved substantially by the availability of reliable Arrhenius parameters, with which one can calculate the values of rate coefficients at other temperatures for use in experimental or modelling studies.

4.1 MEASUREMENT OF ABSOLUTE RATE COEFFICIENTS

The evaluation of absolute rate coefficients of elementary reactions (hereafter referred to only as rate coefficients) is one of the most important steps in the kinetic analysis. Comparison of such values with our general chemical knowledge of radical reactions serves first as a check on the kinetic analysis and second, if shown to be reliable, they may be used in the kinetic analysis of other systems. It is often possible and useful to evaluate the rate coefficients directly in oxidation reactions as well as in much more simplified systems where many of the competing steps have been eliminated.

In most free radical processes, a rate coefficient of interest is measured relative to a second rate coefficient which hopefully has been accurately evaluated. Thus the rate coefficient for reactions of $RO_2·$ are measured, in most cases, relative, directly or indirectly, to the bimolecular termination of $RO_2·$. Rate coefficients for reactions involving either alkoxy or carbon radicals do not have any well-measured competing reaction as a reference point; however, fragmentation of the alkoxy radical and oxygenation of the carbon radical are useful semi-quantitative reference points.

4.1.1 Overall kinetics of oxygen consumption

The loss of oxygen during an oxidation reaction is the result of a complex set of reactions. In the simplest situation where radicals are introduced into the system at a constant rate, such as by thermal decomposition of a free radical source, and where loss of radicals occurs upon every interaction of two $RO_2\cdot$, the overall scheme is

$$\text{Initiator} \xrightarrow{k_i} 2e \ R\cdot \qquad \text{Rate} = R_i = 2ek_i[\text{initiator}] \tag{59}$$

$$R\cdot + O_2 \xrightarrow{k_o} RO_2\cdot \tag{27}$$

$$RO_2\cdot + RH \xrightarrow{k_p} RO_2H + R\cdot \tag{28}$$

$$2 \ RO_2\cdot \xrightarrow{k_t} O_2 + \text{stable products} \tag{29}$$

where e is the efficiency of free radical production.

The rate of oxygen consumption is given by

$$-\frac{dO_2}{dt} = R_o = k_o[R\cdot][O_2] - k_t[RO_2\cdot]^2 \tag{60}$$

Since at high oxygen pressure

$$k_o[R\cdot][O_2] = k_p[RO_2\cdot][RH] + R_i \tag{61}$$

and since

$$R_i = 2k_t[RO_2\cdot]^2 \tag{62}$$

$$[RO_2\cdot] = \left(\frac{R_i}{2k_t}\right)^{1/2} \tag{63}$$

we have

$$R_o = k_p[RH]\left(\frac{R_i}{2k_t}\right)^{1/2} + \frac{R_i}{2} \tag{64}$$

When R_i is small compared with the first term, the simple and familiar relation

$$R_o = k_p[RH]\left(\frac{R_i}{2k_t}\right)^{1/2} \tag{65}$$

holds. In cases where the reaction rate is followed by pressure drop or by a decrease in non-condensible gas and the initiator forms nitrogen, the rate of gas consumption, R_g, is given by

$$R_g = R_o - k_d[\text{initiator}] = R_o - \frac{R_i}{2e} \tag{66}$$

$$= k_p[\text{RH}]\left(\frac{R_i}{2k_t}\right)^{1/2} + \frac{R_i(e-1)}{2e} \tag{67}$$

In cases where $e \sim 1$ or the chain length is long ($R_o/R_i > 20$), the second term in the equation may be neglected.

For many hydrocarbons, a significant fraction of the interactions of $RO_2\cdot$ produce $RO\cdot$ [step (29)] and $RO\cdot$ also propagate [step (31)] with consumption of oxygen

$$RO\cdot + RH \rightarrow ROH + R\cdot \tag{31}$$

$$R\cdot + O_2 \rightarrow RO_2\cdot \tag{27}$$

Thus the termination rate coefficient is ak_x and where a is <0.2, much of RH can be converted to alcohol rather than hydroperoxide. In this more general case

$$R_o = k_p[\text{RH}]\left(\frac{R_i}{2ak_x}\right)^{1/2} + \frac{R_i}{2a} \tag{68}$$

which reduces to the simple case [eqn. (65)] when every interaction of $RO_2\cdot$ results in termination; i.e. $a = 1$ and $k_x = k_t$.

Another variation in the overall kinetic scheme occurs when some peroxy radicals terminate by a first-order process involving an inhibitor such as a phenol

$$RO_2\cdot + \text{inhibitor} \rightarrow \text{termination products} \tag{69}$$

In some cases, the products of termination may be more effective inhibitors than the initial inhibitor so that more than one $RO_2\cdot$ is rapidly consumed by each inhibitor

$$n\,RO_2\cdot + \text{inhibitor} \xrightarrow{k_{inh}} \text{termination products} \tag{70}$$

Thus, at a constant rate of formation of radicals (R_i)

$$R_i = nk_{inh}[RO_2\cdot][\text{inhibitor}] \tag{71}$$

$$[RO_2\cdot] = \frac{R_i}{nk_{inh}[\text{inhibitor}]} \tag{72}$$

and

$$R_o = \frac{k_p[\text{RH}]R_i}{nk_{inh}[\text{inhibitor}]} \tag{73}$$

All of the above rate expressions for oxygen consumption yield ratios of rate coefficients rather than individual coefficients; therefore in order to obtain individual (absolute) rate coefficients some other techniques must be used instead of, or in addition to, measuring oxygen consumption at a constant rate of radical formation.

4.1.2 Evaluation of k_t independent of k_p

There are both steady state and non-steady state approaches to the evaluation of k_t. In order to evaluate k_t independently of k_p, the measured parameter must be related directly to the termination process, e.g. peroxy radical disappearance by ESR. If some measure of the overall reaction, such as O_2 absorption, is also obtained, then both k_p/k_t and $k_p/k_t^{1/2}$ are obtained from which absolute values of both k_p and k_t can be calculated.

(a) Peroxy radical disappearance

Conceptually, the simplest method of measuring of k_t is to generate a relatively high concentration of peroxy radicals and follow their disappearance by a suitable spectrometric method such as ESR or UV. Some of the techniques by which peroxy radicals may be generated in high concentrations are:

(i) photolysis of a peroxide [45] in the presence of hydrocarbon and oxygen

$$R'-O_2-R' \xrightarrow{hv} 2\,R'O\cdot \tag{74}$$

$$R'O\cdot + RH \rightarrow R'OH + R\cdot \tag{75}$$

$$R\cdot + O_2 \rightarrow RO_2\cdot \tag{27}$$

(ii) photolysis of azo-compounds [46] in the presence of oxygen

$$R-N_2-R \xrightarrow{hv} 2\,R\cdot + N_2 \tag{76}$$

$$R\cdot + O_2 \rightarrow RO_2\cdot \tag{27}$$

(iii) photolysis of hydroperoxides

$$RO_2H \xrightarrow{hv} RO\cdot + \cdot OH \tag{77}$$

$$RO\cdot(HO) + RO_2H \rightarrow ROH(HOH) + RO_2\cdot \tag{78}$$

(iv) reaction of hydroperoxides with a stable free radical or metal ion [47,48]

$$RO_2H + Ce(IV) = RO_2\cdot + Ce(III) + H^+ \tag{79}$$

(v) pulse radiolysis of hydrocarbon in the presence of oxygen [49]

$$RH \xrightarrow{(e^-)} R\cdot + H\cdot \tag{80}$$

$$R \cdot (H \cdot) + O_2 \rightarrow RO_2 \cdot (HO_2 \cdot) \tag{27}$$

Ideally, the source of peroxy radicals must be controllable so that their formation may be stopped quickly to allow their disappearance to be followed spectrometrically. This condition is easily met when photolysis or radiolysis is used; the use of metal ions [reaction (79)] requires rapid mixing and stop flow techniques [50].

The kinetic process for loss of peroxy radicals is bimolecular and may be expressed simply as

$$\frac{d[RO_2 \cdot]}{dt} = 2k_t [RO_2 \cdot]^2 \tag{81}$$

$$\frac{d[RO_2 \cdot]}{[RO_2 \cdot]^2} = 2k_t \, dt \tag{82}$$

which upon integration in the limits of $t = 0$ to t gives

$$\frac{1}{[RO_2 \cdot]_t} - \frac{1}{[RO_2 \cdot]_o} = 2k_t t \tag{83}$$

A plot of $1/[RO_2 \cdot]_t$ verse time gives a line with slope $2k_t$. The value of k_t will depend on the precise mechanism of interaction as well as on the reactants present. If the interaction gives stable products or a fraction of alkoxy radicals are formed, such as in reactions (29) and (30)

$$2 \, RO_2 \xrightarrow{ak_x} \text{termination products} \tag{29}$$

$$\xrightarrow{(1-a)k_x} 2 \, RO \cdot + O_2 \tag{30}$$

$$RO \cdot + RH \rightarrow ROH + R \cdot \tag{31}$$

$$R \cdot + O_2 \rightarrow RO_2 \cdot \tag{27}$$

where the alkoxy radical regenerates the original peroxy radical, then the measured k will equal ak_x which in turn equals k_t, the termination coefficient. However, in the absence of reactant to convert the $RO \cdot$ back to $RO_2 \cdot$, then k_t will equal k_x.

(b) Steady-state peroxy radical concentration

In the oxidation of a hydrocarbon or in the induced decomposition of a hydroperoxide, the peroxy radical concentration will reach a dynamic equilibrium (steady-state) concentration if the rate of generation of radicals is constant (R_i) according to the equation

$$R_i = 2k_t [RO_2 \cdot]_{ss}^2 \tag{84}$$

Thus, if the radicals are formed at a known rate, such as by the thermal decomposition of a peroxide or azo compound, measurement of the radical concentration gives $2k_t$ directly.

The use of ESR for evaluating rate coefficients for peroxy radical disappearance, while apparently simple enough, does not appear to have the desired accuracy that some other techniques have. The reason for the inaccuracy is not totally clear but is probably related to the difficulty of integrating the ESR signal for $RO_2\cdot$, which is considerably broader than the typical ESR signal, and in particular is broader than that of the standards such as pitch or DPPH which are used to calibrate the peroxy radical signal. An additional problem is that the line shape varies with temperature so that, for meaningful temperature-dependent measurements, calibration must be carried out at each temperature [51]. But even with laborious calibration at each temperature, the question remains of how precisely the broad peroxy radical signal can be calibrated.

One approach to solving this problem, which has not been reported to our knowledge but which holds considerable promise, involves combining the steady-state and decay measurements. By inspection of eqn. (84), one may see that, in the decay method, any constant error, f, in $[RO_2\cdot]$ will show up as fk_t. However, in the steady-state technique, eqn. (84), the same error in $[RO_2\cdot]$ shows up as f^2k_t. If both measurements can be done under comparable conditions so that the error f is the same in each case, the ratio of the apparent termination constants is given by

$$\frac{k_{\text{decay}}}{k^{\text{SS}}} = \frac{fk_t}{f^2 k_t} = \frac{1}{f} \tag{85}$$

The two methods will give the error which can be used to correct the observed rate coefficients.

(c) Photoemission decay

A third approach to measuring k_t independently of k_p requires monitoring the light that is generated by the oxidation reaction [52,53]. If the light intensity (I) can be demonstrated to be proportional to the rate of termination (R_t), then

$$I = CR_i = C2k_t[RO_2\cdot]^2 \tag{86}$$

where C is the proportionality constant. Solving this expression for $RO_2\cdot$ and substituting into eqn. (83) followed by combination with eqn. (87), which relates the initial steady-state light intensity to the rate of initiation

$$I_o = CR_i \tag{87}$$

gives the following expression from which k_t may be evaluated,

$$\left(\frac{I_o}{I}\right)^{1/2} = t(2k_t R_i)^{1/2} \tag{88}$$

Thus the proportionality constant conveniently cancels out. The validity of this technique rests on the requirement that the proportionality in eqn. (86) holds for each hydrocarbon. In oxidation of primary or secondary C—H bonds, the light emitted apparently comes from excited carbonyl produced in the termination step (89) [54] following the mechanism originally suggested by Russell [55]

$$2 \ R_2CHO_2\cdot \rightarrow \qquad \rightarrow \tag{89}$$

Spin conservation rules require that either the carbonyl be in a triplet state or the oxygen molecule be in a singlet state but not both.

In the oxidation of tertiary carbon—hydrogen bonds, where termination cannot involve this mechanism, the rate of light emission cannot be directly proportional to the rate of termination [34,56]. However, it is known that in oxidation of such hydrocarbons, termination can involve at least some primary or secondary peroxy radicals which are formed indirectly from fragmentation of the alkoxy radical corresponding to the parent hydrocarbon (see Sect. 3.4).

4.1.3 Measurement of k_t and k_p dependently

(a) Decay of reaction rate

Any quantitative measure of the overall reaction rate may be used to monitor the decay of the rate after the source of initiation is stopped. However, the decay of such a parameter actually produces the ratio of k_p/k_t. But by introducing the overall steady-state rate expression which gives a measure of $k_p/k_t^{1/2}$, it is then possible to separate k_t from k_p.

The rate of reaction as measured by oxygen consumption (R_o) is directly related to [RO$_2$·] in an oxidizing hydrocarbon by the equation

$$R_o = k_p[RO_2\cdot][RH] - ak_x[RO_2]^2 = k_o[R\cdot][O_2] - ak_x[RO_2\cdot]^2 \tag{60}$$

If radical interaction and termination produce a small fraction of oxygen compared with the total oxygen consumed, then

$$[RO_2\cdot] = \frac{R_o}{k_p[RH]} \tag{90}$$

This expression may be substituted in eqn. (83)

$$\frac{1}{R_o} - \frac{1}{R_o^i} = \frac{(k_t/k_p)t}{[RH]} \tag{91}$$

where R_o^i is the initial rate at the time the decay begins. If the rate of initiation (R_i) is known at the initial point of decay of $[RO_2 \cdot]$, eqn. (91) may be combined with the steady-state equation

$$R_o^i = k_p[RH] \left(\frac{R_i}{2k_t} \right)^{1/2} \tag{92}$$

thereby giving

$$\frac{R_o^i}{R_o} - 1 = (k_t R_i)^{1/2} t \tag{93}$$

This technique requires monitoring the oxygen consumption during the decay process which is difficult because diffusion of oxygen into the liquid is not instantaneous [57]. However, the technique could be used if the consumption of oxygen in a static solution was monitored by following the oxygen—hydrocarbon charge transfer UV absorption [58]. This approach should be the most accurate, precise, and sensitive of the methods available.

Because of the lack of reliable techniques to follow the decay of the rate of oxidation under non steady-state conditions, a number of approaches have been used to bypass this problem as discussed below.

(b) Photochemical pre- and after effect

In cases where the radical life times are sufficiently long, a change in initiation rate will produce a gradual change in rate until the new, steady-state rate is reached. As originally shown by Bateman and Gee [57], the quantities of oxygen, ΔO_2 (growth), obtained by extrapolating the new rate back to the time the rate of initiation was increased and ΔO_2 (decay) obtained by extrapolating the rate back to the time the rate of initiation was decreased, can be related to the rate of reaction, R, which at steady-state in the dark and light, are R_D and R_L, respectively, viz.

$$\Delta O_2 \text{ (growth)} = -\int_0^\infty (R - R_L) \, dt \qquad \Delta O_2 \text{ (decay)} = \int_0^\infty (R - R_D) \, dt \tag{94}$$

The rate of reaction and its derivative may be expressed accordingly as

$$R = k_p[RH][RO \cdot_2] \quad \text{and} \quad \frac{dR}{dt} = \frac{k_p[RH] d[RO_2 \cdot]}{dt} \tag{95}$$

The disappearance of $RO_2 \cdot$ may be expressed as

$$\frac{d[RO_2 \cdot]}{dt} = -2k_t[RO_2 \cdot]^2 + R_i \tag{96}$$

where R_i is the rate of formation of radicals. Thus the expression for change in the reaction rate in going from light off, where there is only thermal initiation, $(R_i)_D$, to light on is

$$\frac{dR}{dt} = k_p[RH]\{-2k_t[RO_2 \cdot]^2 + (R_i)_L\} \tag{97}$$

Since $[RO_2 \cdot] = R/k_p[RH]$ and $R_i = (R^2\, 2k_t)/(k_p^2[RH]^2)$

$$\left(\frac{dR}{dt}\right)_{D \to L} = -\frac{2k_t R^2}{k_p[RH]} = \frac{2k_t}{k_p[RH]}\{-R^2 + R_L^2\} \tag{98}$$

while for the transition from light rate to dark rate the expression is

$$\left(\frac{dR}{dt}\right)_{L \to D} = \frac{2k_t}{k_p[RH]}\{-R^2 + R_D^2\} \tag{99}$$

Solving each of these expressions and substituting into the equation to be integrated gives

$$\Delta O_2 \text{ (growth)} = -\frac{k_p[RH]}{2k_t} \int_{R=R_D}^{R=R_L} \frac{R-R_L}{R_L^2 - R^2}\, dR \tag{100}$$

and

$$\Delta O_2 \text{ (decay)} = \frac{k_p[RH]}{2k_t} \int_{R=R_L}^{R=R_D} \frac{R-R_D}{R_D^2 - R^2}\, dR \tag{101}$$

Upon integration they become

$$\Delta O_2 \text{ (growth)} = -\frac{k_p[RH]}{2k_t} \ln\left(\frac{2R_L}{R_L + R_D}\right) \tag{102}$$

and

$$\Delta O_2 \text{ (decay)} = \frac{k_p[RH]}{2k_t} \ln\left(\frac{R_L + R_D}{2R_D}\right) \tag{103}$$

If ΔO_2 (growth) and ΔO_2 (decay) are sufficiently large to be accurately measured, then the ratio $k_p/2k_t$ is easily obtained and with a value of $k_p/(2k_t)^{1/2}$ from direct, initiated oxidation, yields k_p and k_t separately. The limiting factor on this method is the sensitivity of the oxygen consumption measurement. If a pressure detector can detect $\pm 2 \times 10^{-3}$ torr, $k_p/2k_t$ must be greater than 10^{-8} to be measured with a precision greater than

±10%. A more practical lower limit for this technique is in the range 10^{-6} to 10^{-7} for $k_p/2k_t$. The technique has been applied most successfully to reactive tertiary systems where, for example, with cumene [34] $k_p/2k_t \simeq 10^{-5}$.

One source of error arises from slow equilibration of oxygen with solvent, which depends on a dynamic equilibrium governed both by the oxygen consumption rate and the diffusion rate [59]. Thus this error should be subtracted from ΔO_2 (growth) and added to ΔO_2 (decay). The difference between ΔO_2 (growth) and ΔO_2 (decay) is independent of this error. Under conditions where the error is significant, the combined equation

$$\Delta O_2 \text{ (decay)} - \Delta O_2 \text{ (growth)} = \frac{k_p[\text{RH}]}{k_t} \ln \left\{ \frac{(R_L + R_D)^2}{4R_L R_D} \right\} \tag{104}$$

should be used to measure k_p/k_t. Whether or not this correction needs to be applied depends on the magnitude of k_p/k_t as well as the efficiency of gas—liquid mixing in the apparatus. The correction was not necessary in the oxidation of cumene [34], where k_p/k_t is relatively large (10^{-5}), but was necessary in the study of hydrocarbons where the reactive C—H bonds were secondary and k_p/k_t was small (10^{-7}) [34,59].

(c) Intermittent illumination technique

The success of this technique, commonly referred to as the rotating sector method, is due to the dependence of radical lifetimes in a photo-initiated reaction on the frequency of the light pulse [61—64]. The radical lifetime (λ) is

$$\lambda = \frac{[\text{RO}_2 \cdot]}{R_t} = \frac{[\text{RO}_2 \cdot]}{k_t[\text{RO}_2 \cdot]^2} = \frac{1}{k_t[\text{RO}_2 \cdot]} \tag{105}$$

where R_t is the rate of termination of the radical and k_t is the rate coefficient for that process. If the on—off cycle of the light pulse is much shorter than the lifetime of the radical, which generally is of the order of seconds, pulsing does not cause any fluctuation in the radical concentration. However, the rate of initiation is reduced compared with the rate in the full light. Thus the ratio of rate of oxidation in the pulse light (R_o^p) to rate in full light (R_o^f) is

$$\frac{R_o^p}{R_o^f} = f^{1/2} \tag{106}$$

where f is fraction of time the light is on during the pulsing. If the pulsing cycle is much longer, the reaction rate will have distinct periods, one where the light is completely on and one where it is off. The overall rate of oxidation is then

$$R_o^p = fR_o^f + (1 - f)R_o^D \tag{107}$$

where R_o^D is the rate in the dark. If there is no dark initiation ($R_o^D = 0$), then

$$\frac{R_o^p}{R_o^f} = f \tag{108}$$

A more quantitative description involves consideration of the decay of the radical concentration in going from light on to light off

$$\frac{d[RO_2\cdot]}{dt} = R_i^D - \alpha k_t[RO_2\cdot] \tag{109}$$

and from light off to light on

$$\frac{[RO_2\cdot]}{dt} = R_i^L + R_i^D - \alpha k_t[RO_2\cdot]^2 \tag{110}$$

Integrating these expressions and combining them, taking into account the ratio of the lengths of time of dark and light periods, gives the following expression for the average rate of oxidation, R_o^p, in pulse light

$$R_o^p = \frac{1}{(q + 1)(\tau_2 - \tau_1)}$$
$$\times \ln \frac{\cosh \tau_2}{\cosh \tau_1} \cdot \frac{R_o^m \sinh q\gamma(\tau_2 - \tau_1)}{\gamma} + \cosh q\gamma(\tau_2 - \tau_1) \tag{111}$$

where q is the ratio of dark to light periods in the intermediate light cycle, $\lambda = (R_i^D/R_i^L)^{1/2}$, τ_2 and τ_1 are the durations of illumination in multiples of the lifetime required to obtain the maximum rate (R_o^m) and minimum rate of the cycle, respectively, and thus $\tau_2 - \tau_1$ is the duration in units of lifetime over which the reaction is illuminated in the intermittent cycle. This equation has been evaluated [63,34] for the average rate of reaction as a function of $\tau_2 - \tau_1$ for various values of q and γ. The time interval of the intermittent cycle, $t_2 - t_1$, for each point equals some $\tau_2 - \tau_1$ or multiple of lifetimes from the theoretical curve with the corresponding average rate of reaction. Thus the lifetime (λ) equals the ratio $(t_2 - t_1)/(\tau_2 - \tau_1)$. The relation of λ to other parameters is

$$\lambda = \tfrac{1}{2}k_t[RO_2\cdot] = \frac{1}{(2k_tR_i)^{1/2}} \tag{112}$$

or

$$k_t = (2\lambda_i^2)^{-1} \tag{113}$$

4.1.4 Evaluation of k_p using added hydroperoxide

In many cases autoxidation processes are sufficiently complex that all of the above techniques fail to give meaningful data. However, addition of hydroperoxide to the reaction simplifies many of the systems by the following mechanism.

Initiation.

$$\text{Initiator} \xrightarrow{k_i} 2\,\text{X} \cdot \tag{24}$$

$$\text{X} \cdot + \text{RH} \rightarrow \text{XH} + \text{R} \cdot \tag{26}$$

$$\text{R} \cdot + \text{O}_2 \xrightarrow{\text{fast}} \text{RO}_2 \cdot \tag{27}$$

Propagation.

$$\text{RO}_2 \cdot + \text{RH} \xrightarrow{k_p} \text{RO}_2\text{H} + \text{R} \cdot \tag{28}$$

$$\text{RO}_2 \cdot + \text{R}'\text{O}_2\text{H} \rightarrow \text{RO}_2\text{H} + \text{R}'\text{O}_2 \cdot \tag{114}$$

$$\text{R}'\text{O}_2 \cdot + \text{RH} \xrightarrow{k'_p} \text{R}'\text{O}_2\text{H} + \text{R} \cdot \tag{115}$$

Termination.

$$2\,\text{R}'\text{O}_2 \cdot \xrightarrow{k'_t} \text{termination} \tag{30}$$

$$2\,\text{RO}_2 \cdot \xrightarrow{k_t} \text{termination} \tag{47}$$

$$\text{RO}_2 \cdot + \text{R}'\text{O}_2 \cdot \xrightarrow{k''_t} \text{termination} \tag{116}$$

If enough $\text{R}'\text{O}_2\text{H}$ is added to ensure that all $\text{RO}_2 \cdot$ and $\text{RO} \cdot$ are converted to $\text{R}'\text{O}_2 \cdot$, the kinetic expression for oxidation becomes (with long kinetic chains)

$$R_o = k'_p[\text{RH}] \left(\frac{R_i}{2k'_t} \right)^{1/2} \tag{117}$$

The presence of $\text{R}'\text{OOH}$ thus eliminates the complications due to different radicals formed by fragmentation of $\text{RO} \cdot$ [34,65]. The principle of this procedure rests on the rapid chain transfer between $\text{RO}_2 \cdot$ (and $\text{RO} \cdot$) and added $\text{R}'\text{OOH}$ ($k_{114} \sim 10^3$ l mole^{-1} s^{-1}) [66]. Use of a stable t-RO_2H, such as t-BuO_2H, gives a single $\text{RO}_2 \cdot$ radical which has a well-known self-termination rate coefficient of 1.2×10^3 l mole^{-1} s^{-1}. Thus by measuring oxygen uptake at a constant rate of initiation in the presence of 0.1—1.0 M t-BuO_2H, eqn. (117) can be used to solve for k'_p.

Other advantages of this procedure over autoxidation are

(i) The reactivity of a series of substrates toward one peroxy radical can be determined, free from differences in the reactivity of the peroxy radical.

(ii) Since the termination rate coefficient (k_t') does not change, the ratio $k_p/(2k_t')^{1/2}$ gives much more reliable relative reactivities than does $k_p/(2k_t)^{1/2}$.

(iii) Addition of a hydroperoxide, which gives a peroxy radical with a low value of k_t' (e.g. t-butyl hydroperoxide), can significantly increase the chain length of an autoxidation. Thus reliable values of k_p' can be determined for substrates that normally oxidize too slowly to give reliable values of k_p by a non-stationary state method.

4.2 THE REACTION OF CARBON RADICALS WITH OXYGEN

4.2.1 Formation of peroxy radicals

The oxidation of hydrocarbons is the result of the rapid reactions of the carbon radicals with oxygen, viz.

$$-\overset{|}{\underset{|}{C}}\cdot + O_2 \rightarrow -\overset{|}{\underset{|}{C}}-O_2\cdot \qquad (27)$$

The rapidity of the reaction can be seen by the large effect low pressures (~1 torr) of oxygen can have on the free radical polymerization of a reactive olefin such as styrene [22]. The reaction rate coefficients are expected to be typical for exothermic radical—radical reactions with essentially no activation energy. Thus, if R· is alkyl, $\log(k_o/\mathrm{l\ mole^{-1}\ s^{-1}})$ would be 9.0 ± 0.5, and be independent of temperature. For simple resonance-stabilized radicals, $\log(k_o/\mathrm{l\ mole^{-1}\ s^{-1}})$ would be 8.5 ± 0.5.

(a) Thermochemistry of radical—oxygen reactions

The thermochemistry of the reaction of oxygen with carbon radicals has been evaluated for the gas phase [31] and these data would be expected to be valid also for the liquid phase. Table 1 contains averages of the thermodynamic values estimated by Benson [31].

Values of ΔH^0 correspond to the $C-O_2\cdot$ bond strength and indicate that this bond can dissociate readily to the corresponding carbon radical and O_2. Comparison of the equilibrium constants for both alkyl and allyl/benzyl systems show that the value for the latter system is considerably smaller although both decrease with temperature. The bond strengths of the allyl and benzyl peroxy radicals are weaker than those of the alkyl peroxy radicals by approximately the resonance stabilization associated with the carbon radical [67]. The fraction of carbon radicals which would be oxygenated if other loss mechanisms for the radicals were not impor-

TABLE I
Thermodynamics and equilibrium constants for reactions of oxygen with carbon radicals [31]
All thermodynamic values for a standard state of 1 atm at 25°C

Quantity	Alkyl	Allyl/benzyl
$-\Delta H^\circ$ (kcal mole^{-1})	28 ± 2	14 ± 2
$-\Delta S^\circ$ (cal mole^{-1} K^{-1})	32 ± 2	29 ± 2
$-\Delta G^\circ$ (kcal mole^{-1})	18.5 ± 2.5	5.4
$-\Delta G^{500}$ (kcal mole^{-1})	16.0 ± 2.5	0.5
K^{298} (atm^{-1})	3.5×10^{13}	8.5×10^3
K^{500} (atm^{-1})	1.8×10^5	0.4

tant can be expressed as

$$\frac{[RO_2 \cdot]}{[R \cdot]} = K[O_2] \tag{118}$$

Thus for alkyl radicals, the equilibrium lies essentially completely on the side of $RO_2\cdot$ at typical oxygen pressures (>0.01 atm) up to 500 K. However, as the temperature increases still further, the ratio continues to decrease and reaches unity at about 700 K and 0.01 atm oxygen. These values are considerably lower for the allyl and benzyl systems. For example, at 0.01 atm oxygen, the ratio equals 10 at 300 K and 10^{-2} at 500 K. Thus a significant fraction of carbon radical is present under conditions that are *typically* used for studying oxidation reactions. The greater tendency of alkenes and aralkanes to show oxygen pressure dependence of the rate compared with alkanes can be attributed to the difference in these equilibrium constants, although a clear demonstration of the importance of these equilibria in solution has not been made for these simple systems. However, in cases where highly stabilized carbon radicals are involved, oxygen pressure dependence as high as 1 atm of oxygen has been observed [68]. In fact, hydrocarbons such as fluorene and triphenylmethane will retard the oxidation of cumene and cyclohexene at moderate oxygen pressures [68]. As expected if an equilibrium is involved, the effect is reduced by increasing the oxygen pressure. Howard and Ingold [69] have demonstrated that triphenylmethyl hydroperoxide retards the oxidation of cumene, tetralin, and 9,10-dihydroanthracene. The kinetics are best explained by the reactions

$$RO_2 \cdot + Ph_3CO_2H \rightarrow RO_2H + Ph_3CO_2 \cdot \tag{119}$$

$$Ph_3CO_2 \cdot \overset{1/K}{\rightleftharpoons} Ph_3C \cdot + O_2 \tag{120}$$

where K was estimated to be 8×10^3 l mole^{-1} (60 atm^{-1}) at 30°C which is reasonably consistent with the value of Janzen and coworkers [70] who obtained $K = 2.5$ atm^{-1} at $27°$C by observing the effect of air and temperature on the triphenylmethyl peroxy radical concentration in a crystal matrix. Janzen obtained $\Delta H^0_{298} = 9.0$ kcal mole^{-1}, which is consistent with the data in Table 1 considering the greater resonance stabilization of the triphenylmethyl radical.

(b) Kinetic data for radical—oxygen reactions

There are very few data on rate coefficients for reactions of oxygen with carbon radicals in solution. Table 2 summarizes the available data for the liquid phase; data for the ethyl radical in the gas phase are included for comparison. Because of difficulties in measuring k_o, these values are only approximate. There are no data for the reaction of an alkyl radical plus oxygen in solution because the reaction is so fast that the oxygen pressure must be maintained at a very low value. On the basis that the reactions of the resonance-stabilized radicals are really reversible, we have re-examined the data and calculated experimental values of K. In evaluating k_o, the assumption was originally made that

$$\frac{[RO_2\cdot]}{[R\cdot]} = \frac{k_o[O_2]}{k_p[RH]} \tag{121}$$

However, if the equilibrium is important, then

$$\frac{[RO_2\cdot]}{[R\cdot]} = K[O_2] \tag{122}$$

Therefore K may be calculated according to the expression

$$K = \frac{k_o}{k_p[RH]} \tag{123}$$

where k_o, k_p, and [RH] are derived initially without assuming reversibility. The agreement of these experimental values of K with those in Table 1 is very good considering the uncertainties. Thus the reported values of k_o are probably experimental artifacts and the apparent oxygen pressure dependence is a reflection of the rapid reversibility of reaction (27) for allyl and benzyl radicals. The effect of low oxygen pressures on the rate of oxidation has been discussed by Bolland [17].

4.2.2 Olefin formation

In cases where there is a hydrogen atom beta to the radical site, the reaction

$$-\underset{\underset{H}{|}}{\overset{|}{C}}-\overset{|}{\underset{\cdot}{C}}- + O_2 \rightarrow \overset{|}{\underset{/}{C}}=\overset{|}{\underset{|}{C}} + HO_2\cdot \tag{124}$$

TABLE 2
Reported rate coefficients for the reaction
$R\cdot + O_2 \rightarrow RO_2\cdot$

Radical	Phase [a]	Temp. (°C)	$10^{-6} k_0$ (l mole⁻¹ s⁻¹)	$10^{-3} K$ (calc.) [b] (atm⁻¹)	Ref.
$CH_3CH_2\cdot$	GP		4200		71
$-O_2CH_2CH{=}CH\dot{C}H_2$	LP	50	9	0.56	72
$-O_2CH_2\dot{C}H(Ph)$	LP	50	100	3.8	22
$\left\{\begin{array}{l}-CH{=}CH-CH{=}CH-\dot{C}H_2-\\-CH{=}CH_2-CH_2-CH{=}CH_2-\dot{C}H-\end{array}\right\}$	LP	25	10	2.8	60
$-CH{=}CH-\dot{C}H-$	LP	25	1	3.2	60
(naphthalenyl)	LP	25	68	28	59

[a] GP = gas phase; LP = liquid phase.
[b] See text.

is possible. The rate of this reaction is generally slower than the formation of the organic peroxy radical. However, if the organic peroxy radical formation step is reversible, then this reaction can become important. This appears to be the case in high temperature alkane oxidation (see Sect. 3.3). In the liquid phase near room temperature, the best example of olefin formation is the oxidation of 1,4-cyclohexadiene to benzene and H_2O_2 [73]. The important reactions are

$$X \cdot (HO_2 \cdot) + \underset{}{\bigcirc} \overset{|k_p'}{\to} HX(HO_2H) + \underset{}{\bigcirc} \cdot \tag{125}$$

$$\underset{}{\bigcirc} \cdot + O_2 \underset{k_{-0}}{\overset{k_0}{\rightleftharpoons}} \underset{O_2 \cdot}{\bigcirc} \tag{126}$$

$$\underset{O_2 \cdot}{\bigcirc} + \underset{}{\bigcirc} \overset{k_p}{\to} \underset{O_2H}{\bigcirc} + \underset{}{\bigcirc} \cdot \tag{127}$$

$$\underset{}{\bigcirc} \cdot + O_2 \overset{k_h}{\to} HO_2 \cdot + \underset{}{\bigcirc} \tag{128}$$

The ratio of formation of organic peroxide (RO_2H) to H_2O_2 is

$$\frac{\Delta[RO_2H]}{\Delta[HO_2H]} = \left(\frac{k_p}{k_h}\right) K_{126}[RH] \tag{129}$$

The cyclohexadienyl peroxy radical has an extremely weak $C-O_2 \cdot$ bond (~ 5 kcal mole^{-1}) and K_{126} has been estimated to be 0.1 l mole^{-1} at 50°C [73]. The removal of an H-atom by O_2 (23 kcal mole^{-1} exothermic) to form an olefinic bond, reaction (128), is undoubtedly slower than the formation of the peroxy radical, reaction (126), but it is not reversible and therefore it can be the major product-producing step.

The dissociation energy of the β-C—H bond is critical to the rate of this process since it affects the value of k_h. Howard and Ingold [74] found that 1,4-dihydronaphthalene gave organic peroxide as well as naphthalene depending on the hydrocarbon concentration, viz.

$$\underset{}{\bigcirc\bigcirc} + O_2 \xrightarrow{\text{high RH}} \underset{}{\bigcirc\bigcirc}^{OOH} \tag{130}$$

$$\xrightarrow{\text{low RH}} \underset{}{\bigcirc\bigcirc} + H_2O_2 \tag{131}$$

while 9,10-dihydroanthracene formed the corresponding organic peroxide independent of the concentration of dihydroanthracene, viz.

(132)

The reason for this trend in going from 1,4-cyclohexadiene to 1,4-dihydronaphthalene to 9,10-dihydroanthracene is that, while the ease of removal of the initial H-atom is expected to be about the same for each compound, ease of removal of the β-H-atom in the resulting radical decreases with the increase in benzo substitution as shown in Table 3. Thus the removal of an H-atom from the hydroanthracene radical by oxygen is thermoneutral and the reaction is sufficiently slow so that the organic hydroperoxide route dominates. The reaction of 1,4-dihydronaphthalene is intermediate. The reaction of the 1-hydronaphthyl radical with oxygen to form naphthalene is exothermic by 10 kcal mole^{-1}; however, it is sufficiently slow so that, as the hydrocarbon concentration is increased, the organic peroxide product increases.

Data are included in Table 3 for cyclohexane and cyclohexene. For cyclohexyl and other alkyl radicals, the β-C—H bond is sufficiently weak so that, as the temperature increases, addition of O_2 becomes reversible and olefin formation is facilitated, viz.

$$-CH_2CH_2O_2 \cdot \xrightarrow{k-o} -CH_2CH_2 \cdot + O_2 \qquad (133)$$

$$-CH_2CH_2 \cdot + O_2 \xrightarrow{k_h} -CH=CH_2 + HO_2 \cdot \qquad (134)$$

In the oxidation of olefins such as cyclohexene, the β-C—H bond in an

TABLE 3
Thermochemistry of dihydro and related compounds [73]

Dihydro compd. (ArH$_2$)	Enthalpies of formation (kcal mole^{-1})					
	Dihydro compd. (ArH$_2$)	Dehydro compd. (Ar)	Radical (ArH)	Bond dissociation energy (kcal mole^{-1})		
				H—ArH	Ar—H	HAr—O$_2$·
1,4-Cyclohexadiene	26.3	19.3	45.3	71	27	5
1,4-Dihydronaphthalene	30.6	36.1	51.6	71	37	5
9,10-Dihydroanthracene	42.0	55.2	61.0	71	46	5
Cyclohexane	−29.4	−0.8	13.0	94	38	27
Cyclohexene	−0.8	26.0	29.0	82	49	15

TABLE 4

Estimates of rate parameters for the reaction
$R\cdot + O_2 \rightarrow$ olefin [a]

Compound	Conc. (M)	Temp. (°C)	$\dfrac{[ROOH]}{[HOOH]}$	K [b] (l mole⁻¹)	k_p [c] (l mole⁻¹ s⁻¹)	k_h [d] (l mole⁻¹ s⁻¹)	E_h [e] (kcal mole⁻¹)
1,4-Cyclohexadiene	10.2	50	0.01 [f]	0.1	50	1×10^4	7.4 ± 2.0
	10.6	30	0.14	0.16	20		
	0.2	30	0.01 [f]	0.16	20	1.3×10^2	
1,4-Dihydronaphthalene	6.0	30	31	0.16	20	1.3	10.7 ± 2.0
	0.1	30	0.28	0.16	20	4.6	
9,10-Dihydroanthracene	1.0	30	>0.99 [g]	0.16	20	<6	$>11.4 \pm 2.0$
	0.1	30	>0.99 [g]	0.16	20	<0.6	

[a] Data from refs. 73 and 74.
[b] Estimate $\log K = -4.4 + (5000/4.57T)$ (ref. 73).
[c] Estimate from data in ref. 43 for H-atom transfer to t-BuO$_2\cdot$; values per hydrogen.
[d] Value of k_h (per hydrogen), estimated using eqn. (129).
[e] Weighted average value assuming $\log(A/\text{l mole}^{-1}\text{ s}^{-1}) = 8.0$.
[f] No organic peroxide observed, but assumed to be 1%.
[g] No HO$_2$H detected, assumed to be less than 1%.

allyl radical is as strong as the H—$O_2\cdot$ bond; thus there is less driving force to remove the β-H by O_2 to form a diene even though the equilibrium ($R\cdot + O_2 \rightleftharpoons RO_2\cdot$) is more favorable for olefin formation than in the case of alkane. Therefore one may expect the formation of dienes from olefin to occur only at elevated temperatures, possibly as high as for the formation of olefin from alkanes.

Using eqn. (129) for the ratio of organic peroxide to hydrogen peroxide, it is possible to estimate k_h for the three systems discussed under the various conditions. The necessary data are compiled in Table 4.

4.3 PROPAGATION

4.3.1 Hydrogen atom transfer to $RO_2\cdot$ radicals

(a) Relative reactivities

Oxidations of mixtures of organic compounds or of two or more different CH bonds in a single molecule is the most common procedure for evaluating reactivity of one hydrocarbon relative to a second toward $RO_2\cdot$ [75—77]. Under most conditions, the values for relative reactivity are independent of the values for k_t; however, as measured in co-oxidation, k(rel) is the ratio of the two rate coefficients for H-atom transfer by one peroxy radical. Although capable of giving valuable information concerning the reactivity of different organic compounds towards $RO_2\cdot$ radicals, co-oxidations require exceptional care in analysis to avoid very large errors [75] (see Sect. 6.1) and have been all but superseded by the technique of added hydroperoxide discussed in Sect. 4.1. Relative reactivity data may be used to estimate absolute rate coefficients if there is a suitable absolute value for comparison. However, care must be taken in selecting absolute standards as the absolute reactivities of differently substituted peroxy radicals with the same substrate vary roughly in the order tert : sec : primary = 1 : 5 : 10 [78]. Because of the large number of absolute coefficients available [43], no relative data are specifically included in this review, although the general techniques are discussed in Sect. 4.1.

(b) Estimation of rate parameters log A and E from transition state theory

One approach to estimating rate parameters for elementary steps rests on the relationship between the Arrhenius parameters defined by

$$k = A \exp[-E/RT] \tag{135}$$

and the activation parameters $\Delta S^{\ddagger}$ and $\Delta H^{\ddagger}$ defined by

$$k = \frac{kT_m}{h} \exp[(\Delta S^{\ddagger}/R) - (\Delta H^{\ddagger}/RT)] \tag{136}$$

Thus

$$A = \frac{ekT_m}{h} \exp \frac{\Delta S^{\ddagger}}{R} \tag{137}$$

where T_m is the mean temperature and $E = \Delta H^{\ddagger} + RT_m$. Since there is always a loss of entropy in going to the transition state for hydrogen atom transfer or for addition

$$RO_2 \cdot + R'H \rightleftharpoons RO_2 \cdots H \cdots R' \tag{138}$$

$$RO_2 \cdot + \overset{|}{\underset{|}{C}} = \overset{|}{\underset{|}{C}} \rightleftharpoons RO_2 C - C \cdot \tag{139}$$

$\Delta S^{\ddagger}$ is negative. For H-atom transfers involving polyatomic $RO_2 \cdot$, $\log A$ should be no lower than 8—9 (l mole^{-1} s^{-1}) for alkanes, and no larger than 7—8 (l mole^{-1} s^{-1}) for open-chain allylic or benzylic systems where greater stiffening in the transition state for these latter reactants will increase $\Delta S^{\ddagger}$, and lower $\log A$ by about one log unit [13]. Since there is good experimental evidence that $\log A \geqslant 9$ (l mole^{-1} s^{-1}) for *sec*-H-atom transfer from alkanes to $RO_2 \cdot$ [39,79,80], we have assigned all values of $\log A$ from this benchmark because these H-atom transition states should have the largest degree of bond breaking of any of the common H-atom transfers owing to the high activation energies. Therefore these reactions should show the largest effects of resonance on $\Delta S^{\ddagger}$.

Table 5 summarizes the assigned values of $\log A$ for H-atom transfers

TABLE 5

Assigned value of $\log A$ for H-atom transfer to alkylperoxy radicals

Reactant	$\log A$ (l mole^{-1} s^{-1}) [a]
Alkyl primary	8.8
sec	9.0
tert	9.2
Allyl primary	7.8
sec	8.0
tert	8.2
Benzyl primary	7.8
sec	8.0
tert	8.2
Cycloalkyl sec	9.2
tert	9.6
Cycloallyl sec	8.7
tert	9.1
Cyclobenzyl sec	8.7
tert	9.1
Heterosubstituted [b]	8.5

[a] Values ± 0.5.
[b] Includes ethers, alcohols, sulfur analogs, amines, and carbonyls.

References pp. 83—87

with an additional refinement (perhaps unjustified) that, for reactions of secondary (*sec*) and tertiary (*tert*) H-atom transfers, values of log A are larger by 0.2 and 0.4, respectively, than for primary H-atom. Several kinds of H-atom transfers exhibit this structural effect in gas phase reactions [81] and some support for these differences also comes from reactions of

TABLE 6

Summary of absolute rate coefficients k for $RO_2 \cdot + RH$, k' for $t\text{-}BuO_2 \cdot + RH$ and rate parameters at $30°C$ [a]

Substrate and position	k $(l \, mole^{-1} \, s^{-1})$	k' $(l \, mole^{-1} \, s^{-1})$	$\log A'$ [b,d] $(l \, mole^{-1} \, s^{-1})$	E' [c,d] $(kcal \, mole^{-1})$
Alkyl				
primary			8.8	
sec		0.00027	9.0 (9.4)	17.40 (17.0)
tert	0.0048	0.0048	9.2	15.95 (15.5)
cyclo (*sec*)		0.00087	9.0 (8.8)	16.70 (16.3)
Allyl				
primary	0.14			
sec	0.50	0.084	8.0	12.57
tert	1.2			
cyclo	1.6	0.80	8.7	12.18
Benzyl				
primary	0.08	0.012	7.8 (6.1)	13.46 (11.0)
sec	0.05	0.10	8.0	12.46
tert	0.18	0.16	8.2	12.46
cyclo	1.6	0.50	8.7 (8.6)	12.46 (12.3)
Acetylenic	0.7			
Dienes				
1,4	7.0	0.23	7.3	10.99
1,3	31.0			
cyclo 1,3	55	1.4	7.8	10.60
cyclo 1,4	370	20	8.3 (8.7)	9.69 (11.3)
Alcohols				
sec		0.009	9.0	15.29
cyclo	0.036			
benzyl	2.4	0.065	7.8	12.44
Ethers				
sec	0.3	0.016	8.8	14.67
tert	0.02	0.02	8.8	14.54
benzyl	5.8	0.55	7.8	11.16
cyclo *sec* 5 ring	1.1	0.085	8.5	13.25
cyclo *tert* 5 ring	2.4	0.4	8.5	12.32
cyclo 5 ring	0.14	0.006	8.5	14.85

[a] Data from ref. 43; updated for alkanes from ref. 39.

[b] Assigned; see text and Table 5.

[c] Values in parentheses are experimental values from ref. 39 and 80.

[d] Calculated from k' and log A'.

the methyl radical in the liquid phase [82]. Recent data of Howard et al. [39,80] on *sec-* and *tert-*CH H-atom transfers to $RO_2\cdot$ also point in this direction. Experimental $\log A$ values for cycloalkanes range from 8.8 to 10.4 [39] depending on whether the ring is C_5 or C_6.

Intermediate values of $\log A$ were assigned to reactions involving cyclic allyl and benzyl systems and to heteroatom-substituted systems on two bases: for cyclic allylic systems, changes in $\Delta S^{\ddagger}$ should be smaller than for open chain systems since resonance stabilization cannot additionally stiffen these structures very much. In heteroatom systems where resonance effects involving p-electrons appear to be small as indicated by only small changes in CH bond strengths on substituting oxygen or nitrogen for carbon [83], we have decreased $\log A$ to 8.5 to reflect a lessened degree of resonance interaction but have neglected differences between primary, *sec*, and *tert* C—H bonds. We believe this procedure, with a probable error of ±0.5 log unit, is at least as reliable as most rate measurements over limited temperature spans and, for some radical systems such as peroxy or alkoxy, one of the most reliable methods of obtaining rate parameters.

Table 6 summarizes the absolute rate coefficients and rate parameters for representative types of organic structures. Rate coefficients for the reaction of the H-atom donor with its own $RO_2\cdot$ radical (k) and with t-$BuO_2\cdot$ (k') are reported at 30°C. Values of $\log A'$ and E' are reported for H-atom transfer to the t-$BuO_2\cdot$ radical only. E' was calculated from the value of k' and the assigned value of $\log A$ with sufficient accuracy to recalculate k' at 30°C. The probable error in E calculated in this way is about ±1 kcal mole^{-1}, or a factor of six in rate at 30°C. Where reliable experimental evidence indicates that some other value of $\log A$ is applicable, we have also listed the experimental value. However, in no case do the estimated and experimental values differ by more than one log unit and larger differences, which are often reported, should be viewed with considerable skepticism.

(c) Structure—reactivity relationships for H-atoms transfers to $RO_2\cdot$

A major goal for chemists is understanding how the structure of molecules affects their chemical reactivity. We can categorize structure—reactivity relationships for reactions of $RO_2\cdot$ radicals as (i) thermochemical, (ii) steric, and (iii) electronic, recognizing that this division is somewhat arbitrary, but still a useful point of departure for discussion.

The dissociation energy of the first C—H bond in methane is 33 kcal mole^{-1} greater than that for the weakest C—H bond in the highly reactive 1,4-cyclohexadiene [13]; this is equivalent to a rate factor of 10^{24} at 25°C. The usual correlation between activation energy and bond dissociation energy or reaction enthalpy is found in the form of the Polyani equation

$$E_p = C + \alpha \Delta H$$

where C is a constant, $0 < \alpha < 1$ and ΔH is the exothermicity of the reaction. Early efforts by Bolland [84] to apply this relation to oxidation was less than satisfactory owing to a series of unjustified assumptions concerning activation energies for individual chain steps. Nonetheless, he did show some correlation with $\alpha = 0.4$.

The seminal paper in this area is that of Korcek et al. [79] in 1972. These investigators measured values of k_p for the reactions of over fifty hydrocarbons and heteroatom-substituted compounds with secondary and tertiary $RO_2\cdot$ radicals and showed that the rate coefficients for H-atom transfer to tertiary $RO_2\cdot$ fit the relationships

$$\log k_p^{30°}(t\text{-}RO_2\cdot) = 15.4 - 0.2(D[R\text{—}H])/C\text{—}H \text{ bond} \tag{140}$$

$$\log k_p^{30°}(sec\text{-}RO_2\cdot) = 16.4 - 0.2(D[R\text{—}H])/C\text{—}H \text{ bond} \tag{141}$$

These relations indicate that secondary peroxy radicals are generally about two to ten times as reactive as tertiary peroxy radicals.

A few experimental data are available for absolute rate parameters (A_p and E_p) for elementary oxidation reactions; measurements have been made for propagation rate coefficients for $t\text{-}BuO_2\cdot$ + RH over a 50° temperature span using seven hydrocarbons [79,80]. Activation energies correlated well with bond strengths for the reactive C—H bond using the relation

$$E_p = 0.55(D[R\text{—}H] - 62.5) \tag{142}$$

Another form of this equation

$$E_p = 14 + 0.55\Delta H_r \tag{143}$$

can be written on the basis of the assumption that $D[O\text{—}H]$ for $t\text{-}BuO_2H$ is 88 kcal mole^{-1}.

These equations provide a way of calculating values of k_p for $t\text{-}BuO_2\cdot$ from $D[R\text{—}H]$ which should provide estimates of propagation rate coefficients at any reasonable temperature to within a power of ten for most hydrocarbons. Marked variations are found in the reactions of peroxy radicals attached to electronegative centers, such as acylperoxy, α-acyloxyalkylperoxy and α-chloroalkylperoxy, where the k_p's are as much as four orders of magnitude as large as k_p for $t\text{-}BuO_2\cdot$ [85].

Table 7 summarizes values of $D[R\text{—}H]$ and E_p calculated from eqn. (142) for selected alkanes and olefins. For about half the examples, measured values of E_p are also given in parentheses; the average deviation is ±1.1 kcal mole^{-1} or a factor of six in rate at 25°C.

These semi-empirical procedures for estimating rate coefficients and parameters, if used carefully with due regard for their limitations, are a valuable alternative method to transition-state procedures described above. The latter methods have a firm theoretical foundation but the former methods rest on solid experimental bases. Used together, the two

TABLE 7

Carbon—hydrogen bond strengths for selected hydrocarbons and activation energies for H-atom transfer to the t-BuO$_2$ · radical

RH	$D[R—H]$ (kcal mole^{-1})	E_a [a] (kcal mole^{-1})	Ref. [b]
Pentane CH$_3$	98	19.5 (18.7)	41
2-Methylpentane CH$_2$(4)	95	17.9 (17.5)	39
n-Butane CH$_2$	95	17.9 (16.4)	35
Cyclohexane	95	17.9 (19.5)	39
Methylcyclohexane CH	92	16.2 (16.3)	39
i-Butane CH	92	16.2 (15.5)	29
Toluene CH$_3$	85	12.4	
p-Xylene CH$_3$	85	12.4 (11.0)	80
2,3-Dimethylbutene-2	84	11.8	
Cyclohexene allylic H	83	11.3	
Tetralin α-CH$_2$	82	10.7 (12.3)	80
Ethylbenzene CH$_2$	82	10.7	
Diphenylmethane CH$_2$	81	10.2	
Cumene CH	79	9.1 (10.0)	34
9,10-Dihydroanthracene CH$_2$	71	4.7	
Cyclo-1,4-hexadiene CH$_2$	71	4.7 (11.3)	80

[a] Values are calculated from eqn. (142); the values in parentheses are measured.
[b] Reference to measured values.

procedures can provide a valuable check on unsuspected experimental complications or unwarranted theoretical assumptions. For the most part, values of E_p calculated by the two methods in Tables 6 and 7 agree within ±1 kcal mole^{-1}. In a few cases where the difference is larger, we prefer first the measured values in Tables 6 and 7, second the TS values in Table 6, and last the thermochemical values in Table 7.

Correlations of bond strengths with E_p or k_p are necessarily restricted to those cases where transition states for the reactions are very similar and probably have only small contributions from extreme charge separated forms such as

$$RO_2 \cdot H—R \leftrightarrow RO_2^-H \cdot^+ R \leftrightarrow RO_2^- H^+ \cdot R \qquad (144)$$

For transition states with significant charge separation, we may expect to find marked departures from simple relations between bond strengths and rate coefficients or parameters. Polar effects in H-atom transfer to RO$_2$· radicals were first evaluated quantitatively by Russell and Williamson [8] for a group of p-substituted arylalkanes at 60°C. Propagation rate coefficients for substituted styrenes and cumenes follow a $\sigma\rho$ relationship with $\rho \sim -0.3$ to -0.4, respectively. These are small polar effects representing only a factor of about 2 in rate at 60°C in going from a p-t-butyl to a p-nitro substituent. Compared with other less electrophilic radicals such as bromine or trichloromethyl [86], the effect is surprisingly small and suggests that the transition states for these other radical processes are quite different and that, for the case of simple $RO_2\cdot$, charge separation in the transition state is not very large.

However, Howard and Korchek [85] have shown that polar effects are much larger in peroxy radicals than in H-atom donors. Electron-withdrawing substituents on the peroxy radical, such as chloro, acetate and carbonyl, give values of k_p that are several hundred to several thousand times as large as k_p for t-$BuO_2\cdot$ with these same H-atom donors. For example, k_p for $PhCO(O_2)\cdot$ + $PhCHO$ is 33,000 at $30°C$ compared with a value of 0.85 for t-$BuO_2\cdot$ + $PhCHO$ [86]. Similarly, but not as dramatic, the $HO_2\cdot$ radical has been found to be more reactive than various $RO_2\cdot$ radicals with the same hydrocarbons [87].

Also of direct concern in oxidation of hydrocarbons are the special rate effects found for H-atom transfer to $RO_2\cdot$ from heteroatoms in molecules such as $PhOH$, RO_2H, or $PhNH_2$. Based on bond strengths, these reactions have exceptionally low A-factors and activation energies [88]. This effect is attributed to the formation of a hydrogen-bonded radical complex [89]. Typical rate parameters for such reactions are in the region of $\log A = 4$ and $E_p = 1$ kcal mole^{-1}; from the bond dissociation energy [88] and eqn. (142), we calculate $E_p = 15$. Thus one must exercise great care in applying any of the estimation techniques to reactions of substituted $RO_2\cdot$ involving H-atom transfer from any donor, and to $RO_2\cdot$ reactions involving H-atom transfer from —OH, —NH or —SH.

Steric effects in oxidation are found primarily in the H-atom donor as, for example, isobutane versus 2,4-dimethylpentane where both co-oxidation and added hydroperoxide measurements indicate isobutane to be 1.4 times as reactive toward either $RO_2\cdot$ radical [38,39]. For a series of branched alkanes, the value of k_p was more sensitive to changes in steric bulk on the adjacent carbon than on the same carbon but all values were within a factor three [39]. Increasing size in t-$RO_2\cdot$ radicals has little effect on the value of k_p for H-atom transfer from primary, secondary or tertiary sites [69]. However, k_p for a given hydrocarbon does increase by a factor of about 5—10 as $RO_2\cdot$ is changed from a tertiary to secondary or primary RO_2 [78]. Moreover, in the oxidation of aromatic compounds, the ratio of k_p's for H-atom transfer to parent $RO_2\cdot$ and to t-$BuO_2\cdot$ correlate with *meta* substituent constants which suggests that this difference in reactivity is due mainly to polar effects [86].

To summarize: the value of k_p for the reaction

$$RO_2\cdot + R'H \rightarrow RO_2H + R'\cdot$$

is affected mainly by polar effects in $RO_2\cdot$ (which give rise to rate variations as large as 30,000), by the C—H bond dissociation energy of the donor which causes rate variations as high as 10^{10}, and by a change in donor from C—H to O—H or N—H which, for the same bond strength,

TABLE 8

Summary of thermochemical, steric and electronic effects for H-atom transfer to $RO_2\cdot$ radicals ar 30°C

RH	k_p (per reactive hydrogen) (l mole^{-1} s^{-1})						
	$t\text{-BuO}_2\cdot$	$sec\text{-BuO}_2\cdot$	prim. $\text{BuO}_2\cdot$	$RO_2\cdot$ [a]	$Me_3CCH_2CMe_2O_2\cdot$	$CH_3CPhCH_2CH_3$—$O_2\cdot$	$C_6H_5C(CH_3)_2O_2\cdot$
Effect of the structure of $RO_2\cdot$:							
Toluene	0.05	0.1 [b]	0.1 [b]	0.08 [b]			
p-Methyltoluene	0.015			0.14 [b]			
Cumene (isopropyl-benzene)	0.22	0.4 [b]	0.45 [b]	0.18 [b]			
Effect of the bulk of $t\text{-}RO_2\cdot$ [69]							
Toluene	0.012				0.013	0.014	
Effect of the polarity of $RO_2\cdot$ [85]							
Toluene				0.11			
Benzaldehyde				3.4×10^4			
Benzyl alcohol				2.4			
Benzyl chloride				1.5			
Benzyl acetate				2.3			
Effect of the polarity of RH [86]							
Cumene-α-d_1							0.2
p-Methoxycumene							0.33
p-Isopropylcumene							0.24
Cumene							0.18
p-Chlorocumene							0.15
p-Cyanocumene							0.13
p-Nitrocumene							0.12

[a] $RO_2\cdot$ is the peroxy radical derived from the H-atom donor. [b] Ref. 78.

References pp. 83—87

gives a rate variation of about 1000. Table 8 provides data for k_p for substituted toluenes toward several $RO_2 \cdot$ radicals and illustrates the range of steric and polar effects to be expected in these reactions.

4.3.2 Addition of $RO_2 \cdot$ radicals to carbon double bonds: formation of di- and polyperoxides

Rate coefficients and rate parameters for the addition of $RO_2 \cdot$ to double bonds (k_a) are much more limited in number than for H-atom transfer (k_p). Howard [90] lists about twenty values of k_a for the addition of t-$BuO_2 \cdot$ to eleven activated vinyl compounds at $30°C$. Van Sickle et al. [18] in a careful study of a variety of unactivated alicyclic [18a] and cyclic olefins [18b], measured $k_a/(2k_t)^{1/2}$ at $60—110°C$ for nineteen olefins, separating the addition from abstraction mechanisms by product analyses. For these olefins, addition accounted for $2—100\%$ of the oxygen consumed.

From the measurements of the rates of addition of t-$BuO_2 \cdot$ and other $RO_2 \cdot$ radicals to activated olefins, Howard [90] concludes that (i) rate coefficients for addition to these olefins correlate well with the stabilization energies of the adduct radicals, (ii) the reactivity of the olefins toward addition by t-$BuO_2 \cdot$ decreases as the electron-withdrawing capacity of the substituent α to the forming radical increases, and (iii) the reactivity of different $RO_2 \cdot$ radicals towards addition increases with increasing electron-withdrawing capacity of R. All of these effects parallel those found for H-atom transfer (Table 8). For these activated olefins, addition of parent $RO_2 \cdot$ was as much as 1258 times as fast as for addition of t-$BuO_2 \cdot$ to unreactive vinyl acetate and as little as 3.4 times as fast in the case of very reactive α-methylstyrene.

Van Sickle et al. [18] developed some generalizations for the reactivity of cyclic olefins based on the composite term $k_a/(2k_t)^{1/2}$ which suggested that (i) rates of addition for most olefins are closely similar and (ii) rates of abstraction vary much more widely among cyclic olefins. Thus the decrease in rate of abstraction is mostly responsible for the larger proportion of addition products found with cyclooctene than with smaller ring alkanes.

In Table 9, we have used $k_a/(2k_t)^{1/2}$ for the oxidation of simple olefins [18] to calculate the values of k_a, the rate coefficient for the addition of $RO_2 \cdot$ to the olefin; values of $2k_t$ were assumed to be similar to those for olefins oxidized by Howard [90], corrected to $60—90°C$ by increasing Howard's values by a factor of three. This procedure introduces errors in addition to those inherent in the original measurement and values of k_a calculated in this way probably are not accurate to better than a factor of ten. The values of k_a in Table 9 calculated in this way are remarkably constant, so much so that one suspects that, even allowing for substantial errors, several factors work to compensate for differences in radical stabilization and steric requirements.

TABLE 9

Calculated rate coefficients for addition of $RO_2 \cdot$ to unactivated olefins at 60—90°C

Olefin	Temp. (°C)	$[k_a/(2k_t)^{1/2}] \times 10^4$ [a]	$2k_t \times 10^{-6}$ [b]	k_a	% Addition
Ethylene	110	3.7	600 [c]	4.5	100
Propene	110	2.3	500 [d]	5.1	50
Butene-1	70	2.8	500 [d]	6.3	26
trans-Butene-2	70	5.5	12 [e]	1.9	62
Isobutene	80	8.5	20 [f]	3.8	81
3-Me-Butene-1	70	1.0	1.0 [g]	0.1	6
Trimethylethylene	60	18	1.0 [g]	1.8	52
2,3-Dimethylbutene-2	50	42	1.0	4.2	42
Hexene-1	90	1.8	500 [d]	4.0	33
t-Butylethylene	90	3.7	600 [c]	9.1	100
Cyclopentene	50	13	14 [h]	4.8	11.2
Cyclohexene	60	7.83	14	2.6	4.4
Cyclooctene	70	9.2	20 [i]	6.4	71

[a] From ref. 18; in $(1 \text{ mole}^{-1} \text{ s}^{-1})^{1/2}$.

[b] From ref. 90, increased by 3× for closest corresponding olefin (footnotes c—i); in $1 \text{ mole}^{-1} \text{ s}^{-1}$.

[c] $RO_2CH_2CH_2O_2 \cdot$. The same as for toluene.

[d] The same as for octene-1.

[e] Same as for heptene-3.

[f] Cross-termination between $RO_2CH_2\overset{\displaystyle |}{\underset{\displaystyle O_2 \cdot}{C}Me_2}$ and $CH_2=C(CH_3)CH_2O_2 \cdot$.

[g] Same as for 2,3-dimethylbutene-2.

[h] Same as for cyclohexene.

[i] Cross-termination between $RO_2\overset{\diagdown}{C}H—\overset{\diagup}{C}HO_2 \cdot$ and $—CH=CH—\overset{\displaystyle |}{C}HO_2 \cdot$.

Activation parameters (A_a and E_a) for the addition of $RO_2 \cdot$ radicals to olefins have not been measured except in the case of styrene [91] and α-methylstyrene [92]. Therefore, we have calculated the value of E_a for the addition of $RO_2 \cdot$ radicals, using the same procedure as described previously for H-atom transfer (Sect. 4.3.1). First we assign a value of A_a and, from that value and k_a, calculate the value of E_a with sufficient accuracy to recalculate k_a at the temperature cited.

The correct assignment of values of A_a is the main problem. In this regard it is instructive to examine the results from studies of alkyl radical addition to olefins in the gas phase [93]. Although data are limited, addition of $CH_3 \cdot$, $Et \cdot$, and $i\text{-}Pr \cdot$ give progressively smaller values of A_a in adding to ethylene; for both $Me \cdot$ and $Et \cdot$ radicals values of A_a decrease as steric bulk around the olefinic center increases. Electronic factors are reflected both in E_a and A_a but stabilization of the adduct radical is reflected mostly in E_a. If, to a first approximation, we can consider $RO_2 \cdot$ similar to an $Et \cdot$, the steric bulk of R can be disregarded and we may then

TABLE 10

A-Factors assigned for addition of $RO_2\cdot$ to olefins

Olefin type	$\log(A_a/1\,\text{mole}^{-1}\,\text{s}^{-1})$ [a]
$CH_2=CH_2$	8.5
$RCH=CH_2$	8.2
$R-CH=CH-R$	8.0
$R_2C=CH_2$	8.0
$R_2C=CHR$	7.7
$R_2C=CR_2$	7.5
1,3-Diene	7.5
Vinylbenzene	7.7

[a] Cyclic olefins have values of $\log A_a$ 0.3 units higher.

assign values of A solely on the basis of steric bulk in the olefin. The paucity of reliable experimental data for $RO_2\cdot$ additions to simple olefins is a major reason why we use values similar to those measured for addition of alkyl radicals in the gas phase [93] rather than values assigned for H-atom transfer to $RO_2\cdot$ (Table 5) despite the fact that both reactions have somewhat similar transition states. However, the effect of the structure of the olefin on the value of $\log A$ is similar in the two cases with lower values of $\log A$ found for more conjugated systems. But, contrary to effects of structure noted for H-atom transfer, increasing bulk in the olefin leads to a decrease in the value of $\log A$. Thus it becomes clearer why the rate of addition of $RO_2\cdot$ to bulky olefins is not much different from addition to terminal olefins: the lower A-factor is compensated by a lower E_a reflecting greater electron availability in the transition state from α and β substituents, viz.

$$\overset{\oplus}{\underset{/}{\text{C}}}\overset{}{\underset{\backslash}{-\text{C}}}\quad \overset{\ominus}{:}O_2R \tag{145}$$

Van Sickle et al. reached the same conclusion from correlations between $k_a/(2k_t)^{1/2}$ and excitation energy [18b,94].

Addition of $RO_2\cdot$ to a cyclic olefin leads to a gain of fewer rotational modes than addition to acyclic olefins. Therefore we have assigned values of $\log A$ 0.3 units higher for cyclic olefins. Table 10 summarizes assignments of $\log A_a$ for the addition of $RO_2\cdot$ and Table 11 summarizes rate parameters for the addition of $RO_2\cdot$ and t-$BuO_2\cdot$ to a variety of olefins using values of k_a from Table 9 and from Howard [90], and values of $\log A_a$ from Table 10.

4.3.3 Intramolecular H-atom transfer to peroxy radicals

In preceding sections, we have noted that styrene gives some epoxide during oxidation (Sect. 3.2) and that 2,4-dimethylpentane gives mostly

TABLE 11

Rate coefficients and parameters for addition of $RO_2\cdot$ and t-$BuO_2\cdot$ to activated and non-activated olefins at 30—90°C.

Olefin	Temp. (°C)	k_a [a] (1 mole^{-1} s^{-1})	log A (1 mole^{-1} s^{-1})	E (kcal mol^{-1})
Butene-1	70	6.3	8.2	11.62
trans-Butene-2	70	1.9	8.0	12.12
Isobutene	80	3.8	8.0	11.98
3-Methyl-butene-2	70	0.1	8.2	14.44
Trimethylethylene	60	1.8	7.7	11.35
Tetramethylethylene	50	4.2	7.5	10.16
t-Butylethylene	90	9.1	8.2	12.03
Cyclopentene	50	4.8	8.3	11.26
Cyclohexene	60	2.6	8.3	12.02
Cyclooctene	70	6.4	8.3	11.77
Vinylcyclohexene	70	0.22	8.0	13.59
Vinyl acetate	30	0.002	8.2	15.1
Methyl acrylate	30	0.02 (0.6)	8.2	13.7
Methyl methacrylate	30	0.08 (1.0)	8.0	12.6
Acrylonitrile	30	0.01 (3.3)	8.2	14.1
Methyl acrylonitrile	30	0.094 (4.5)	8.0	12.51
Styrene	30	1.3 (41)	7.5 (7.67) [b]	10.24 (8.39) [b]
α-Methylstyrene	30	2.9 (10)	7.3 (6.82) [c]	9.48 (8.07) [c]

[a] Values for unactivated olefins from Table 9; values for vinyl esters and styrenes from ref. 90.
[b] Measured value from ref. 91.
[c] Measured value from ref. 92.

dihydroperoxide and, at low oxygen pressures, some oxetane (Sect. 3.5). In gas phase oxidations, high proportions of cyclic ethers have been reported from the oxidation of simple alkanes [42]. These products have their origin in intramolecular processes that usually become important only when the competing intermolecular processes, notably H-atom transfer and addition of oxygen, are slow. Reliable measurements of rate coefficients for these processes are few, although Benson [31], Mill et al. [38,41], and Fish [95,96] have considered the kinetic features of the reactions.

Estimation of rate parameters (A_{ip} and E_{ip}) for intramolecular H-atom transfer

$$\begin{array}{c}\diagdown\\C-(C)_n-C\\\diagup\;|\quad\quad|\diagdown\\O_2\cdot\quad\quad H\end{array}\xrightarrow{k_{ip}}\begin{array}{c}\diagup\overset{\diagdown(C)_n\diagup}{C}\diagdown\\\diagdown C\quad\quad C\diagup\\\diagup\;|\quad\quad\diagdown\\O-O\cdots H\end{array}\rightarrow\begin{array}{c}\diagdown\quad\quad\diagup\\C-(C)_n-C\\\diagup\;|\quad\quad\cdot\diagdown\\O_2H\end{array}\qquad(146)$$

must take into account the effect of transition-state ring size on both log A and E. In general, log A increases as ring size decreases owing to loss of fewer rotational modes on going from ground state to transition state for smaller rings. However, since E includes a ring strain term, it should be larger in rings both smaller and larger than six atoms [13]. On this basis, we would expect that the internal H-atom transfer would be fastest in a six-center process and slower in other internal transfers with the ring size order $5 > 7 > 8$. In fact, the few experimental data on 5- and 7-center reactions using 2,3-dimethylbutane [38] and 2,5-dimethylhexane [37] show the opposite order. Therefore we have adopted a pragmatic approach to estimating rate parameters for these reactions based on the relative yields of dihydroperoxides formed in the oxidations of 2,3-dimethylbutane, 2,4-dimethylpentane and 2,5-dimethylhexane at 100—120°C [37,38]. To a first approximation, the ratio of yields of dihydroperoxide may be set equal to the ratio of rate coefficients (k_{ip}) for the corresponding 5-, 6- and 7-center intramolecular processes (see Sect. 3.5). Since we know the value of k_{ip} for 2,4-dimethylpentane [38], we can estimate from the ratio of hydroperoxides values of k_{ip} for other cyclic H-atom transfers. We also can estimate how A_{ip} varies with ring size [13] and therefore we can calculate E_{ip} directly. The difference between the E_{ip} and E_p for the same C—H bond by similar $RO_2\cdot$ radicals can be interpreted in terms of the strain energy E_s that must be added to E_p when internal H-atom transfer takes place. The data and calculated values of E_{ip} and E_s are summarized in Table 12.

The one citation for a n-alkane shows a very large value for E_s; possibly n-alkanes will not exhibit much difference in E_s between 5-, 6-, 7- or 8-center processes owing to lack of side-chain interactions; in fact, data cited below in connection with ring closures suggest that very little, if any,

TABLE 12

Intramolecular H-atom transfer in branched alkanes at 100°C

Carbon chain	Ring size	k_{ip} (s)	log(A/s)	E_{ip} [a] (kcal mole^{-1})	E_s [b] (kcal mole^{-1})
2	5	<0.2 [c]	12.5	>22.5	6.5 (6.5)
3	6	17.7 [d]	11.5	17.5	1.5 (0.6)
4	7	8	11.0	17	1.2 (6.5)
5	8	<1 [c]	11.0	>19	2.8 (10)
3	6 (n-alkane)	0.007 [e]	11.5	23.3	5.5 (0.6)

[a] $E_{ip} = 1.72(\log A - \log k) = (E_p + E_s)$ where $E_p = 16$ kcal mole^{-1} for intermolecular H-atom transfer from tertiary C—H bond (Table 6).

[b] $E_s = E_{ip}$ (calc.) $- E_p$; values in parentheses are the actual strain energies in the corresponding n-membered carbocycles (ref. 13).

[c] From ref. 37 based on the yield of hydroperoxide relative to 2,4-dimethylpentane.

[d] From ref. 38.

[e] From ref. 41.

ring strain shows up in the activation energy term even for closure to 3- and 4-membered rings.

4.3.4 Ring closures of peroxy radicals

Significant amounts of epoxides, oxetanes and tetrahydrofurans can be formed during oxidation of olefins and alkanes through the intermediate alkylperoxyalkyl or hydroperoxyalkyl radicals formed by addition of $RO_2\cdot$ or by internal H-atom transfer followed by C—O ring closure [96]

$$RO_2\cdot + \ \overset{\diagdown}{\underset{\diagup}{C}}{=}\overset{\diagup}{\underset{\diagdown}{C}} \ \rightarrow ROO\overset{\diagdown}{\underset{\diagup}{C}}{-}\overset{\diagup}{\underset{\diagdown}{C}}{\cdot} \ \rightarrow \ \overset{\diagdown}{\underset{\diagup}{C}}\overset{}{\underset{O}{-}}\overset{\diagup}{\underset{\diagdown}{C}} + RO\cdot \qquad (147)$$

$$\overset{\diagdown}{\underset{\diagup}{C}}\overset{(C)_n}{\underset{OOH}{}}\overset{\diagup}{\underset{\diagdown}{C}} \rightarrow \overset{\diagdown}{\underset{\diagup}{C}}\overset{(C)_n}{\underset{\overset{O}{\underset{OH}{}}}{}}\overset{\diagup}{\underset{\diagdown}{C}} \rightarrow \overset{\diagdown}{\underset{\diagup}{C}}\overset{(C)_n}{\underset{O}{}}\overset{\diagup}{\underset{\diagdown}{C}} + HO\cdot \qquad (148)$$

Unlike H-atom transfer, C—O ring closures are highly exothermic processes because of stronger C—O bonds formed from cleavage of weaker O—O bonds, less the ring strain of the ether. For the simplest cases of epoxides or oxetanes, in which ring strain is 26—28 kcal mole^{-1} [13]

$$RO_2CH_2(CH_2)_nCH_2\cdot \rightarrow RO\cdot + CH_2(CH_2)_nCH_2O \qquad (149)$$

$\Delta H_r \sim -23$ kcal mole^{-1}. Values for ΔH_r for 5- and 6-center rings in which ring strain is only ~ 6 and ~ 1 kcal mole^{-1}, respectively, should be much larger. Addition of $RO_2\cdot$ to olefins to give epoxides is commonly observed in liquid phase oxidations at 1 atm O_2 and 40—150°C. Twigg [97] was the first to suggest that epoxides arose by ring closure of the $RO_2\cdot$—olefin adduct [reaction (147)].

The kinetics of these reactions have been examined solely in the context of the competition between ring closure [reaction (150)] and addition of oxygen to the peroxyalkyl radical [reaction (151)]

$$ROO\overset{\diagdown}{\underset{\diagup}{C}}(\overset{|}{\underset{|}{C}})_n\overset{\cdot}{\underset{|}{C}}\overset{/\ k_r}{\underset{\diagdown k_O}{\diagup}} \quad \overset{\diagdown}{\underset{\diagup}{C}}\overset{(C)_n}{\underset{O}{}}\overset{\diagup}{\underset{\diagdown}{C}} + RO\cdot \qquad (150)$$

$$O_2\diagdown \qquad ROO\overset{|}{\underset{|}{C}}(\overset{|}{\underset{|}{C}})_n\overset{|}{\underset{|}{C}}O_2\cdot \qquad (151)$$

Reaction (151) leads to a mixture of alkylperoxyalkylhydroperoxide and polyperoxide depending on the relative ease of H-atom transfer or addition to the parent olefin.

TABLE 13
Ring closure of $RO_2 \cdot$ to cyclic ethers

Olefin or alkane	Temp. (°C)	k_r/k_o [a]	k_r (s^{-1})	log(A_r s^{-1})	E_r [b] (kcal mole^{-1})	Ref.
Epoxides						
Styrene	50	1.9×10^{-5}	7.6×10^3	12.5	12.81	91
α-Methylstyrene	50	1.7×10^{-4}	6.8×10^4	12.5	11.39	92
Cyclopentene	50	4.5×10^{-3}	4.5×10^6	12.8	9.13	18
Cyloheptene	60	2.9×10^{-2}	2.9×10^7	12.8	8.18	18
Cyclooctene	70	4.5×10^{-1}	4.5×10^8	12.8	6.54	18
2-Butene	90	9.1×10^{-3}	9.1×10^6	12.1	8.11	18
2-Methyl-1-pentene	70	9.1×10^{-3}	9.1×10^6	12.1	8.11	18
Trimethylethylene	60	6.7×10^{-3}	6.7×10^6	12.1	8.08	18
Oxetanes and furans						
2,4-Dimethylpentane	100	8.2×10^{-5}	5.4×10^5	11.5	8.00	38
Butadiene	50	1.0×10^{-6}	4.0×10^2	11.0	>12.40	72

[a] The value for k_o was assigned as 10^9 l mole^{-1} s^{-1} in all cases except for styrene, methylstyrene and butadiene where $10^{8.6}$ was used.
[b] $E_r = (\log A - \log k) 4.6T \times 10^{-3}$.

Van Sickle et al. [18,94] were able to separate addition from H-atom transfer and ring closure from O_2 addition for a series of simple olefins at 50—90°C. Mayo and Miller [22,23] had earlier examined the effect of O_2 pressure on the formation of styrene oxide. Their data on k_o/k_r together with a value for the ring closure in the 2-hydroperoxy-2,4-dimethyl-4-pentyl radical [38] and in the polyperoxybutadienyl radical [72] comprise most of the reliable data base from which to estimate absolute rate coefficients and parameters.

Estimation of k_r is done readily by assuming that addition of O_2 to non-stabilized carbon radicals is diffusion-controlled, i.e. $k_o \sim 10^9$ l mole^{-1} s^{-1}, and for resonance-stabilized radicals $k_o \sim 10^{8.6}$ l mole^{-1} s^{-1}. These assumptions seem justified on the grounds that for most unhindered complex radicals, rates of mutual interaction generally are close to the diffusion limit (see Sect. 4.2).

Table 13 summarizes the values for k_r together with values of log A_r and E_r calculated in the usual way. In this case, we can assign values for A_r based on the thermochemical analysis of the cyclic transition states for 3-, 4- and 5-centered ring closures [13]. These estimates take into account losses and gains in rotational entropy on going from the ground state to the transition state, by assigning 2—3 e u for each locked rotation. For example, a tight 3-center transition state loses, at most, 2 rotations

(152)

when $\Delta S^{\ddagger} = -5$ eu and since

$$A = \frac{ekT}{h} \, 10^{\Delta S^{\ddagger}/R} \tag{153}$$

$$A = 10^{13.2} \, 10^{-1.1} = 10^{12.1} \tag{154}$$

Cyclic and conjugated olefins lose fewer rotational modes during ring closure, as reflected in larger values of $\log A$. However, Table 13 shows that ring closure to epoxide by the peroxybenzyl radical is appreciably slower than by peroxyalkyl radicals, despite a more favorable A-factor. A high value of E may be viewed as a penalty paid for loss of resonance stabilization, which amounts to 3—4 kcal mole^{-1}. Among alkylperoxy radicals, 6- and 7-membered cyclic radicals close most rapidly owing both to a more favorable A-factor and a lower value of E.

4.3.5 H-Atom transfer to alkoxy radicals

(a) General considerations

Alkoxy radicals (RO·) are important chain carriers in many oxidation reactions especially at higher temperatures [36,98]. They are formed in several reactions including self-reaction of tertiary RO$_2$· (Sect. 4.4)

$$2 \, RO_2 \cdot \rightarrow 2 \, RO \cdot + O_2 \tag{29}$$

ring closures of β-peroxyalkyl radicals (Sect. 4.3.2)

$$RO_2 \overset{|}{\underset{|}{C}} \overset{|}{\underset{|}{C}} \cdot \rightarrow RO \cdot + \overset{\backslash}{\underset{/}{C}} \overset{/}{\underset{\backslash}{\underset{O}{\longrightarrow}}} \overset{/}{C} \tag{147}$$

and homolysis of hydroperoxides, induced thermally or by metal ions [99]

$$RO_2 H \overset{\Delta}{\rightarrow} RO \cdot + HO \cdot \tag{155}$$

$$RO_2 H + M^{n+1} \rightarrow RO \cdot + HOM^n \tag{156}$$

Much of the chemistry of RO· in the liquid phase is associated with the competition between H-atom transfer to give alcohol and β-cleavage to give carbonyl and a new alkyl radical, viz.

$$R' \overset{|}{\underset{|}{C}} O \cdot + RH \overset{k_a}{\rightarrow} R' \overset{|}{\underset{|}{C}} OH + R \cdot \tag{157}$$

$$R' \overset{|}{\underset{|}{C}} O \cdot \overset{k_d}{\rightarrow} R' \cdot + \overset{\backslash}{\underset{/}{C}} = O \tag{158}$$

Addition of RO· to double bonds is not observed with most olefins despite the fact that the reaction is exothermic by about 20 kcal mole^{-1}.

The following sections review the thermochemistry of RO· reactions, absolute rate coefficients for H-atom transfer, β-cleavage and intramolecular reactions.

(b) Thermochemistry of RO· reactions

H-Atom transfer to RO· results in the formation of a 104 kcal mole^{-1} O—H bond with the result that all H-atom transfers from organic H-atom donors are exothermic to at least 6 kcal mole^{-1} and as much as 32 kcal mole^{-1} (see Table 4). For most alkanes and olefins, H-atom transfer reactions are exothermic by 9—16 kcal mole^{-1} and activation energies are low.

Heats of formation of RO· $[\Delta H_f(\text{RO·})]$ have been calculated from the gas phase pyrolysis of RONO [100], by group additivity [101] and from the heats of formation of the corresponding alcohols [102] using the assumption that $\Delta H_f(\text{RO·}) = \Delta H_f(\text{ROH}) + 52$ [101]. On this basis, ΔH_f for RO· ranges from +4 kcal mole^{-1} for MeO· to —40.4 kcal mole^{-1} for t-BuCMe$_2$O·. Calculation of the heats of reaction (ΔH_r) for β-cleavage of RO·, viz.

$$R'-\overset{|}{\underset{|}{C}}-O\cdot \rightarrow R'\cdot + {>}C{=}O \qquad (158)$$

may be made in a straightforward manner

$$\Delta H_r = \Delta H_f(R\cdot) + \Delta H_f({>}C{=}O) - \Delta H_f(\text{RO·}) \qquad (159)$$

Values of ΔH_r for β-cleavage reactions of some RO· are given in Table 14.

TABLE 14
Estimated ΔH_r for β-cleavage of the RO· radical
All values in kcal mole^{-1}.

RO·	$\Delta H_f(\text{RO·})$	$\Delta H_f(R_1\cdot)$	$\Delta H_f(R_2R_3CO)$	ΔH_r
MeO	4.2 [a]	52 (H)	—28 (CH$_2$O)	19.8
EtO	—4.1 [a]	34 (Me)	—28 (CH$_2$O)	10
n-PrO	—9.9 [a]	26 (Et)	—40 (EtCHO)	—4.1
i-PrO	—12.5 [a]	34 (Me)	—40 (MeCHO)	6.5
sec-BuO	—17.0 [a]	34 (Me)	—46 (EtCHO)	5
sec-BuO	—17.0 [a]	26 (Et)	—40 (MeCHO)	3
t-BuO	—21.7 [a]	34 (Me)	—51.7 (Me$_2$CO)	4
t-AmO	—24.7 [b]	26 (Et)	—51.7 (Me$_2$CO)	—1
i-PrCMe$_2$O	—31.7 [b]	17.6 (i-Pr)	—51.7 (Me$_2$CO)	--2.4
t-BuCMe$_2$O	—40.4 [b]	6.7 (t-Bu)	—51.7 (Me$_2$CO)	—4.6
C$_6$H$_{11}$O	—18 [c]	—14(C$_6$H$_{11}$O)		4
C$_6$H$_5$CMe$_2$O	+17.0 [b]	34 (Me)	—22 (C$_6$H$_5$COMe)	—5
C$_6$H$_5$CH$_2$CMe$_2$O	+14.0 [b]	45 (C$_6$H$_5$CH$_2$)	—51.7 (Me$_2$CO)	—2
ClCH$_2$CMe$_2$O	—28[b]	22.4 (ClCH$_2$) [b]	—51.7 (Me$_2$CO)	—1.3

[a] From RONO (ref. 100).
[b] From group additivity (ref. 101).
[c] From ROH (refs. 101 and 102).

Most of these reactions are endothermic suggesting that, for many $RO\cdot$, the activation energy for cleavage (E_d) will be substantially greater than 5 kcal mole^{-1}, and often greater than E_a, the activation energy for H-atom transfer to $RO\cdot$. Thus more stable $RO\cdot$ will usually react by H-atom transfer [reaction (157)] rather than β-cleavage [reaction (158)].

(c) Structure—reactivity relationships

The electrophilic character of $RO\cdot$ is manifested in substituent effects on H-atom transfer from toluenes

$$XC_6H_4CH_3 + t\text{-}BuO\cdot \rightarrow XC_6H_4CH_2\cdot + t\text{-}BuOH \qquad (160)$$

where the rate coefficients for a series of X follows the relation [103]

$$\log(k \text{ l mole}^{-1} \text{ s}^{-1}) = -(0.32 - 0.39)\sigma^+ + 4.78 \qquad (161)$$

Unlike the case of $RO_2\cdot$, there are few data on the effect of the structure of $RO\cdot$ on the rate of H-atom transfer to $RO\cdot$ mainly because of the lack of absolute rate coefficients for H-atom transfer for a series of $RO\cdot$. For certain $RO\cdot$, relative reactivity data such as k_a/k_d are readily available, but are unreliable except for a series of closely related H-atom donors because values of both k_a and k_d may change on changing from one donor to another due to changes in both donor reactivity toward $RO\cdot$ and solvent effects on the β-cleavage of $RO\cdot$ [104]. Much of the relative reactivity data is competently summarized in the review by Gray et al. [105] in 1967 and to some extent by Howard [10] in his 1972 review. For both aliphatic and aromatic hydrocarbons, the range of reactivity for H-atom transfer to $RO\cdot$ is only about 75 at 135°C on going from unreactive primary (t-butylbenzene) to very reactive cyclic benzylic (tetralin) [106]. At 100°C, the range for primary to tertiary C—H in alkanes is about 50 [107]. Under the same conditions and toward the same alkanes, the range for $RO_2\cdot$ is nearly 1000 (Table 4). One important consequence of this low selectivity toward C—H bonds is that $RO\cdot$ chain carriers will create many more primary $R\cdot$ by H-atom transfer than will $RO_2\cdot$. Thus the importance

TABLE 15
Relative reactivity of $RO\cdot$ and $RO_2\cdot$ in H-atom transfer at 100°C

CH bond	$RO\cdot$ [a]	$RO_2\cdot$ [b]
Primary	1.0	1.0
Secondary	10	50
Tertiary	50	1000
Allylic	30	3000

[a] From data in ref. 107.
[b] From data in Table 4.

of RO· in an oxidation process may be evaluated by careful product analysis. Table 15 shows how RO· and RO_2· will contribute to propagation by H-atom transfer at tertiary, secondary and primary C—H bonds at 100°C, based on data in Table 4 and ref. 107.

(d) Absolute rate coefficients and parameters for H-atom transfer to RO·

Although several different RO· have been investigated in the gas phase [105] the bulk of kinetic investigations in solution involving RO· have been with t-BuO· [10,43]. Relative reactivities of organic compounds toward t-BuO· (k/k') may be measured in competitive experiments where two substrates (R_1H and R_2H) react with t-BuO· and the alkyl radicals formed then react with CCl_4 to form alkyl chlorides.

$$R_1· + CCl_4 \xrightarrow{k} R_1Cl + CCl_3·$$ (162)

$$R_2· + CCl_4 \xrightarrow{k'} R_2Cl + CCl_3·$$ (163)

The ratio of rate coefficients, k/k', may be determined indirectly by comparing ROH/ketone ratios on reaction with each substrate separately or by determining the relative yields of R_1Cl and R_2Cl or the consumption of reactants in competitive experiments.

Both competitive methods give fairly reliable relative rate coefficients in most cases. However, discrepancies between them have been found when t-butyl hypochlorite was used as the source of alkoxy radicals and when aralkanes (e.g. toluene) were the substrates because of the incursion of a chlorine atom chain, and relative reactivities to Cl· rather than to t-BuO· were determined. Absolute rate coefficients reported in this review do not include this suspect·data.

Recently, Scaiano and coworkers [108] reported a series of absolute rate coefficients for H-atom transfer to t-BuO· at 25°C. Their measurement technique utilized nanosecond laser flash photolysis and optical spectroscopy. These data supersede older relative data of Walling and Kurkov [109] and Zavitsas and Blank [110], who used the photo-initiated reactions of t-BuOCl and toluene, and which now appear to be too low by a factor of three.

The laser kinetic method depends on the fact that a large concentration of t-BuO· can be generated by photolysis in a few nanoseconds followed by reaction of t-BuO· with Ph_2CHOH, viz.

$$t\text{-BuOOBu-}t \xrightarrow{h\nu} 2\ t\text{-BuO·} \quad (6\ \text{ns})$$ (164)

$$t\text{-BuO·} + Ph_2CHOH \xrightarrow{k_A} t\text{-BuOH} + Ph_2\overset{·}{C}OH$$ (165)

$$t\text{-BuO·} \xrightarrow{\tau^{-1}} (\text{first-order loss})$$ (166)

TABLE 16

Summary of absolute rate coefficients and parameters for H-atom transfer to t-BuO·
at 40°C (per active hydrogen).

RH bond	$k_A \times 10^{-4}$ [a] (l mole^{-1} s^{-1})	log(A/l mole^{-1} s^{-1}) (± 0.5)	E_A [c] (kcal mole^{-1})
Alkane			
primary	0.84	9.0	7.25
sec	7.2	9.2	6.20
tert	37.8	9.4	5.46
cyclic-sec	6.0	9.2	6.32
Alkene			
primary	15.9	8.0	4.00
sec	72	8.2	3.35
tert	156	8.4	3.16
cyclic-sec	333	8.7	3.11
Phenylalkyl			
primary	6.0 [d]	8.0	4.61
sec	31.5	8.2	3.86
tert	61.5	8.4	3.73
cyclic-sec	195	8.7	3.46
	45	8.7	4.36
Diphenylmethane	42.3	8.5	4.11
Triphenylmethane	87	8.5	3.66
Ethers (α-CH) [e]			
primary	30	8.7	4.61
sec	93	8.7	3.91
cyclic sec 3-ring [f]	82.5	9.0	4.41
cyclic sec 4-ring	176	8.7	3.51
cyclic sec 5-ring	128	8.7	3.71
cyclic sec 6-ring	66.9	8.7	4.11
Alkyl-X (α-CH)			
chloro	4.2	9.0	6.26
cyano	0.54	9.0	7.53
acetoxy	2.4	9.0	6.61
Ketone (α-CH)			
primary	2.2 [g]	8.7	6.23
sec	1.8 [g]	8.5	6.07
cyclo	3.0 [g]	8.7	6.04

[a] Recalculated from relative reactivity data in ref. 43 compared with toluene.
[b] Assigned: see Sect. 4.2.
[c] Calculated from k and log A.
[d] Absolute value measured (ref. 108).
[e] Extrapolated from 273 K.
[f] Ring size.
[g] At 0°C.

On the time scale of the reaction, $Ph_2\dot{C}OH$ is stable and its rate of formation can be monitored by optical spectroscopy. Under these conditions

$$\ln\left\{\frac{[Ph_2\dot{C}OH]}{[Ph_2\dot{C}OH]_o - [Ph_2\dot{C}OH]_t}\right\} = (\tau^{-1} + k_A[Ph_2CHOH])t \qquad (167)$$

A plot of $\ln[A/(A - A_t)]$ versus t gives a slope $= (\tau^{-1} + k_A Ph_2[CHOH])$ from which k_A can be evaluated by using different initial concentrations of Ph_2CHOH. Once k_A for Ph_2CHOH is evaluated, k'_A for other compounds can be measured by photolyzing binary mixtures with Ph_2CHOH. Scaiano and coworkers [108] measured a value for toluene $k_A = 6 \times 10^4$ l mole^{-1} s^{-1} or a factor of about 3 larger (per H) than estimated earlier by Walling and Kurkov [109]. From this value of k_a together with other values for other H-atom donors relative to toluene, a series of absolute values of k_a for t-BuO$\cdot$ can be developed. Values of A for H-atom transfer have been assigned on the same basis as for H-atom transfer to $RO_2\cdot$ (Table 4) and thus provide the basis for the data shown in Table 16. Since we have no reason to expect that other t-RO$\cdot$ values should be very different in reactivity from t-BuO$\cdot$, the values in Table 16 should, to a first approximation, be applicable to all t-RO$\cdot$. Very likely *sec*- and prim-RO$\cdot$ will have k_a values that are larger by a factor of (at most) five owing to smaller steric requirements.

4.3.6 Absolute rate coefficients and parameters for C—C cleavage of alkoxy radicals (RO·)

Cleavage of C—C bonds in RO$\cdot$ (β-cleavage) by reaction (158) is the most important process for chain scission in oxidation reactions and is responsible for the great majority of lower molecular weight products found in oxidation reactions. β-Cleavage also exhibits one of the largest solvent effects known for a free radical reaction, proceeding faster in more polar solvents by factors as large as twenty on going from alkane to acetic acid [104].

Values for absolute rate coefficients for this reaction (k_d) can be calculated with fair reliability from the ratio k_a/k_d with an H-atom donor for which k_a is known. The most reliable values of the ratio k_a/k_d for t-BuO$\cdot$, *sec*-BuO$\cdot$ and, t-AmO$\cdot$ toward hydrocarbon donors are those measured by Allara et al. [107] at 50 and 100°C. Ratios of k_a/k_d for several other t-RO$\cdot$ estimated by Walling and Padwa [111] using CCl_4 solvent at 0, 40, and 70°C are very useful and, where comparison between the two sets of data is possible, the agreement is quite good (a factor of 2) considering the difference in solvents. For consistency, we have used the data of Walling and Padwa [111] at 40°C and their values of $E_d - E_a$ to calculate values of k_d, log A_d, and E_d. All of these values rest on the assumption that k_a per C—H bond for the reaction of t-BuO$\cdot$ and cyclo-

TABLE 17

Rate coefficients and parameters for β-scission of RO· at 40°C

RO·	k_d/k_a [a]	k_d [b] $\times 10^{-4}$	$\log A_d$ [c]	E_d [d]
t-BuO	0.021	0.27	15.5	17.4
$ClCH_2CMe_2O$	0.121	1.57	11.4	10.5
$PhCMe_2O$	0.477	6.20	12.4	11.0
$EtCMe_2O$	2.09 (1.55) [e]	27.2 (5.12) [f]	11.8	9.2
i-$PrCMe_2O$	76	988	12.3	7.6
t-$BuCMe_2O$	>300	>3900		
$PhCH_2CMe_2O$	1.98	25.7	10.7	7.6
$(CH_2)_4CMeO$	97 [e]	3200 [f]	13.0	7.0
$(CH_2)_5CMeO$	6.6 [e]	21.8 [f]	11.6	8.0

[a] In mole l^{-1}; measured toward cyclohexane in CCl_4 (ref. 111).
[b] In s^{-1}; calculated by assuming that $k_a = 1.3 \times 10^5$ l mole^{-1} s^{-1} at 40°C.
[c] In s^{-1}, calculated from $(\log k + E_a/4.6T \times 10^{-3})$.
[d] In kcal mole^{-1}; estimated from measured value of $E_d - E_a$ with the assumption that $E_a = 6.32$ kcal mole^{-1} (Table 16).
[e] Value at 0°C.
[f] Value calculated at 0°C using $k_a = 3.3 \times 10^4$ l mole^{-1} s^{-1}.

hexane is 6×10^4 l mole^{-1} s^{-1} at 40°C and that the same value holds for other t-RO·. The data are summarized in Table 17.

In general, there is a fair correlation between increasing stability of the carbon radical formed and increasing ease of β-cleavage [112]; loss of an ethyl group from the t-amyloxy radical is about four times faster than loss of ethyl from the less-hindered sec-butoxy radical [35,107], both at 100°C.

Despite the relatively good agreement between sets of data for liquid phase reactions of RO·, agreement with the best value for k_d estimated for gas phase reactions is poor. Recently, Baldwin et al. [113] have concluded that for t-BuO· in the gas phase

$$\log k_d = 15.2 - \frac{15.9}{4.6\,T \times 10^{-3}} \tag{168}$$

At 40°C, this relation predicts that $k_d = 1.2 \times 10^4$ s^{-1} compared with a value of 1.3×10^3 calculated from liquid phase data. Close scrutiny of the different sets of data reveals no obvious reason for the large discrepancy but the disagreement should serve as a cautionary note against the use of either liquid or gas phase data under conditions much different from those used in their original measurement.

4.3.7 Intramolecular H-atom transfer by RO·

RO· can transfer an H-atom intramolecularly in much the same way as RO_2·, viz.

$$\underset{\substack{| \\ R_2}}{\overset{\substack{R_1 \\ |}}{RCH_2 CH_2 CO}} \cdot \xrightarrow{k_{ai}} \underset{\substack{| \\ R_2}}{\overset{\substack{R_1 \\ |}}{R\dot{C}CH_2 CH_2 COH}} \qquad (169)$$

but the process exhibits greater sensitivity to the size of the transition state. Some semi-quantitative data are available from Walling and Padwa's investigations of intramolecular H-atom transfer for a series of t-RO· generated from hypochlorites [114]. The marked preference for a 6-center transition state for this process is evident from the result with II where the

(II)

ratio of chloroalcohols resulting from competition between 6- and 7-center processes is 10 : 1 even though the 7-center process is activated by a factor of 5 by reactive benzyl C—H bonds (see Table 18). The result is especially surprising when viewed in the context that a low activation energy transition state should have a relatively loose configuration in which steric strain should not be important. A possible explanation for this selectivity in ring size lies in the rigid geometric requirements for H-atom transfer to RO· in which colinearity between C—O—H, required to consummate the process, is achieved only in a six-center configuration [115]. It should be noted that the specific steric requirements for internal H-atom transfer to RO· forms the basis for the very successful achieve-

TABLE 18
Intramolecular H-atom transfer in t-RO· at $0°C$

CH bond type	$\dfrac{[ClROH]}{[Me_2CO]}$ [a]	k_{ai} [b] $\times 10^4$	$\log A_{ai}$ [c]	E_{ai} [d]
6-Center process				
Primary	0.48	2.4	11.8	9.2
Secondary	6.2	32	12.9	0.2
7-Center process				
Secondary	0.37	1.9	11.7	9.2
Benzyl	0.40	2.0	11.7	9.2

[a] Averaged values at $0°C$ from ref. 114; where no value was given for Me_2CO, the value was assumed from $100 - \%ClROH$.
[b] From relation $k_{ai} = k_d([ClROH]/[Me_2CO])$; $k_d = 5.1 \times 10^4$ s^{-1}.
[c] Calculated from $\log A = (\log k_{ai} + E/RT)$.
[d] Assigned; see text.

ments of Barton et al. [116] in steroid synthesis (the "Barton reaction") whereby selected, remote C—H bonds in the steroid rings are functionalized.

Values of k_{ai} at $0°C$ for primary and secondary C—H bonds involved in 6- and 7-centered processes are summarized in Table 18. The values are averages from those calculated from the ratio of yields of chloroalcohols (internal H-atom transfer k_a), and acetone (cleavage k_d) reported by Walling and Padwa [114] for the series $RCMe_2O\cdot$, where $R = C_3—C_6$ and $C_6H_5C_4$. Values of k_{ai} are calculated from the relation

$$k_{ai} = k_d \frac{[ClROH]}{[Me_2CO]} \tag{170}$$

where k_d at $0°C$ is $5.12 \times 10^4 \text{ s}^{-1}$, the value for cleavage of $Et\cdot$ from $EtCMe_2O\cdot$. For purposes of the calculation, we have assumed that other primary alkyl groups cleave from $RCMe_2O\cdot$ with similar rate coefficients. Values of k_{ai} calculated in this way are probably reliable within a factor of five.

The lack of change in product composition observed over a change of $70°C$ for several $t\text{-}RCMe_2O\cdot$, indicates that $E_{ai} \sim E_d$ for both 6- and 7-center processes. Since E_D is 9.2 kcal mole^{-1} for $EtCMe_2\cdot$ (Table 17), E_{ai} must also be 9.2 kcal mole^{-1}. The activation energies for analogous intermolecular H-atom transfers are $3—5$ kcal mole^{-1} (Table 16). The values of $\log A_{ai}$ and E_{ai} calculated from these data and listed in Table 18 are larger than expected and indicate some consistent error in the measurements or in the assigned values of $\log A_a$ for H-atom transfer to $RO\cdot$.

4.4 PEROXY RADICAL INTERACTIONS

4.4.1 Chemistry of $RO_2\cdot$ radical interactions

The major termination process in most oxidations involves interactions of $RO_2\cdot$ with like (self-reaction) or unlike $RO_2\cdot$ to form stable products through the intermediacy of RO_4R. Other possible interactions such as $RO\cdot$ and $RO_2\cdot$ or $2RO\cdot$ are not observed ordinarily because the high reactivity of $RO\cdot$ keeps the concentration of $RO\cdot$ too low to permit a significant contribution to the total rate of termination. Reactions of $RO_2\cdot$ and $R\cdot$ are important only at low oxygen pressures and were extensively investigated by Bateman and his coworkers [6].

During the past thirty years, a considerable effort has been made to measure accurately values of the termination rate coefficient, $2k_t$, for a wide variety of hydrocarbons in order to provide a more accurate picture of the effects of structure on rates of oxidation. Howard [117] has summarized the recent advances in this area.

Our present understanding of termination reactions of $RO_2\cdot$ comes from a variety of studies of hydrocarbon oxidations [56,65,69,117—119], chain decompositions of hydroperoxides [120—122], and low temperature reactions of $RO_2\cdot$ in inert solvents [46,123—125]. Together, these studies provide the following picture of the elementary reactions in termination by $RO_2\cdot$ [117].

$$2\ RO_2\cdot\ \rightleftharpoons RO_4R \tag{171}$$

$$RO_4R \rightarrow (2\ RO\cdot)_{cage} + O_2 \tag{172}$$

$$RO_4R \rightarrow R'CHO + R'CH_2OH + O_2 \text{ (concerted)} \tag{173}$$

$$(2\ RO\cdot)_{cage} \rightarrow ROOR \text{ or } R'CH_2OH + R'CHO \tag{174}$$

$$(2\ RO\cdot)_{cage} \rightarrow 2\ RO\cdot \tag{175}$$

$$RO\cdot + RO_2\cdot \rightleftharpoons RO_3R \tag{176}$$

Dramatic differences in values of $2k_t$ are found for prim- or sec-$RO_2\cdot$ compared with t-$RO_2\cdot$ [10,55]. Increased rate factors of 10^2—10^4 for prim- or sec-$RO_2\cdot$ arise from the intervention of reaction (173), a rapid concerted cleavage of prim- or sec-tetroxides ("Russell termination") to give carbonyl, alcohol, and oxygen

$$R_1R_2C\underset{\substack{\diagdown \\ O\text{--}O}}{\overset{\substack{H\text{--}O\text{--}CHR_1R_2 \\ \diagup \quad \diagdown}}{\diagup \qquad O}} \rightarrow R_1R_2C{=}O + HOCHR_1R_2 + O_2 \tag{177}$$

in preference to the slower cleavage of tetroxide to $RO\cdot$ and oxygen, reaction (172), the only reaction available to t-RO_4R [55,126]. For prim- or sec-$RO_2\cdot$, where $k_{173} < k_{-171}$, the measured termination rate coefficient, $2k_t$, is simply the product of the equilibrium constant for reaction (171) and the rate coefficient for reaction (173)

$$2k_t = k_{171}k_{173} \tag{178}$$

If $k_{173} > k_{-171}$ then $2k_t = k_{171}$. For t-$RO_2\cdot$, however, the measured rate coefficient, $2k_t$, is related to the elementary rate coefficients in more complex ways depending on the fraction of, and the fate of, those $RO\cdot$ that escape the cage [reaction (175)]. In the simplest case, where no $RO\cdot$ escape ($k_{175} = 0$), $2k_t = 2k_{171}k_{172}$. In the usual case, where some large fraction of $RO\cdot$ escape the solvent cage but are completely scavenged by added ROOH [119] through H-atom transfer

$$RO\cdot + ROOH \rightarrow ROH + RO_2\cdot \tag{179}$$

$$2k_t = 2k_{172}K_{171}\frac{k_{174}}{k_{174} + k_{175}} = k_{180} \tag{180}$$

If t-$RO_2\cdot$ are generated at low temperatures in the absence of H-atom donors [46], then combination of $RO\cdot$ with $RO_2\cdot$ to form RO_3R, reaction (176), is efficient and the product is stable. Under these conditions

$$2k_t = 2k_{172}K_{171} \frac{1 + k_{175}}{k_{175} + k_{174}} \tag{181}$$

A more complex situation arises when $RO\cdot$ cleaves to give $prim\text{-}R'O_2\cdot$, which is more reactive in both propagation and termination reactions. Under these conditions, where termination is mainly via

$$prim\text{-}R'O_2\cdot + RO_2\cdot \rightarrow R'{=}O + ROH + O_2 \tag{182}$$

the absolute termination rate coefficient is [65]

$$2k_t = 2k_{173} + 2k_{175} + k_{182}[RO_2\cdot] \tag{183}$$

where k_d and k_a are coefficients for cleavage (158) and abstraction (157) by $RO\cdot$. Usually, the net effect of cleavage is to retard the rate of oxidation [56,65] and the effect of added hydroperoxide is to accelerate the reaction by scavenging $R'O_2\cdot$ [56,65,118].

4.4.2 Structure—reactivity relationships in radical interactions

Howard and Ingold [10,69,119] have carefully measured values of $2k_t$ for a variety of $RO_2\cdot$. The $RO_2\cdot$ radicals fall in the following order with respect to increasing values of $2k_t$: t-alkyl $<$ acrylic $<$ allylic $<$ cyclic secondary $<$ acrylic benzylic $<$ primary. Rate coefficients span a range from 10^3 to 10^8 l mole^{-1} s^{-1} [10].

Substituent effects in $RO_2\cdot$, where R is benzyl or substituted benzyl, have only small effects on $2k_t$, largely unrelated in direction to mesomeric or inductive effects of the substituents: ring-substituted styrenes [127] have $2k_t = (3.88 \pm 1.24) \times 10^7$ l mole^{-1} s^{-1}; α-substituted toluenes have $2k_t = (10.3 \pm 10.1) \times 10^7$ l mole^{-1} s^{-1} [128]. Among $t\text{-}RO_2\cdot$, increasing bulk in R leads to higher values for $2k_t$ which range from 1.3×10^3 l mole^{-1} s^{-1} for t-butyl to 6×10^4 l mole^{-1} s^{-1} for 1,1-diphenylethyl [129].

Acylperoxy radicals arising from the oxidation of aldehydes apparently terminate via a tetroxide which then cleaves to form, first, primary carbon radicals and second, primary alkylperoxy radicals which terminate rapidly [130,131]. Termination rate coefficients for aliphatic aldehydes have values ranging from 0.7×10^7 to 10×10^7 l mole^{-1} s^{-1} [10].

4.4.3 Thermochemistry of radical interactions

Benson and Shaw [101] have calculated heats for formation and bond strengths for many peroxides, polyoxides, and their precursor radicals, from which we can estimate the heats of reaction of the important interaction (termination) steps. For terminations of $MeO_2\cdot$, the important

TABLE 19
Bond strengths in peroxides and polyoxides [101]

RO_nR [a]	DC—O) (kcal mole^{-1})	D(O—O) (kcal mole^{-1})
RO_2H	70	44
ROOR	70	38
ROOOR	70	21
ROOOOR	70	5(2,3) [b] 21(1,2) [b]

[a] Bond strengths are the same for R=Me and t-Bu.
[b] Refers to 1,2 or 2,3 O—O bond.

reactions are

$$2\,MeO_2\cdot \rightarrow 2\,MeO\cdot + O_2 \qquad \Delta H_r = -6 \text{ kcal mole}^{-1} \qquad (184)$$

$$2\,MeO\cdot \rightarrow MeOOMe \qquad \Delta H_r = -37 \text{ kcal mole}^{-1} \qquad (185)$$

$$2\,MeO_2\cdot \rightarrow CH_3OH + CH_2O + O_2 \qquad \Delta H_r = -81 \text{ kcal mole}^{-1} \qquad (186)$$

Formation of RO· from $RO_2\cdot$ is only slightly exothermic, in contrast to the concerted process which is sufficiently exothermic to generate a small population of excited carbonyl, or singlet oxygen, both of which have been detected in oxidation systems [56,124]. Bond strengths for several intermediates, peroxides and polyoxides are summarized in Table 19.

Equilibrium enthalpies and entropies for reactions (171) and (−171), the reversible dissociation of tetroxide to $RO_2\cdot$, have been measured with good precision between −140 and −80°C for several t-RO$_4$R using low temperature ESR to monitor changes in the concentration of $RO_2\cdot$ [11, 117,128]. Values for ΔH_{171} and ΔS_{171} and calculated values of K_{171} are summarized in Table 20.

TABLE 20
Equilibrium values for the reversible dissociation of t-RO$_4$R at 30°C [a]

t-RO$_4$R	K_{171} [b] (l mole^{-1})	ΔH_{171} (kcal mole^{-1})	ΔS_{171} (cal mole^{-1} K^{-1})
t-Butyl	12.2	8.8	34
2-Ethyl-2-propyl	8.5	7.5	29
2-Isopropyl-2-propyl	10.3	8.6	33
Cumyl [c]	2.2	9.2	32
1-Methylcyclopentyl	6.2	8.0	30
2-Phenyl-2-butyl		11	

[a] From data of refs. 46 and 117.
[b] Calculated from relation $K = \exp(\Delta S/R)\exp(-\Delta H/RT) = k_{-171}/k_{171}$.
[c] Cumyl = $C_6H_5C(CH_3)_2$.

The relatively small differences in K_{171} suggest that any differences in the observed values of $2k_t$ must originate in the irreversible dissociation of RO_4R to $RO\cdot$, reaction (173); that is kinetic, not thermochemical, properties govern the overall rate of termination.

4.4.4 Absolute rate coefficients for termination

The kinetic aspects of termination are most usefully discussed in the context of the assembly of elementary steps (171)—(175).

(a) Rate coefficients k_{171} and k_{-171}

The value for k_{-171} and rate parameters for k_{171} and k_{-171} can be calculated from the value of K_{171}, some assumptions concerning the value of k_{171}, the rate coefficient for recombination of two $RO_2\cdot$, and the relationships

$$\log \frac{A_{-171}}{A_{171}} = \frac{\Delta S_{171}}{2.3R} \tag{187}$$

$$\log A_{-171} = \frac{\Delta S_{171}}{2.3R} + \log A_{171} \tag{188}$$

If we assume that the radical combination coefficient, k_{171} has the Arrhenius form

$$\log k_{171} = 9.5 - \frac{2}{2.3RT \times 10^{-3}}$$

then $\log A_{-171}$ ranges from 16 to 17 s^{-1} for those t-$RO_2\cdot$ investigated (Table 20), values expected for simple O—O fission. Values for E_{-171} follow from the relation

$$E_{171} - E_{-171} = \Delta H_{171} \tag{189}$$

$$E_{-171} = \Delta H_{171} - 2 \tag{190}$$

The values of E_{-171} are equal to 5—9 kcal mole^{-1}, in reasonably good agreement with the estimated bond strength of the 2,3 O—O bond in t-RO_4R plus E_{171}.

(b) Rate coefficients k_{172} and k_{174}/k_{175}

Values of k_{172} can be calculated from known values of k_{171}, k_{174}/k_{175}, and eqn. (180). Independent estimates of k_{174}/k_{175} are possible via the induced decomposition of the hydroperoxide corresponding to the t-$RO_2\cdot$ [120—122], viz.

$$In \overset{R_i}{\rightarrow} 2 X\cdot \tag{24}$$

$$X\cdot + ROOH \to H + RO_2\cdot \qquad (191)$$

$$2\ RO_2\cdot \to (2\ RO\cdot)_{cage} + O_2 \qquad (171,172)$$

$$(2\ RO\cdot)_{cage} \to ROOR \qquad (174)$$

$$\searrow 2\ RO\cdot \qquad (175)$$

$$RO\cdot + ROOH \to ROH + RO_2\cdot \qquad (179)$$

The rates of loss of ROOH and evolution of O_2 obey the relations [120]

$$\frac{d(O_2)}{dt} = \frac{d(ROOH)}{dt} = R_i \left(1 + \frac{k_{175}}{k_{174}}\right) \qquad (192)$$

Several workers [120,121] have shown that, for t-BuOOH, k_{175}/k_{174} is close to 10 at 45°C in benzene, while Howard et al. [122] found values of 5—7 for six other hydroperoxides in CCl_4 at 30°C. These data coupled with values of K_{171} enabled Howard et al. to calculate values of k_{172} for two t-$RO_2\cdot$ using eqn. (180).

The activation parameters $\log A_{172}$ and E_{172} calculated for reaction (172) are in fair agreement with values expected for single bond fission of a 20—22 kcal mole^{-1} bond: $\log (A_{172}/s^{-1}) \sim 17$, $E_{172} \sim 17$ kcal mole^{-1}.

(c) Rate coefficient for reaction (173): competition between concerted and stepwise cleavage of RO_4R

The much larger values of $2k_t$ for prim- and *sec*-$RO_2\cdot$ compared with t-$RO_2\cdot$, together with the failure to observe either $RO_2\cdot$ or RO_4R from *sec*-$RO_2\cdot$ at -140°C [124,125], indicate that concerted collapse of RO_4R

TABLE 21

Absolute rate coefficients and parameters for radical interactions (171)—(175) at 30°C [a]

Reaction number (n)	k_n [a,b]	$\log A_n$	E_n (kcal mole^{-1})
171	1(9)	9.5	2
-171	6(9)	16	8.6
172	2.8(5)	17	16
173	6.8(7)	10	3
174 [c]	1(15)	15	0
175 [c]	6(15)	17	2
174/175	0.16		

[a] Units are s^{-1} l mole^{-1} s^{-1}.

[b] Numbers in parentheses are exponents of 10.

[c] Estimated on the basis that k_{174}/k_{175} decreases with temperature corresponding to an activation energy of -2 kcal mol^{-1} (ref. 133).

TABLE 22

Measured values of $2k_t$ for selected peroxy radicals at $30°C$ [10]

$RO_2\cdot$	$2k_t$ ($l\ mol^{-1}\ s^{-1}$)
t-Bu	1.5×10^4
$PhCMe_2$	6×10^3
C_5H_8Me	$(1-20) \times 10^4$
Ph_2CMe	6.4×10^4
$MeC(O)$	10.4×10^7
n-C_4H_9	3×10^8
Me_2CH	3×10^6

[reaction (173)] to products is much faster than fission of the 2,3 O—O bond to give $RO_2\cdot$ or fission of the 1,2 O—O bond to give $RO\cdot + O_2$. If we assume that $k_{172} \cong k_{-171}$ and that $\log A_{-171}$, $\log A_{173}$ and E_{-171} have reasonable values of 16.5 and 11 (6-center process) and 6 kcal mole^{-1}, respectively, then $E_{173} \sim 3$ kcal mole^{-1}, a value consistent with many direct measurements of $2k_t$ for sec-$RO_2\cdot$ [117,125].

Using estimated and measured [125] values of the rate parameters for reactions (172) and (173) the isokinetic temperature, where the rate coefficients k_{172} and k_{173} are equal, is about $237°C$, a value much higher than predicted from oxidation experiments with n-butane at $100-125°C$ [35] or $MeO_2\cdot$ in the gas phase at $25°C$ [132] where significant fractions of prim- or sec-$RO_2\cdot$ appear to give prim- or sec-$RO\cdot$ products rather than only carbonyl and alcohol as expected if $k_{173} \gg k_{172}$. We can accommodate the experimental observations if we adjust somewhat the values of the rate parameters for reactions (172) and (173). If $\log k_{172} = 17 - (16/4.6T \times 10^{-3})$ and $\log k_{173} = 10 - (3/4.6T \times 10^{-3})$ then the isokinetic temperature is 406 K or $133°C$, a value in reasonable agreement with data for sec-$BuO_2\cdot$ though still too high for agreement with the gas phase data [132].

Table 21 summarizes preferred values of rate parameters for reactions (171)—(175) and Table 22 summarizes some selected values of $2k_t$ for several kinds of hydrocarbon.

5. Special features of initiation

Oxidation as a free radical chain process requires a continuous source of free radicals to maintain a steady rate of reaction. Free radical sources, such as azo compounds, are often added in order to ensure a consant rate of initiation which then allows one to measure the ratio of the rates of the propagation and termination reactions [4,18,38]. However, without added initiators, relatively slow "spontaneous" initiation processes often occur and these can be a significant source of radicals in some systems [26].

5.1 AUTOCATALYSIS

Autocatalysis is the term applied to initiation resulting from hydroperoxide build-up during a reaction. The rate of initiation and the rate of oxidation increase as the hydroperoxide concentration builds up, thus producing the characteristic autocatalytic rate curve. If reactions producing radicals are both first and second order in hydroperoxide, then the rate of initiation (R_i) may be written

$$R_i = 2k'[ROOH] + 2k''[ROOH]^2 \tag{193}$$

and, assuming that termination is bimolecular in $[RO_2 \cdot]$

$$R_t = 2k_t[RO_2 \cdot]^2 \tag{194}$$

Then, equating R_t and R_i yields

$$[RO_2 \cdot] = \frac{1}{2k_t} (2k'[RO_2H] + 2k''[RO_2H]^2)^{1/2} \tag{195}$$

Substituting this expression in the usual rate expression

$$R_o = k_p[RO_2 \cdot][RH] \tag{196}$$

leads to the often observed [60] rate expression

$$R_o = \frac{k_p}{(2k_t)^{1/2}} (2k'[RO_2H] + 2k''[RO_2H]^2)^{1/2}[RH] \tag{197}$$

Generally, the bimolecular term predominates, except at very low [ROOH], in which case

$$R_o = \frac{k_p}{(2k_t)^{1/2}} (2k'')^{1/2}[RO_2H][RH] \tag{198}$$

This expression qualitatively fits a number of systems [60]. Van Sickle et al. [18] have made a careful study of the oxidation of cyclopentene and its autocatalysis at 50°C. This reaction yields about 75% 3-cyclopentenylhydroperoxide and 25% of the dimer hydroperoxide III

(III)

They found that the rate of oxidation is linear with respect to [ROOH] up to 10% conversion (1 M ROOH), but at higher conversions the rate gradually falls off, even when corrected for the decrease in cyclopentene concentration, so that at 20% conversion (2 M ROOH) the rate dropped off by 12%. The fit to eqn. (198) is quite good in view of the change in the reaction medium, deviation of the rate of decomposition of ROOH from second order with concentration [18], the presence of a small proportion of dimer hydroperoxide which may decompose at a different rate from the simple hydroperoxide [18], the reaction of the initial products [134], and the possibility of retarders [18] being formed.

The decomposition of 3-cyclopentenyl hydroperoxide in cyclopentene in the absence of oxygen is approximately second order, which is consistent with the first-order dependence of the rate of oxidation on [ROOH]. Similarly, in the oxidation of cyclohexene, the rate depends on the 0.75—0.85 power of the hydroperoxide concentration [135], in good agreement with the observed 1.7 power dependence for hydroperoxide decomposition in absence of oxygen [136]. The reaction that generally has been proposed [137] for the bimolecular decomposition of RO_2H is

$$2 \, ROOH \rightarrow RO\cdot + RO_2\cdot + H_2O \tag{199}$$

which is endothermic by ~16 kcal mole^{-1}. The $RO\cdot$ radical can then react with ROOH or RH to generate $RO_2\cdot$ (or $R\cdot$ which in turn forms $RO_2\cdot$).

$$RO\cdot + RO_2H \rightarrow ROH + RO_2\cdot \tag{200}$$

$$RO\cdot + RH \rightarrow ROH + R\cdot \tag{31}$$

Initially, however, $RO\cdot$ and $RO_2\cdot$ from reaction (199) would be formed in a solvent cage, and some proportion of radical pairs will disproportionate

$$2 \, RO_2H \rightarrow (RO\cdot + RO_2\cdot)_{cage} + H_2O \tag{201}$$

$$(RO\cdot + RO_2\cdot)_{cage} \overset{f}{\rightarrow} RO\cdot + RO_2\cdot \tag{202}$$

$$(RO\cdot + RO_2\cdot)_{cage} \overset{(1-f)}{\longrightarrow} {\setminus \atop /}{=}O + RO_2H \tag{203}$$

where f is the fraction of caged radicals that diffuse apart and initiate the oxidation chain and $(1 - f)$ the fraction that are lost by termination. Van Sickle et al. [18] have estimated f to be 0.36—0.42 from $k_p/(2k_t)^{1/2}$ for cyclopentene and the measured second-order decomposition rate coefficient for the peroxide.

The autocatalytic effect can be affected by the solvent. If the oxidation of cyclopentene is carried out in benzene solution, no evidence of autocatalysis is observed [18]. Consistent with this result is the fact that the decomposition of 3-cyclopentenyl hydroperoxide in benzene is much slower than in cyclopentene and is first order in hydroperoxide [18].

Autocatalyzed oxidation reactions can, in principle, reach a maximum rate at the point where the rate of loss of RO_2H by decomposition equals the formation rate of RO_2H, as first discussed by Tobolsky et al. [138], i.e.

$$2k_2 [RO_2 H]_{ss}^2 = k_p [RO_2 \cdot] [RH] \tag{204}$$

Since rates of radical formation and termination must be equal at the limit

$$2fk_2 [RO_2 H]_{ss}^2 = 2k_t [RO_2 \cdot]^2 \tag{205}$$

Then, from eqn. (204)

$$2k_t [RO_2 \cdot]^2 = fk_p [RO_2 \cdot] [RH] \tag{206}$$

and

$$[RO_2 \cdot]_{ss} = \frac{fk_p [RH]}{2k_t} \tag{207}$$

Since, at short chain lengths, the rate of oxidation (R_o) is

$$R_o = k_p [RO_2 \cdot] [RH] - \frac{R_i}{2a} \tag{208}$$

then

$$R_o^{ss} = \left(f - \frac{f^2}{2a} \right) \frac{k_p^2}{2k_t} [RH]^2 \tag{209}$$

The value of $[RO_2 H]$ at which this will occur may be calculated from

$$R_o^{ss} = 2fk_2 [RO_2 H]^2 \tag{210}$$

Assuming all hydrocarbon is converted to ROOH

$$\left(f - \frac{f^2}{2a} \right) \frac{R_p^2}{2k_t} ([RH]_i - [ROOH])^2 = 2fk_2 [ROOH]^2 \tag{211}$$

According to this expression, at 50°C the value of [ROOH] for cyclo-pentene oxidation where this maximum rate is obtained is 9.0 M or at about 82% conversion of the cyclopentene. Such a conversion is practically unattainable because other reactions of the hydroperoxide become important as its concentration increases. However, for less reactive hydrocarbons where $k_p/(2k_t)^{1/2}$ is only 0.01—0.1 that for cyclopentene, the maximum hydroperoxide concentration (0.5—3.0 M) is lower and the limiting rate is obtainable.

6. Co-oxidation

6.1 KINETIC RELATIONS

Oxidation of a mixture of two hydrocarbons simultaneously introduces cross-propagation reactions, where the peroxy radical from one hydrocarbon reacts with the other hydrocarbon and vice versa. Thus, for the co-oxidation of two hydrocarbons, R_1H and R_2H four propagation reactions may be important, viz.

$$R_1OO\cdot + R_1H \xrightarrow{k^p_{11}} R_1OOH + R_1\cdot \tag{212}$$

$$R_1OO\cdot + R_2H \xrightarrow{k^p_{12}} R_1OOH + R_2\cdot \tag{213}$$

$$R_2OO\cdot + R_2H \xrightarrow{k^p_{22}} R_2OOH + R_2\cdot \tag{214}$$

$$R_2OO\cdot + R_1H \xrightarrow{k^p_{21}} R_2OOH + R_1\cdot \tag{215}$$

In addition to the cross-propagation reactions, one cross-termination reaction is introduced in addition to the two self-termination reactions, viz.

$$R_1OO\cdot + R_1OO\cdot \xrightarrow{k^t_{11}} \text{termination} \tag{216}$$

$$R_1OO\cdot + R_2OO\cdot \xrightarrow{k^t_{12}} \text{termination} \tag{217}$$

$$R_2OO\cdot + R_2OO\cdot \xrightarrow{k^t_{22}} \text{termination} \tag{218}$$

Under conditions where kinetic chain lengths are high, the rate of total oxygen consumption is

$$-\frac{d[O_2]}{dt} = \frac{(r_1[R_1H]^2 + 2[R_1H][R_2H] + r_2[R_2H]^2)R_i^{1/2}}{(r_1^2\delta_1^2[R_1H]^2 + \phi r_1 r_2 \delta_1 \delta_2[R_1H][R_2H] + r_2^2\delta_2^2[R_2H]^2)^{1/2}} \tag{219}$$

where

$$r_1 = \frac{k^p_{11}}{k^p_{12}} \qquad r_2 = \frac{k^p_{22}}{k^p_{21}} \tag{220}$$

$$\delta_1 = \frac{(2k^t_{11})^{1/2}}{k^p_{11}} \qquad \delta_2 = \frac{(2k^t_{22})^{1/2}}{k^p_{22}} \tag{221}$$

and the cross-termination is expressed as

$$k^t_{12} = \phi(k^t_{11}k^t_{22})^{1/2} \tag{222}$$

In early studies of the co-oxidation of hydrocarbons, Russell [76,139] showed the importance of the cross-termination reaction on the rate of oxygen consumption. Although the different propagation rate coefficients do not vary significantly, the rate of the two self-termination reactions can vary by as much as 10^4. Thus, as the composition of the mixture is varied from 100% of one hydrocarbon to 100% of the other, the importance of the various termination reactions changes accordingly. The effect of the cross-termination is most dramatic when $\phi \gg 2$. Small amounts of a hydrocarbon such as tetralin, which has a large self-termination rate

66

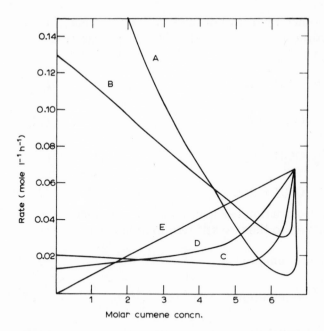

Fig. 1. Oxidation of mixtures of cumene and aralkyl hydrocarbons at 90°C, 0.02 M
t-butyl perbenzoate. A, Dibenzyl ether; B, indan; C, diphenylmethane; D, ethyl-
benzene; E, theoretical for an inert diluent. Reprinted with permission from ref. 139.
Copyright by the American Chemical Society.

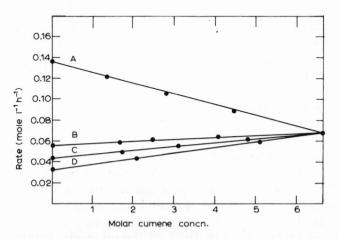

Fig. 2. Oxidation of mixtures of cumene and substituted cumenes at 90°C, 0.02 M
t-butyl perbenzoate. A, *p*-diisopropylbenzene; B, *p-t*-butylcumene; C, *p*-bromocumene;
D, *p*-nitrocumene. Reprinted with permission from ref. 139. Copyright by the American
Chemical Society.

coefficient, can drastically reduce the rate of oxidation of cumene, which has a very small termination rate coefficient [77,139]. The effect is accentuated because the cross-termination rate coefficient is also large ($\phi \gg 2$). Figures 1 and 2 illustrate the effect of a number of hydrocarbons on the oxidation of cumene (isopropylbenzene). Figure 1 shows the results of adding to cumene several compounds which form secondary peroxy radicals that can terminate much faster than can cumyl peroxy. In all cases, small amounts of the secondary hydrocarbon reduce the rate of oxidation of cumene more than does an inert material. In the case of compounds that oxidize faster than cumene, a distinct minimum rate is observed upon addition of a few percent of these compounds. Figure 2 shows the effect of adding hydrocarbons that form predominantly tertiary peroxy radicals as does cumene. The variation in oxidation rate with the mixture composition is linear.

By carrying out a number of co-oxidations with various hydrocarbons, it is possible to compare the termination rate coefficient of these hydrocarbons and thereby group them accordingly [140,141]. Although more direct and more precise methods of measuring termination rate coefficients are available, this technique is an effective qualitative method for estimating these coefficients.

6.2 REACTIVITY RATIOS r_1 AND r_2

From the measured consumption of the two hydrocarbons, the values of r_1 and r_2, the ratios of the propagation coefficients as defined earlier, may be determined. At low conversions

$$\frac{\Delta[R_1H]}{\Delta[R_2H]} = \frac{k_{11}^p[R_1O_2\cdot][R_1H] + k_{21}^p[R_2O_2\cdot][R_1H]}{k_{22}^p[R_2O_2\cdot][R_2H] + k_{12}^p[R_1O_2\cdot][R_2H]} \tag{223}$$

The ratio of concentrations of the two peroxy radicals is expected to remain constant at low conversions of the hydrocarbons, i.e.

$$k_{21}^p[R_2O_2\cdot][R_1H] = k_{12}^p[R_1O_2\cdot][R_2H] \tag{224}$$

Thus, eqn. (223) may be simplified to

$$\frac{\Delta[R_1H]}{\Delta[R_2H]} = \frac{(r_1[R_1H]/[R_2H]) + 1}{(r_2[R_2H]/[R_1H]) + 1} \tag{225}$$

Fineman and Ross [142] have used $\rho = \Delta[R_1H]$ and $R = [R_1H]/[R_2H]$ to show that eqn. (4) may be converted to the form

$$\frac{(\rho - 1)}{R} = r_1 - \frac{\rho}{R^2} r_2 \tag{226}$$

Thus, a plot of the left-hand side of eqn. (226) against ρ/R^2 gives $1/r_2$ as the slope and r_1 as the intercept.

The values of r_1 and r_2 are measures of the relative reactivity of the two hydrocarbons toward each peroxy radical. Thus, if one or both of the self-propagation rate coefficients is known, the corresponding cross-propagation rate coefficients can be determined. The differences in r_1 and $1/r_2$ are the results of the differences in the organic structure of the two peroxy radicals because differences in hydrocarbons should cancel out. Typically, the quantity

$$r_1 r_2 = \frac{k_{11}^p}{k_{12}} \frac{k_{22}^p}{k_{21}}$$
(227)

is a measure of the differences in selectivity. The quantity reduces to unity if the selectivities are the same. However, differences in selectivity between peroxy radicals are never large and are generally less than a factor of 2 [75]; hence, it is difficult to distinguish small effects from the experimental uncertainties that can occur in the analyses.

By studying the co-oxidation of a series of hydrocarbons with one standard hydrocarbon, it is possible to determine the reactivity of the series toward the peroxy radical of this standard. However, in some cases alternative methods such as the hydroperoxide method discussed in Sect. 4.1.4, can be used. The hydroperoxide method is preferred for determining the reactivity of peroxy radicals formed from readily obtainable hydroperoxides. It is not satisfactory if the hydroperoxide is not stable, if it is not appreciably soluble in the reaction mixture, as is the case for HO_2H,

TABLE 23

Relative reactivity of hydrocarbons towards $H(O_2C_4H_6)_x O_2 \cdot$, $t\text{-}C_4H_9O_2 \cdot$, and $HO_2 \cdot$

Hydrocarbon	$H(O_2C_4H_6)_x O_2 \cdot$ [a] (50°C)	$t\text{-}C_4H_9O_2 \cdot$ [b] (30°C)	$HO_2 \cdot$ [c] (50°C)
Butadiene	3.3		1.7
Cumene	0.14	0.08	
sec-Butylbenzene	0.13		
Tetralin	1.00	1.00	1.00
Styrene	1.5		
Cyclohexadiene	80	40	7.2
Cyclopentene	1.7	1.7	
Cyclohexene	0.8		
Cycloheptene	1.1		
Cyclooctene	0.4		
Tetramethylethylene			1.8

[a] Reactivities taken from ref. 143. To convert to absolute rate coefficients, multiply all values by 80 l mole^{-1} s^{-1} (refs. 119 and 127); this assumes that the styrenyl and butadienyl peroxy radicals have identical reactities.

[b] Reactivities taken from ref. 79. To convert to absolute rate coefficients, multiply all values by 2.0 l mole^{-1} s^{-1}.

[c] Reactivities taken from ref. 87. To convert to absolute rate coefficients, multiply all values by 580 l mole^{-1} s^{-1} (ref. 69).

or if the hydroperoxide is not readily prepared, as is the case for the butadiene peroxy radical.

Table 23 summarizes data obtained by co-oxidation techniques for the relative reactivity of a number of hydrocarbons towards $HO_2\cdot$ and the butadiene polyperoxy radical, $H(O_2C_4H_6)_xO_2\cdot$ [143]. For comparison, Table 23 gives the relative reactivities of the same hydrocarbons towards the t-butylperoxy radical [79]. These were obtained by carrying out the oxidation in the presence of t-butyl hydroperoxide (Sect. 4.1.4). The data for $H(O_2C_4H_6)_xO_2\cdot$ and t-$C_4H_9O_2\cdot$ are quite parallel. However, for $HO_2\cdot$ the data show a much smaller spread in the reactivity, which indicates a lower selectivity. Consistent with this lower selectivity is the higher propagation rate coefficient for $HO_2\cdot$. The high reactivity is consistent with the steric simplicity of $HO_2\cdot$ as well as the difference in electron-withdrawing ability of H compared with alkyl groups [87]; these factors have been observed for other systems [78,144]. Complexing of $HO_2\cdot$ with H_2O and H_2O_2 may also affect the reactivity of $HO_2\cdot$ [145].

6.3 EVALUATION OF THE CROSS-TERMINATION PARAMETER ϕ

Once r_1, r_2, δ_1, and δ_2 are determined, it is then possible to determine ϕ by substitution into eqn. (219). Table 24 summarizes some values obtained in the co-oxidation of cumene with various hydrocarbons. The term ϕ has been defined such that it is expected to equal 2 because of the statistical advantage of the bimolecular reaction between different species. Thus, in the reaction of cumene with α-methylstyrene, where the self-termination rate coefficients are nearly identical, the value of ϕ is 2. However, for the other hydrocarbons, all of which form peroxy radicals that terminate much faster than cumene, ϕ is much greater than 2.

The cross-termination rate coefficients between secondary and tertiary peroxy radicals are expected to approximate the self-termination rate

TABLE 24
Cross-termination parameters, ϕ, for co-oxidation of hydrocarbons with cumene

Hydrocarbon	Temp. (°C)	ϕ	Ref.
Tetralin	90	12	1
	70	5	146
	30	12	66
Ethylbenzene	70	13	146
Diphenylmethane	70	13	146
Styrene	70	26	146
	60	21	147
α-Methylstyrene	70	2	146

coefficients of the secondary peroxy radicals, because both react by a similar mechanism (Sect. 4.4.1). Thus, to at least the first approximation, $k_{12}^t \cong k_{22}^t$. We know that in all cases included in Table 24, except for α-methylstyrene, k_{22}^t/k_{11}^t is 100—1000, and thus ϕ should equal 10—30, which is in good agreement with the values reported in Table 24.

7. Inhibition of oxidation

7.1 CHEMISTRY OF INHIBITION

No account of liquid phase oxidations would be complete without some discussion of the inhibition of oxidation by chain-breaking anti-oxidants. For well over one hundred years, antioxidants have been used in a variety of commercial products to slow deterioration in air, rubber being among the first to receive attention [148]. Excellent reviews of the practical aspects of antioxidant use and development are given by Lundberg [149] and Scott [150].

Progress in understanding the role of antioxidants has paralleled the understanding of oxidation kinetics; the first real insight into antioxidant mechanisms occurred roughly at the time that Backstrom [2] defined the radical chain character of benzaldehyde oxidation.

Modern kinetic investigations of antioxidant action began with the investigations of Bolland and ten Haave [151,152] on inhibited oxidation of ethyl linoleate and with the broad theoretical and experimental studies of Waters and his coworkers [153—155]. Bolland and ten Haave proposed that inhibition resulted from chain-breaking by the faster reaction of $RO_2\cdot$ with antioxidant, AH, than with hydrocarbon RH to give an unreactive radical $A\cdot$ which then terminates with $RO_2\cdot$ or $A\cdot$, viz.

$$RO_2\cdot + RH \xrightarrow{k_p} RO_2H + R\cdot \tag{28}$$

$$RO_2\cdot + AH \xrightarrow{k_{inh}} RO_2H + A\cdot \tag{228}$$

$$2\,A\cdot \rightarrow products \tag{229}$$

$$RO_2\cdot + A\cdot \rightarrow products \tag{230}$$

Under conditions where reaction (228) is much faster than reaction (28), no oxygen uptake by RH is noted and oxidation of RH is inhibited until nearly all the AH is consumed, at which time oxygen uptake begins rather abruptly. In many oxidations, the actual fate of $A\cdot$ depends on several factors, including the reactivity of $A\cdot$, RH, ROOH, and the concentration of $RO_2\cdot$. Thus with simple unhindered phenols, chain-transfer by $A\cdot$ with RH leads to propagation, albeit at a slower rate, via reaction (231)

$$A\cdot + RH \rightarrow AH + R\cdot \tag{231}$$

If reaction (231) is important, the oxidation process is only retarded and some oxygen uptake is found even in the initial stages. With many hindered phenols, reaction (231) is very slow and only coupling between radicals occurs [reactions (229) and (230)].

Chemical evidence for the importance of reaction (229) was first obtained by Waters and Wickenham-Jones [153,154] for reactions in oxidizing benzaldehyde inhibited by 2,6-dimethylphenol. They found that the phenol was converted to the corresponding diphenoquinone

$$(232)$$

The exact mode of self-reaction of other phenoxy radicals depends on the structure of the phenol; coupling generally takes place at the site most remote from side-chain substitution [155,156]

$$(233)$$

Proof that coupling of A· with $RO_2·$, reaction (230), can also be important in inhibited oxidation reactions, was provided by several workers [157—159] in the 1950s using mixtures of phenols and high concentrations of azo initiators in oxygen. Thus Hammond and his coworkers [159] found that the peroxycyclohexadienone (III) resulted from coupling of the 2,6-di-t-butyl-4-methylphenoxy radical with the peroxy radical from azobis(isobutyronitrile)

$$(234)$$

Competition between reactions (229) and (230) occurs even under conditions where the concentration of $RO_2·$ is high; Ingold and Horswill [156] showed that for 2,4-di-t-butylphenol and t-BuO$_2$·, over ten products, corresponding to self-reaction or coupling of $RO_2\cdot$ and $A\cdot$ and further oxidation, were formed in the reaction mixture. Aromatic amines react in ways similar to phenols yielding quinoimine coupling products with $RO_2\cdot$ [159] and complex dimers [156] on self-reaction. Scott [150] and Ingold [160] have provided comprehensive accounts of the progress in phenolic antioxidant chemistry through 1964 and 1970, respectively. A review by Howard in 1974 covers a broader area in less detail [161].

7.2 KINETICS OF INHIBITION BY PHENOLS AND AMINES

Bolland and ten Haave [151] found that the oxidation of ethyl linoleate inhibited by hydroquinone was described by the relation

$$\text{rate of oxygen uptake } (R_{O_2}) \propto R_i \frac{[\text{RH}]}{[\text{AH}]} \tag{235}$$

A similar relation was found by Howard and Ingold [162] for the oxidation of styrene inhibited by a variety of phenols. Both of these systems apparently involve some combination of reactions (229) and (230), but their relative importance is indistinguishable on the basis of the kinetic relationship alone except for the stoichiometric ratio of $RO_2\cdot$ consumed for each AH. Thus the steady-state concentration of $RO_2\cdot$ in the inhibited system is

$$[RO_2\cdot] = \frac{R_i}{nk_{\text{inh}}[\text{AH}]} \tag{236}$$

and the rate of oxidation of RH, on substitution for $[RO_2\cdot]$ in the rate expression $k_p[RO_2\cdot][\text{RH}]$, is

$$R_{O_2} = \frac{R_i k_p [\text{RH}]}{nk_{\text{inh}}[\text{AH}]} \tag{237}$$

In the oxidations of neat styrene, $n = 2$ for a variety of phenols, indicating that the probable mechanism for inhibition is

$$RO_2\cdot + AH \rightarrow RO_2H + A\cdot \tag{228}$$

$$RO_2\cdot + A\cdot \rightarrow \text{products} \tag{230}$$

Hammond and coworkers [159] also found values of $n \sim 2$ for a variety of amines and phenols in the inhibited oxidation of cumene. However, since the kinetics showed a half-order dependence on cumene and no isotope effect was found in the inhibited oxidation using N-deuterated amines, they proposed a more complex mechanism to account for their findings. Howard and Ingold [162] showed that by using added D_2O to maintain the N- or O-deuteration, large isotope effects are found in the inhibited oxidation. Moreover, Mahoney and Ferris [163] showed that the unusual kinetic dependence on hydrocarbon found by Hammond and coworkers [159] could be accounted for by chain transfer, viz.

$$RO_2 \cdot + RH \overset{k_p}{\rightarrow} RO_2H + R \cdot \tag{28}$$

$$RO_2 \cdot + AH \overset{k_{inh}}{\longrightarrow} RO_2H + A \cdot \tag{228}$$

$$A \cdot + RH \rightarrow AH + R \cdot \tag{238}$$

$$RO_2 \cdot + A \cdot \rightarrow products \tag{230}$$

$$R_{O_2} = \left(\frac{k_{234}R_i}{2k_{inh}k_{230}}\right)^{1/2} k_p \frac{[RH]^{3/2}}{[AH]^{1/2}} \tag{239}$$

Equation (239) also accounts adequately for some unusual results noted by Thomas [164] and Thomas and Tolman [165] for oxidations of cumene inhibited by phenol, diphenylamine, and trimethylamine.

The past ten years have witnessed a significant advance in our understanding of the complex kinetic and equilibrium relationships for inhibited oxidations, thanks largely to the detailed thermochemical and kinetic studies of Mahoney and DaRooge [89,166,167] and the ESR kinetic studies of Howard, Ingold, and their coworkers [10,88,160]. Mahoney and DaRooge [89] have very ably summarized and extended their studies on the kinetic and thermochemical properties of phenoxy radicals in a recent paper.

The kinetic scheme which seems to best represent the currently accepted mechanism for the effects of antioxidants (AH) in autoxidations is

$$Initiator \overset{k_i}{\rightarrow} 2 X \cdot \tag{24}$$

$$X \cdot + RH \rightarrow XH + R \cdot \tag{26}$$

$$R \cdot + O_2 \rightarrow RO_2 \cdot \tag{27}$$

$$RO_2 \cdot + RH \overset{k_p}{\rightarrow} RO_2H + R \cdot \tag{28}$$

$$RO_2 \cdot + AH \overset{k_{inh}}{\rightleftharpoons} RO_2H + A \cdot \tag{228}$$

$$2 A \cdot \rightarrow termination \tag{229}$$

$$A \cdot + RO_2 \cdot \rightarrow termination \tag{230}$$

$$2 RO_2 \cdot \overset{k_t}{\rightarrow} termination \tag{30}$$

$$A \cdot + RH \rightarrow AH + R \cdot \tag{238}$$

Addition of a second more hindered inhibitor (BH) leads to the additional steps

$$A \cdot + BH \rightleftharpoons AH + B \cdot \tag{240}$$

$$RO_2 \cdot + BH \rightleftharpoons RO_2H + B \cdot \tag{241}$$

$$A \cdot + B \cdot \rightarrow \text{termination} \tag{242}$$

$$RO_2 \cdot + B \cdot \rightarrow \text{termination} \tag{243}$$

The kinetic analysis for reactions (24)—(28), (228)—(30), and (238)—(243) has been solved by Mahoney [166] to give a complex rate law for oxygen uptake of the form

$$R_i^{1/2} [RH]^{-1} (R_{O_2} - R_i) = \frac{(1 + \alpha)}{(K_f + K_g \alpha + K_h \alpha^2)} \tag{244}$$

where K_f, K_g and K_h are products and sums of rate coefficients for the above reactions and concentrations of RH and ROOH. From a computer fit of their data, Mahoney and coworkers solved limiting forms of the equation for ratios of values of k_p, k_{inh}, k_{229}, k_{230}, k_t, and k_{238}. The kinetics of inhibition are simplified considerably when reactions (30), (243), and (−228) can be neglected. These conditions are almost always met if (i) BH is hindered enough so that reaction (−241) is very slow and (ii) sufficient $RO_2 \cdot$ are present to scavenge all $A \cdot$ or $B \cdot$ via reactions (230) or (243); under these conditions

$$R_{O_2} = \frac{R_i k_p [RH]}{2k_{inh}[AH]} \tag{245}$$

which is the same form as eqn. (237) found for phenols in styrene. Oddly enough, in more complex situations where two antioxidants AH and BH are present (one unhindered and one hindered) and reactions (−228) and (229) are suppressed, a simple rate expression gives the oxygen uptake as

$$R_{O_2} = R_i + \left(\frac{R_i k_p [RH]}{q k_p [RH] + q k_{241}[BH]} \right) \tag{246}$$

where $q = 1$ if reaction (242) is the only termination process and $q = 2$ if reaction (243) is the only termination process. Much of the temperature-dependent kinetic data are based on ESR studies of the simpler systems of hindered phenols, amines or thiophenols where eqn. (245) holds.

7.3 RATE COEFFICIENTS AND PARAMETERS FOR INHIBITION

The general inhibition scheme shown above has been partly dissected to give absolute rate coefficients and parameters for individual rate steps. Not surprisingly, the bulk of kinetic information is concerned with reactions (228) and (241), and H-Atom transfer from phenols and amines to $RO_2 \cdot$ radicals. However, some data are available for other rate steps involving both unhindered (AH) and hindered (BH) phenols. Some generalized values of these rate coefficients are summarized in Table 25.

TABLE 25
Generalized rate coefficients for inhibited oxidations

Reaction number	Reactants [a]	Rate coefficient (1 mole^{-1} s^{-1})
228	$RO_2 \cdot + AH$ [b]	1×10^4
-228	$A \cdot + RO_2H$	650
229	$2 A \cdot$	$(0.2-20) \times 10^7$
230	$A \cdot + RO_2 \cdot$	$(32 \pm 23) \times 10^7$
231	$A \cdot + RH$	$32-100$
240	$A \cdot + BH$	6×10^5
-240	$B \cdot + AH$	6×10^3
242	$A \cdot + B \cdot$	5×10^8
243	$RO_2 \cdot + B \cdot$	1×10^8
241	$RO_2 \cdot + BH$	1×10^4
-241	$B \cdot + RO_2H$	0.4

[a] In most cases, A is p-MeOC$_6$H$_4$OH, B is $(t$-Bu$)_3$C$_6$H$_2$OH and RH is dihydroanthracene (see ref. 89).
[b] The value of k_{228} for p-methylphenol with the t-BuO$_2 \cdot$ radical (ref. 88).

More extensive listings of rate coefficients are found in Howard's [10] and Denisov's [168] compilations.

H-Atom transfers from hindered phenols to $RO_2 \cdot$, reaction (241), are characterized by moderately strong substituent effects in which rate constants generally fit best to ρ^+ with values ranging from -1.5 to 0.8: electron-donating substituents accelerate transfer from phenol to the phenoxy radical [88]. At 30°C, a 4-cyano-substituted 2,6-dimethylphenol is about one-tenth as reactive as a 4-methoxy-substituted phenol toward t-butylperoxy radical. Rate coefficients for reaction (241) involving a series of 2,4,6-trialkyl phenols all have values close to $(2 \pm 1) \times 10^4$ 1 mole^{-1} s^{-1} at 30—65°C. Similarly, changes in the structure of the alkylperoxy radical seem to have little effect on k_{241} although a steric effect in reaction of a bulky $RO_2 \cdot$ was reported by Mahoney and DaRooge [89]. The most striking kinetic feature of reaction (228) or (241) is the very low values found for log A and E. For a series of 2,6-di-t-butyl-4-substituted phenols, Howard and Furimsky [88] found log $A_{241} = 4.1 \pm 0.4$ 1 mole^{-1} s^{-1} and $E_{241} = 0.8 \pm 0.3$ kcal mole^{-1}, Howard and coworkers [169] also found that, for unhindered phenols, amines and thiols, log A_{228} falls in a similar range of $4-7$ 1 mole^{-1} s^{-1} while E_{228} is in the range 1—5 kcal mole^{-1}. Rate parameters and coefficients for reactions of several phenols and amines with t-BuO$_2 \cdot$ are summarized in Table 26. No single explanation fully accounts for the low values found for log A_{inh} and E_{inh} for these reactions compared with H-atom transfers to $RO_2 \cdot$ from C—H bonds. One explanation offered is that H-bonding of phenols with $RO_2 \cdot$ precedes transfer and has the effect of reducing log A and E by the values

TABLE 26

Rate coefficients and parameters for reactions (228) and (241); reaction of unhindered phenols and amines with the t-BuO$_2$· radical at 30°C [a]

ArXH [b]	log(A/l mole^{-1} s^{-1})	E(kcal mole^{-1})	10^{-3} k_{inh}(l mole^{-1} s^{-1})
PhMe	8.2	11.2	0.00001
PhOH	7.2	5.2	2.8
PhNH$_2$	6.3	5.0	0.5
PhSH	4.5	1.1	5.1
β-NapOH	6.4	2.6	33
β-NapNH$_2$	4.7	2.3	1.1
β-NapSH	4.8	1.5	5.2

[a] From ref. 169.
[b] Ph = phenyl; Nap = naphthyl.

of the enthalpy and entropy for complexation, viz.

$$RO_2\cdot + HOAr \rightleftharpoons [\dot{R}O_2 \cdots HOAr] \qquad (247)$$

$$[RO_2 \cdots HOAr] \rightarrow RO_2H + \cdot OAr \qquad (248)$$

$$k_{inh} = K_{247}k_{248} \qquad (249)$$

Another factor which might also lower E_{inh} is the bond dissociation energy of the additive (XH). Comparison of di-t-butylphenol with phenol suggests that E can be as much as 6 kcal mole^{-1} smaller for the former phenol because of its lower bond dissociation energy [89]. Probably both smaller bond dissociation energies and complexation contribute to lower values of A and E for phenols. Amines and thiophenols do not appear to complex with RO$_2$· and some other explanation must be sought for their low A values. Howard and Furimsky [88] have suggested that, since E is lowered owing to a low D(X—H), A is also lowered because of correlation effects between A and E. Zavitas [170] has calculated quantum effects on E for H-atom transfer from C—H and X—H; this calculation suggests that H-atom transfer from oxygen is faster than from carbon because of more favorable transition state repulsion energies. Differences of as much as 5 kcal mole^{-1} could arise from this effect.

7.4 SYNERGISTIC EFFECTS OF PHENOLS

Perhaps the most important practical applications of inhibitors arise from the use of combinations of phenols that give more effective inhibition than where each is used alone [171]. Mahoney [166] notes that the most striking synergistic effects are found with combinations of a hindered phenol, with one t-butyl in the *ortho* position, and an unhindered phenol. The foregoing kinetic analysis and schemes help in the understanding of this effect: at elevated temperatures, conversion of RO$_2$· to

A· will not inhibit the oxidation process efficiently because A· can also propagate the chain via reaction (234). However, if A· is rapidly converted to B· via reaction (240), then inhibition is much more effective because sterically hindered B· cannot restart chains efficiently but can terminate A· and RO_2· very efficiently. Use of BH alone is not as effective as in a mixture with AH because more BH than AH is needed to compete with RH, that is $k_{inh} > k_{241}$.

7.5 INHIBITION BY METAL COMPLEXES

Howard and coworkers have examined the mode of action of several zinc [172,173], nickel [172,173], and cupric [174,175] complexes of dialkyldithiophosphoric acids and dialkyldithiocarbamic acids. These complexes are good scavengers of RO_2· radicals, having inhibition rate coefficients (k_{inh}) of 10^3—10^4 l mole^{-1} s^{-1} for zinc and nickel and $>10^6$ l mole^{-1} s^{-1} for cupric. Both zinc and nickel complexes exhibit inhibition kinetics in oxidations of cumene or styrene that follow eqn. (245); that is, each complex scavenges two RO_2· by the reaction sequence (228), (230). Cupric complexes exhibit more complex kinetics, which appear to fit a scheme involving stepwise oxidation of the complex through as many as four intermediates. The stoichiometry is therefore high and partly accounts for the high rate coefficients for cupric complexes. Oxidation at phosphorus also occurs in this process.

8. Photooxidations involving singlet molecular oxygen ($^1O_2$)

8.1. INTRODUCTION

Oxygen is a triplet in its electronic ground state ($^3O_2$), but is relatively unreactive with most organic compounds; radical intermediates are required to effect incorporation of $^3O_2$ into the molecule. In contrast, the first electronically excited singlet state of oxygen ($^1O_2$) the so-called $^1\Delta g$, is spin paired [176], has 22 kcal mole^{-1} more energy than $^3O_2$, and exhibits a variety of electrophilic reactions with organic structures unique to this species.

Photooxidations have been known for many years [2] but the identity of $^1O_2$ as the specific oxidant in many reactions was established only about 15 years ago by Foote et al. [178,179] although Kautsky et al. [180] first proposed the idea of $^1O_2$ and Schenck et al. [181,182] made significant contributions to the field in the period from 1930 to 1960. Several recent reviews of the field are those of Schaap [177], Foote [183,184], Ranby and Radek [14], and those edited by Trozzolo [185] and Mayo [186].

8.2 CHEMISTRY OF PHOTOOXIDATIONS WITH $^1O_2$

Two general classes of photooxidation are now recognized: type I and type II. Type I reactions are photooxidations which usually involve formation of an $n-\pi^*$ triplet diradical species by photolysis of carbonyls such as benzophenone or acetophenone.

$$\text{ArC(O)R} \xrightarrow{h\nu} \text{Ar}\dot{\text{C}}(\dot{\text{O}})\text{R} \tag{250}$$

Triplet diradicals react with organic molecules in a very similar manner to RO· radicals (Sect. 4.3.4) and are useful for photoinitiating free radical oxidations; these photooxidations therefore closely resemble oxidations by other oxy radicals and will not be considered further.

Type II photooxidations involve energy transfer from triplet sensitizers to $^3O_2$ to form what is now recognized as $^1O_2$. Dyes such as rose bengal and methylene blue [187] and metal complexes such as ruthenium(bipyr)$_3$ [188—189] are very efficient sensitizers in visible light with quantum efficiencies close to 1. The process is generally described by the reactions

$$\text{Sen.} \xrightarrow{h\nu} {}^1\text{Sen.} * \tag{251}$$

$$^1\text{Sen.} * \to {}^3\text{Sen.} * \tag{252}$$

$$^3\text{Sen.} * + {}^3O_2 \to \text{Sen.} + {}^1O_2 \tag{253}$$

$$^1O_2 \xrightarrow{k_{ST}} {}^3O_2 \tag{254}$$

The reactions of $^1O_2$ with organic molecules invariably involve electron transfer to electrophilic oxygen usually, but not always, accompanied by C—O bond formation. Because organic molecules react as electron donors, alkanes, simple olefins and aromatics are unreactive as are electron-deficient structures such as alcohols, esters, ketones, sulfur(IV) or (VI) and most amines. However, substituted olefins, dienes, polycyclic aromatics, sulfides, imines, and phenols can react with great rapidity producing a bewildering variety of final products, although the intermediates are usually simple dioxetanes or hydroperoxides. Foote [184] has characterized five types of $^1O_2$ reactions with organic molecules:

(i) Ene-reaction with many internal olefins to give allylic hydroperoxides

$$\tag{255}$$

(ii) Cyclo addition to dienes (2 + 4) to give peroxides

$$\tag{256}$$

(iii) Cycloaddition to electron rich olefins (2 + 2)

$$\begin{array}{c} \diagdown \\ \diagup \end{array} C=C \begin{array}{c} X \\ \diagup \\ \diagdown \end{array} + O_2 \rightarrow \qquad \qquad \tag{257}$$

(iv) Oxidation of sulfur in sulfides, disulfides, and mercaptans

$$2\,R_2S + {}^1O_2 \rightarrow 2\,R_2SO \tag{258}$$

(v) Hydrogen transfer from phenols and other hydrogen donors

$$ArOH + {}^1O_2 \rightarrow ArO\cdot + HO_2\cdot \tag{259}$$

Of particular interest for this review are those reactions of $^1O_2$ that closely resemble those found in the autoxidation. Since both $^1O_2$ and $RO_2\cdot$ react with many olefins to form hydroperoxides, some basis for distinguishing between these reactants can be important in understanding the detailed mechanism of oxidation of a specific compound, particularly in photooxidations where both type I and II processes can occur. Many simple substituted olefins give very similar mixtures of hydroperoxides by the two pathways; however, certain structural units do give markedly different hydroperoxides from $RO_2\cdot$ and $^1O_2$ and are useful as criteria for mechanism. Examples are 1,2-dimethylcyclohexene [190]

$$ \xrightarrow{RO_2\cdot} \qquad + \qquad \tag{260} $$

$^1O_2$ 40%   54%

$$ \tag{261} $$

90—96%

and cholesterol (5-cholesten-3β-ol) [191]

$$ \xrightarrow{RO_2\cdot} \qquad OOH(\alpha,\beta) \tag{262} $$

$$ \xrightarrow{{}^1O_2} \qquad + \qquad \tag{263} $$

An important mechanistic distinction between the two processes was illustrated by Golnick [192] using (+)-limonene in which the different and complex mixtures of hydroperoxides formed by both oxidants were racemic from $RO_2\cdot$ oxidation but optically active from $^1O_2$ oxidation.

Other studies using deuterium-labeled allylic sites show that in the ene reaction (255), only hydrogens oriented *cis* to the point of attachment of $^1O_2$ are removed [182]. These lines of evidence support a concerted mechanism in which $^1O_2$ attaches at C-1 with simultaneous transfer of the *cis* H-atom to C-3 and shift of the double bond.

$$---C{=}C \diagup \quad C{-}C \diagdown \quad \diagup \\ O \quad C \longrightarrow O_2H \quad C{-} \\ O{-}H \diagdown \quad \diagup$$

(264)

To suppress the free radical oxidation path in photooxidations, radical inhibitors such as di-*t*-butylphenols are sometimes added to photooxidizing olefins. However, Foote [184] has noted that some phenols react rapidly with $^1O_2$ or quench it. Similarly, the effect of adding a $^1O_2$ quencher [177, 184,185] such as carotene or diazobicyclooctene to oxidizing systems must also be interpreted cautiously since these quenchers may be oxidized via $RO_2\cdot$ or $RO\cdot$ radical chains.

Solvent effects in reactions of $^1O_2$ are quite striking, mostly because of the large solvent effect on the rate coefficient for the singlet—triplet transition [reaction (254)].

Solvent effects on rates of reactions of $^1O_2$ have been investigated in detail by Merkel and Kearns [193], Foote and Denny [194], and Young et al. [195]. Absolute rate measurements of Merkel and Kearns show clearly that solvent mainly affects the value of k_{ST}, the unimolecular rate coefficient for radiationless decay of $^1O_2$ to ground-state triplet oxygen [reaction (254)]. In water, k_{ST} is larger than in any other solvent and the lifetime ($1/k_{ST}$) is the shortest (2 μs). In CS_2 and CCl_4, k_{ST} is much smaller and $1/k_{ST}$ is 200—700 μs. These investigators also showed that deuteration of water or methanol increased the lifetime of $^1O_2$ almost tenfold; deuteration of acetone, however, had little effect on the lifetime. Evidently, reaction (254) involves coupling of electronic to vibrational levels in H—O bonds. This observation is the basis for the use of D_2O or CD_3OD to confirm the role of $^1O_2$, rather than some other oxidant, in a photo-oxygenation process.

Solvent appears to have little effect on the rate of reaction of $^1O_2$ with many classes of chemicals. The rate of reaction with diphenylisobenzo-furan was unchanged in several solvents, excluding water where dimerization may have accelerated its reactivity. However, Young et al. [195] noted that $^1O_2$ oxidation of some furans showed significant solvent effects ($\times 32$), whereas reactions with olefins showed only small effects ($< \times 2$).

Products of $^1O_2$ reactions with some olefins show striking solvent effects on the partition between the ene and $2 + 2$ addition e.g., [196]

Polar solvent (CH_3CN) (265)

Non-polar solvent (C_6H_6) (266)

8.3 KINETIC RELATIONS AND RATE COEFFICIENTS

Kinetic schemes for $^1O_2$ oxidations can be written in the general form

Sen. $\overset{hv}{\to}$ ^{3}Sen. *   (251,252)

^{3}Sen. * $+ {}^3O_2 \to$ Sen. $+ {}^1O_2$ (ϕk_{TS}) (253)

$^1O_2 + S_1 \to S_1O_2$ (k'_{OX}) (261)

$^1O_2 + S_2 \to S_2 + O_2$ or S_2O_2 (k''_{OX}) (267)

$^1O_2 \to {}^3O_2$ (k_{ST}) (254)

where Sen is a triplet sensitizer (usually a dye), S_1 is the chemical of interest; S_2 is a chemical that competes at a known rate with S_1 for $^1O_2$, and k_{ST} is the rate of the radiationless transition from singlet to triplet (ground) state. Of interest here is the value of k'_{OX} where k''_{OX} and k_{ST} are known. Merkel and Kearns [193] measured k'_{OX} directly using a pulsed laser technique in which the loss of the chemical (diphenylisobenzofuran) was measured spectrometrically following pulse irradiation of methylene blue sensitizer in methanol.

Young et al. [195] described a general competitive technique to evaluate relative rate coefficients in which the loss of standard compound, S_2, is followed at different initial concentrations of S_1, the rate expression being

$$-\frac{d[S_2]}{dt} = \frac{\phi k_{TS} k''_{OX}[S_2]}{k''_{OX}[S_2] + k'_{OX}[S_1] + k_{ST}}$$ (268)

If $k''_{OX}[S_2]$ is made small compared with other terms in the denominator, by keeping $[S_2]$ below $<10^{-6}$ M, then the equation simplifies to

$$-\frac{d[S_2]}{dt} = \frac{\phi k_{TS} k''_{OX}[S_2]}{k'_{OX}[S_1] + k_{ST}}$$ (269)

and a simple kinetic analysis gives the ratio k'_{OX}/k_{ST} from which k'_{OX} is readily estimated.

Simple competitive techniques have also been used for estimating relative reactivity toward $^1O_2$. Bartlett et al. [190] evaluated the relative rate coefficients for a series of vinyl ethers in acetone. Gollnick et al. [187] measured the reactivities for a series of olefins and Matsuura et al. [197] measured the relative reactivities for cyclic olefins in methanol. The relative reactivities are expressed as the ratio of two first-order processes

$$\frac{\ln[S_1]_o/[S_1]_t}{\ln[S_2]_o/[S_2]_t} = \frac{k'_{OX}[S_1]}{k''_{OX}[S_2]} \tag{270}$$

where $[S_1]$ and $[S_2]$ refer to concentrations of two different chemicals at times zero and t. This method has great advantages of simplicity over inhibition kinetic methods, but does require more careful analyses for loss of chemicals and selection of a standard chemical that has a reactivity similar to that of the test chemical.

Table 27 summarizes the rate coefficients for several types of $^1O_2$ reactions at or near 25°C. Rate coefficients for electron-rich donors are very large, approaching the diffusion limit for bimolecular rates, and, with significant concentrations of reactants, these reactions can proceed very rapidly.

Kinetic parameters are not available for these reactions but in many cases their activation energies can hardly exceed 2—3 kcal mole^{-1} since the oxidations proceed nearly as rapidly at —70°C as at 25°C [177,181,182]. Indeed, the lack of significant temperature coefficients for these reactions may be the most reliable diagnostic procedure for distinguishing between oxidation by $RO_2\cdot$ and $^1O_2$; the former process generally becomes immeasurably slow below 0°C, whereas the rate of the latter process may be scarcely affected by a change of 50—100°. This distinction may not apply, of course, to type I photooxidation processes where the triplet diradical

TABLE 27
Rate coefficients for oxidation by singlet oxygen at 25°C

Structure	$k'O_2$ [a] (l mole^{-1} s^{-1})
Unsubstituted olefin	3×10^3
Cyclic olefin	2×10^5
Mono-substituted olefin	1×10^6
Dialkyl sulfide	7×10^6
Diene	1×10^7
Imidazole	4×10^7
Disubstituted olefin	5×10^7
Dialkylfuran	1.4×10^8
Trialkyleneamine	8×10^8

[a] From ref. 184

will effect H-atom transfer rapidly even at low temperatures, thus initiating a radical process. In this case, photooxidation at low temperatures will proceed by way of non-chain radical steps or by concerted $^1O_2$ reactions. Since the type I process will lead only to products of radical interactions, rather than hydroperoxides, at low temperature their formation in significant yields from olefins is diagnostic of a $^1O_2$ process.

References

1 Two excellent reviews of the early developments in oxidation chemistry are (a) E.H. Farmer and A. Sundralingham, J. Chem. Soc., (1942) 121; (b) C. Walling, Free Radicals in Solution, Wiley, New York, 1957, Chap. 9.
2 H.L.J. Backstrom, Z. Phys. Chem. Abt. B., 25 (1934) 99.
3 H. Hock and O. Schrader, Naturwissenschaften, 24 (1936) 159.
4 R. Criegee, H. Pilz and H. Flygare, Chem. Ber., B72 (1939) 1799.
5 E.H. Farmer and D.A. Sutton, J. Chem. Soc., (1946) 10 and references therein.
6 D. Barnard, L. Bateman, J.I. Cuneen and J.F. Smith, in L. Bateman (Ed.), The Chemistry and Physics of Rubber-like Substances, McClaren and Sons, London, 1963, Chap. 17.
7 L. Bateman, H. Hughes and A.L. Morris, Discuss. Faraday Soc., 14 (1953) 90.
8 G.A. Russell and R.C. Williamson, J. Am. Chem. Soc., 86 (1964) 2357 and references therein.
9 F.R. Mayo, A.A. Miller and G.A. Russell, J. Am. Chem. Soc., 80 (1958) 2501.
10 J.A. Howard, Adv. Free-Radical Chem., 5 (1972) 49.
11 N.M. Emanuel, E.T. Denisov and Z.K. Maizus, Liquid Phase Oxidation of Hydrocarbons, English translated by B.S. Hazzard, Plenum Press, New York, 1967.
12 F.R. Mayo, Proc. Int. Oxidation Symp., Adv. Chem. Ser., 75—77 (1968).
13 S.W. Benson, Thermochemical Kinetics, Wiley, New York, 2nd edn., 1976.
14 B. Ranby and J.F. Radek (Eds.), Singlet Oxygen Reactions with Organic Compounds, Wiley, New York, 1978.
15 H.N. Stephens, J. Am. Chem. Soc., 50 (1928) 568.
16 J.L. Bolland and G. Gee, Trans. Faraday Soc., 42 (1946) 236.
17 J.L. Bolland, Q. Rev., 3 (1949) 1.
18 (a) D.E. Van Sickle, F.R. Mayo and R.M. Arluck, J. Am. Chem. Soc., 87 (1965) 4824, 4832; (b) D.E. Van Sickle, F.R. Mayo, R.M. Arluck and M.G. Syz, J. Am. Chem. Soc., 89 (1967) 967.
19 C.E. Barnes, R.M. Elofson and G.D. Jones, J. Am. Chem. Soc., 72 (1950) 210.
20 A.A. Miller and F.R. Mayo, J. Am. Chem. Soc., 78 (1956) 1017.
21 F.R. Mayo and A.A. Miller, J. Am. Chem. Soc., 78 (1956) 1023.
22 F.R. Mayo, J. Am. Chem. Soc., 80 (1958) 2465.
23 F.R. Mayo and A.A. Miller, J. Am. Chem. Soc., 80 (1958) 2480.
24 F.R. Mayo, A.A. Miller and G.A. Russell, J. Am. Chem. Soc., 80 (1958) 2501, 6701.
25 W.A. Pryor, Free Radicals, McGraw-Hill, New York, 1966 p. 123.
26 W.A. Pryor, Polym. Prepr. Am. Chem. Soc. Div. Polym. Chem., 12 (1971) 49.
27 J.A. Howard and K.U. Ingold, Can. J. Chem., 43 (1965) 2737.
28 D.E. Winkler and G.W. Hearne, Ind. Eng. Chem., 53 (1961) 655.
29 D. Allara, T. Mill, F.R. Mayo and D.G. Hendry, Adv. Chem. Ser., 76 (1968) 40.
30 G.R. McMillan and J.G. Calvert, in C.F.H. Tipper (Ed.), Oxidation and Combustion Reviews, Elsevier, Amsterdam, 1965, p. 84.
31 S.W. Benson, J. Am. Chem. Soc., 87 (1965) 972.

84

32 J.H. Knox, Combust. Flame, 9 (1965) 297.

33 H.D. Medley and S.D. Cooley, Adv. Pet. Chem. Refin., 3 (1960) 309.

34 D.G. Hendry, J. Am. Chem. Soc., 89 (1967) 5433 and references therein.

35 T. Mill, F.R. Mayo, H. Richardson, K. Irwin and D. Allara, J. Am. Chem. Soc., 94 (1972) 6802.

36 F.R. Mayo, Am. Chem. Soc. Div. Pet. Chem. Prepr., 19 (1974) 627.

37 F.F. Rust, J. Am. Chem. Soc., 79 (1957) 4000.

38 T. Mill and G. Montorsi, Int. J. Chem. Kinet., 5 (1973) 119.

39 J.H.B. Chenier, S.B. Tong and J.A. Howard, Can. J. Chem., 56 (1978) 3047.

40 H.E. O'Neal and S.W. Benson, J. Phys. Chem., 71 (1967) 2903.

41 D.E. Van Sickle, T. Mill, F.R. Mayo, H. Richardson and C.W. Gould, J. Org. Chem., 38 (1973) 4435.

42 T. Berry, C.F. Cullis, M. Saed and D. Trimm, Adv. Chem. Ser., 76 (1968) 86.

43 D.G. Hendry, T. Mill, L. Piszkiewicz, J.A. Howard and H.K. Eigenmann, J. Phys. Chem. Ref. Data, 3 (1974) 937.

44 M. Anbar and P. Neta, Int. J. Appl. Radiat. Isot., 18 (1967) 493.

45 J.E. Bennett, D.M. Brown and B. Miles, Trans. Faraday Soc., 66 (1970) 386, 397.

46 K. Adamic, J.A. Howard and K.U. Ingold, Can. J. Chem., 47 (1969) 3803.

47 J.R. Thomas, J. Am. Chem. Soc., 87 (1965) 3935.

48 J.R. Thomas and K.U. Ingold, Adv. Chem. Ser., 75 (1968) 258.

49 R.L. MacCarthey and A. MacLachlan, J. Chem. Phys., 35 (1961) 1625.

50 W.T. Dixon and R.O.C. Norman, J. Chem. Soc., (1963) 3119.

51 J.R. Thomas, J. Am. Chem. Soc., 88 (1966) 2064.

52 A.A. Vichntinskii, Dokl. Akad. Nauk SSSR, 91 (1969) 7113.

53 A.A. Vichntinskii, A.F. Guk, V.F. Trespalov and V.Ya. Shlyapintokh, Izv. Akad. Nauk SSSR Ser. Khim, (1966) 1672.

54 R.E. Kellog, J. Am. Chem. Soc., 91 (1969) 5433.

55 G.A. Russell, J. Am. Chem. Soc., 79 (1957) 3871.

56 P.D. Bartlett and T.G. Traylor, J. Am. Chem. Soc., 85 (1963) 2407.

57 L. Bateman and G. Gee, Proc. R. Soc. London Ser. A, 195 (1948) 391.

58 J.G. Calvert and J.N. Pitts, Jr., Photochemistry, Wiley, New York, 1967, p. 496.

59 C.H. Bamford and M.J.S. Dewar, Proc. R. Soc. London Ser. A, 198 (1949) 252.

60 L. Bateman, Q. Rev., 8 (1954) 147.

61 F. Briers, D.L. Chapman and E. Walters, J. Chem. Soc., (1926) 562.

62 G.M. Burnett and H.W. Melville, Techniques of Organic Chemistry, Vol. VIII, Part II, Interscience, New York, 1963, Chap. 20.

63 H. Kwart, J. Phys. Chem., 64 (1960) 1250.

64 M.S. Matheson, E.S. Auer, E.B. Bevilacqua and E.J. Hart, J. Am. Chem. Soc., 71 (1949) 497.

65 T.G. Traylor and C.A. Russell, J. Am. Chem. Soc., 87 (1965) 3698.

66 J.A. Howard, W.J. Schwalm and K.U. Ingold, Adv. Chem. Ser., 75 (1968) 6.

67 D.M. Golden and S.W. Benson, Chem. Rev., 69 (1969) 125.

68 D.G. Hendry and G.A. Russell, J. Am. Chem. Soc., 86 (1964) 2371.

69 J.A. Howard and K.U. Ingold, Can. J. Chem., 46 (1968) 2655.

70 C.I. Ayers, E.G. Janzen and F.J. Johnston, J. Am. Chem. Soc., 88 (1966) 2612.

71 W.C. Sleppy and J.C. Calvert, J. Am. Chem. Soc., 81 (1959) 769.

72 D.G. Hendry, F.R. Mayo and D. Schuetzle, Ind. Eng. Chem. Prod. Res. Dev., 7 (1968) 136.

73 D.G. Hendry and D. Schuetzle, J. Am. Chem. Soc., 97 (1975) 7123.

74 J.A. Howard and K.U. Ingold, Can. J. Chem., 45 (1967) 785.

75 F.R. Mayo, M. Syz, T. Mill and J.K. Castleman, Adv. Chem. Ser., 75 (1968) 38.

76 G.A. Russell, J. Am. Chem. Soc., 77 (1955) 4583.

77 L. Sajus, Adv. Chem. Ser., 75 (1968) 59.

78 J.A. Howard and K.U. Ingold, Can. J. Chem., 46 (1968) 2661.

79 S. Korcek, J.H.B. Chenier, J.A. Howard and K.U. Ingold, Can. J. Chem., 50 (1972) 2285.
80 J.A. Howard, J.H.B. Chenier and D.A. Holden, Can. J. Chem., 56 (1978) 170.
81 R.R. Baldwin and R.W. Walker, Fourteenth International Symposium on Combustion, The Combustion Institute, Pittsburgh, 1973, p. 212.
82 A.E. Eachus, J.A. Meyer, J. Pearson and M. Szwarc, J. Am. Chem. Soc., 90 (1968) 3646.
83 R. Walsh and S.W. Benson, J. Am. Chem. Soc., 88 (1966) 3480.
84 J.L. Bolland, Trans. Faraday Soc., 46 (1950) 358.
85 J.A. Howard and S. Korchek, Can. J. Chem., 48 (1970) 2165.
86 J.A. Howard, K.U. Ingold and M. Symonds, Can. J. Chem., 46 (1968) 1017.
87 D.G. Hendry and D. Schuetzle, J. Org. Chem., 4 (1976) 3179.
88 J.A. Howard and E. Furimsky, Can. J. Chem., 51 (1973) 3738.
89 L.R. Mahoney and M.A. DaRooge, J. Am. Chem. Soc., 97 (1975) 4722.
90 J.A. Howard, Can. J. Chem., 50 (1972) 2298.
91 J.A. Howard and K.U. Ingold, Can. J. Chem., 43 (1965) 2729.
92 J.A. Howard and K.U. Ingold, Can. J. Chem., 44 (1966) 1113.
93 J.A. Kerr and M.J. Parsonage, Evaluated Kinetic Data on Gas Phase Addition Reactions, Butterworths, London, 1972, p. 219.
94 D.E. Van Sickle, F.R. Mayo, E.S. Gould and R.M. Arluck, J. Am. Chem. Soc., 89 (1967) 977.
95 A. Fish, Proc. R. Soc. London Ser. A, 298 (1967) 204.
96 A. Fish, Adv. Chem. Ser., 96 (1968) 69.
97 G.A. Twigg, Chem. Eng. Sci. Suppl., 3 (1954) 5.
98 C. Walling, J. Am. Chem. Soc., 91 (1969) 7590.
99 R. Hiatt, in D. Swern (Ed.), Organic Peroxides, Vol. II, Wiley, New York, 1970, pp. 102—106.
100 L. Batt, K. Christie, R.T. Milne and A.J. Summers, Int. J. Chem. Kinet., 6 (1974) 877.
101 S.W. Benson and R. Shaw, in D. Swern (Ed.), Organic Peroxides, Vol. I, Wiley, New York, 1972, pp. 105—139.
102 D.R. Stull, E.F. Westrum and G.C. Sinke, The Chemical Thermodynamics of Organic Compounds, Wiley, New York, 1969.
103 C. Walling and B.B. Jacknow, J. Am. Chem. Soc., 82 (1960) 6113; K.M. Johnson and G.H. Williams, J. Chem. Soc. (1960) 1446.
104 C. Walling and P.J. Wagner, J. Am. Chem. Soc., 86 (1964) 3368.
105 P. Gray, R. Shaw and J.C. Thynne, Prog. React. Kinet., 4 (1967) 63.
106 A.L. Williams, E.A. Oberright and J.W. Brooks, J. Am. Chem. Soc., 78 (1956) 1190.
107 D.L. Allara, T. Mill, K.C. Irwin and H. Richardson, Abstr. 162nd Meeting Am. Chem. Soc., Washington, D.C., Sept. 1971, PHYS 37, 38.
108 H. Paul, R.D. Small and J.C. Scaiano, J. Am. Chem. Soc., 100 (1978) 4520.
109 C. Walling and V.P. Kurkov, J. Am. Chem. Soc., 89 (1967) 4895.
110 A.A. Zavitsas and J.D. Blank, J. Am. Chem. Soc., 94 (1972) 4603.
111 C. Walling and A. Padwa, J. Am. Chem. Soc., 85 (1963) 1593.
112 J.K. Kochi, J. Am. Chem. Soc., 84 (1962) 1183.
113 A. Baldwin, J.R. Barker, D.M. Golden and D.G. Hendry, J. Phys. Chem., 81 (1977) 2483.
114 C. Walling and A. Padwa, J. Am. Chem. Soc., 85 (1963) 1597.
115 P. Kabasakalian, E.R. Townley and M.D. Yudis, J. Am. Chem. Soc., 84 (1962) 2716.
116 D.H.R. Barton, R.P. Budhiraja and J.F. McGhie, J. Chem. Soc. C, (1969) 336.
117 J.A. Howard, Am. Chem. Soc. Symp. Ser., 69 (1978) 413.
118 J.R. Thomas, J. Am. Chem. Soc., 89 (1967) 4872.

86

119 J.A. Howard and K.U. Ingold, Can. J. Chem., 45 (1967) 788, 793.
120 R. Hiatt, J. Clipsham and T. Visser, Can. J. Chem., 42 (1964) 2754.
121 A. Factor, C.A. Russell and T.G. Traylor, J. Am. Chem. Soc., 87 (1965) 3692.
122 J.A. Howard, K.U. Ingold and K. Adamic, Can. J. Chem., 47 (1969) 3797.
123 P.D. Bartlett and G. Guaraldi, J. Am. Chem. Soc., 89 (1967) 4799.
124 T. Mill and R. Stringham, J. Am. Chem. Soc., 90 (1968) 1062.
125 J.A. Howard and J.E. Bennett, Can. J. Chem., 50 (1972) 2374.
126 J.A. Howard and K.U. Ingold, J. Am. Chem. Soc., 90 (1968) 1056.
127 J.A. Howard and K.U. Ingold, Can. J. Chem., 43 (1965) 2737.
128 J.A. Howard and S. Korchek, Can. J. Chem., 48 (1970) 2165.
129 J.A. Howard and K.U. Ingold, Can. J. Chem., 46 (1968) 2366.
130 N.A. Clinton, R.A. Kenley and T.G. Traylor, J. Am. Chem. Soc., 97 (1975) 3746, 3752, 3757.
131 R.A. Kenley and T.G. Traylor, J. Am. Chem. Soc., 97 (1975) 4700.
132 D.A. Parkes, Int. J. Chem. Kinet., 9 (1977) 451; L.J. Kirsch, D.A. Parkes, D.J. Waddington and A. Wooley, J. Chem. Soc. Faraday Trans. 1, (1978) 2293.
133 R. Hiatt, T. Mill, K.C. Irwin and J.K. Castelman, J. Org. Chem., 33 (1968) 1428.
134 D.G. Hendry, C.W. Gould, D. Schuetzle, M.G. Syz and F.R. Mayo, J. Org. Chem., 41 (1976) 1.
135 W.E. Jackson and R.H. Verhoek, report to Air Force Off. Sci. Res. by G. Boord, TR58-82 AD 158384 Armed Services Technical Information Agency, Arlington, VA, 1958.
136 L. Bateman and H. Hughes, J. Chem. Soc., (1952) 4594.
137 R. Hiatt, in D. Swern (Ed.), Organic Peroxides, Vol. II, Wiley, New York, 1970, p. 99.
138 A.V. Tobolsky, D.J. Metz and R.B. Mesrobain, J. Am. Chem. Soc., 72 (1950) 1942.
139 G.A. Russell, J. Am. Chem. Soc., 78 (1956) 1047.
140 J. Alagy, G. Clement and J.C. Balaceanu, Bull. Soc. Chim. Fr., (1959) 1325.
141 L. Sajus, Adv. Chem. Ser., 75 (1968) 59.
142 M. Fineman and S.D. Ross, J. Polym. Sci., 5 (1950) 259.
143 D.G. Hendry, Adv. Chem. Ser., 75 (1968) 24.
144 J.A. Howard, in J.K. Kochi (Ed.), Free Radicals, Wiley-Interscience, New York, 1973, p. 24.
145 E.J. Hamilton, Jr., J. Chem. Phys., 63 (1975) 3682.
146 E. Niki, Y. Kamiya and N. Ohta, Kogyo Kagaku Zasshi, 71 (1968) 1187.
147 C. Chevriau, P. Naffa and J.C. Balaceanu, Bull. Soc. Chim. Fr., (1964) 3002.
148 A.W. Hoffman, J. Chem. Soc., 13 (1861) 81.
149 W.O. Lundberg, Autoxidation and Antioxidants, Vol. I, Interscience, New York, 1962.
150 G. Scott, Atmospheric Oxidation and Antioxidants, Elsevier, Amsterdam, 1965.
151 J.L. Bolland and P. ten Haave, Trans. Faraday Soc., 43 (1947) 201.
152 J.L. Bolland and P. ten Haave, Discuss. Faraday Soc., 2 (1947) 252.
153 W.A. Waters and C. Wickham-Jones, J. Chem. Soc., (1951) 812.
154 W.A. Waters and C. Wickham-Jones, J. Chem. Soc., (1952) 2420.
155 R.F. Moore and W.A. Waters, J. Chem. Soc., (1954) 243.
156 K.U. Ingold and E.C. Horswill, Can. J. Chem., 44 (1966) 263, 269.
157 A.F. Bickel and E.C. Kooyman, J. Chem. Soc., (1953) 3211.
158 T.W. Campbell and G.M. Coppinger, J. Am. Chem. Soc., 74 (1952) 1469.
159 C.E. Boozer, G.S. Hammond, C.E. Hamilton and J.N. Sen, J. Am. Chem. Soc., 77 (1955) 3233.
160 K.U. Ingold, Essays on Free Radical Chemistry, Spec. Publ., No. 24, Chemical Society, London, 1970, pp. 285—293.
161 J.A. Howard, Rubber Chem. Technol., 47 (1974) 976.

162 J.A. Howard and K.U. Ingold, Can. J. Chem., 40 (1962) 1851; 41 (1963) 1744.

163 L.R. Mahoney and F.C. Ferris, J. Am. Chem. Soc., 85 (1963) 2345.

164 J.R. Thomas, J. Am. Chem. Soc., 85 (1963) 2166.

165 J.R. Thomas and C.A. Tolman, J. Am. Chem. Soc., 84 (1962) 2930.

166 L.R. Mahoney, J. Am. Chem. Soc., 89 (1967) 1895.

167 L.R. Mahoney, Angew. Chem. Int. Ed. Engl., 8 (1969) 547.

168 E.T. Denisov, Liquid Phase Reaction Rate Constants, English translation by R.K. Johnson, IFI/Plenum, New York, 1974, pp. 249—279.

169 J.H.B. Chenier, E. Furimsky and J.A. Howard, Can. J. Chem., 52 (1972) 3682.

170 A.A. Zavitsas, J. Am. Chem. Soc., 94 (1972) 2779.

171 J.R. Chipault, in W.O. Lundberg (Ed.), Vol. II, Autoxidations and Antioxidants, Interscience, New York, 1962, p. 477.

172 J.A. Howard, Y. Ohkatsu, J.H.B. Chenier and K.U. Ingold, Can. J. Chem., 51 (1973) 1543.

173 J.A. Howard and J.H.B. Chenier, Can. J. Chem., 54 (1976) 382.

174 J.H.B. Chenier, J.A. Howard and J.C. Tait, Can. J. Chem., 55 (1977) 1644.

175 J.H.B. Chenier, J.A. Howard and J.C. Tait, Can. J. Chem., 56 (1978) 157.

176 A.U. Khan and M. Kasha, Ann. N.Y. Acad. Sci., 171 (1970) 79.

177 A.P. Schaap (Ed.), Singlet Molecular Oxygen, Dowden, Hutchinson and Ross Inc., Stroudsberg, PA, 1976.

178 C.S. Foote and S. Wexler, J. Am. Chem. Soc., 86 (1964) 3879, 3880.

179 C.S. Foote, S. Wexler and W. Ando, Tetrahedron Lett., (1965) 4111.

180 H. Kautsky, H. de Bruijn, R Neuwirth and W. Baumeister, Chem. Ber., 66 (1933) 1588.

181 G.O. Schenck and K. Ziegler, Naturwissenschaften, 32 (1944) 157.

182 K. Gollnick and G.O. Schenck, Pure Appl. Chem., 91 (1964) 507.

183 C.S. Foote, Acc. Chem. Res., (1968) 104.

184 C.S. Foote, in W. Pryor (Ed.), Vol. II, Free Radicals in Biology, Academic Press, New York, 1976, pp. 85—124.

185 A.M. Trozzolo (Ed.), Ann. N.Y. Acad. Sci., 171 (1970).

186 F.R. Mayo (Ed.), Adv. Chem. Ser., 77 (1968).

187 K. Gollnick, T. Franken, G. Schade and G. Dorhofer, Ann. N.Y. Acad. Sci., 171 (1970) 89.

188 (a) J.N. Demas, R.P. McBride and E.W. Harris, J. Phys. Chem., 80 (1976) 2248; J.N. Demas, R.P. McBride and E.W. Harris, J. Am. Chem. Soc., 99 (1977) 3547.

189 V.S. Srinivasan, D. Podolski, N.J. Westrick and D.C. Neckers, J. Am. Chem. Soc., 100 (1978) 6513.

190 P.D. Bartlett, G.D. Mendenhall and A.P. Schaap, Ann. N.Y. Acad. Sci., 171 (1970) 79.

191 L.L. Smith, M.J. Kulig, D. Miller and G.A.S. Ansari, J. Am. Chem. Soc., 100 (1978) 6206.

192 K. Gollnick, Adv. Chem. Ser., 77 (1968) 78.

193 P.B. Merkel and D.R. Kearns, J. Am. Chem. Soc., 94 (1972) 7244.

194 C.S. Foote and R.W. Denny, J. Am. Chem. Soc., 93 (1971) 5168.

195 R.F. Young, K. Wehrly and R.L. Martin, J. Am. Chem. Soc., 93 (1971) 5774.

196 P.D. Bartlett and A.P. Schaap, J. Am. Chem. Soc., 92 (1970) 3223.

197 T. Matsuura, A. Horinaka and R. Nakashima, Chem. Lett., (1973) 887.

Chapter 2

The Liquid Phase Oxidation of Aldehydes

L. SAJUS and I. SÉRÉE DE ROCH

1. Introduction

The liquid phase oxidation of aldehydes by molecular oxygen has been known for a long time. Liebig [1], in 1835, noted that, in the presence of air, aldehydes were transformed into acids. However, in 1897, Bach [2] observed that the primary product of autoxidation is peroxidic. Baeyer and Villiger [3] suggested, in the case of benzaldehyde, that there was a primary formation of perbenzoic acid, which reacts consecutively with the aldehyde to produce benzoic acid. Nevertheless, many doubts have been expressed concerning the capacity of pure aldehydes to be oxidized. Induction periods have been observed by many researchers who have noted that oxidation can be initiated by adding traces of metals such as iron, copper, manganese, and cobalt in the form of salts, or else by light. Likewise, oxidation can be halted by a great many inhibitors.

In 1927, Bäckström [4] succeeded, for the photochemical oxidation of benzaldehyde and heptaldehyde, in making the first quantitative measurement of the quantum yield and found that this is between 560 and 15,000 depending on the conditions. Because a photon can react only with a molecule, this specifically implied a radical chain mechanism.

Since the achievements of these pioneers, the oxidation of aldehydes has been the subject of a lot of work using either thermal, photochemical, or catalytic autooxidation or else catalytic oxidation by silver oxide.

Two general articles [5,6] were written around 1955 and sum up the subject at that time. Maslov and Blyumberg [69] have reviewed aldehyde oxidation in solution in 1976, particularly Russian work, and have calculated values of chain lengths and of kinetic parameters of elementary reactions occurring. Denisov [70] has discussed the mechanism of thermal initiation. The great difficulty in obtaining reproducible results is the reason why work on the topic has only been progressing slowly; at any rate, it is an important reason for making a critical examination of the findings published because they are often contradictory.

Considering the importance that controlling the experimental parameters has on the significance of such findings, we will begin by summing up data concerning experimental conditions, reaction medium purity, the nature of reaction products and the conditions under which they are observed, and the techniques for promoting oxidation.

1.1 ANALYTICAL ASPECTS

The purity of reagents and solvents and both the nature and state of cleanliness of the reactor walls are of great importance in the kinetic study of the liquid phase oxidation of aldehydes. Autoxidation actually occurs via a long chain radical mechanism, and any change made either in initiation or in chain propagation or in chain rupture will have important repercussions on the rate of oxygen absorption. Aldehyde is usually purified by successive distillations in a nitrogen atmosphere. Storage must be in a vessel having walls that are not liable to pollute the aldehyde (glass or teflon) and in an inert atmosphere so as to prevent any prior oxidation capable of causing inhibitors to form. The oxygen used must be thoroughly free from ozone and the reactor must be washed solely with acids.

1.2 OPERATIONAL TECHNIQUES

Liquid phase oxidations of aldehydes are generally carried out at moderate temperatures ($-20°C \leqslant t \leqslant 80°C$) and under partial oxygen pressures ranging from several torr to several atmospheres. The reactor is generally made of glass and has a capacity varying from several cm^3 to several liters.

The stirring system works either by shaking, by oxygen circulation, or by means of an externally powered magnetic bar. Stirring must be sufficient for the kinetic measurements to be meaningful and not dependent on the rate of oxygen dissolution.

The kinetic chains can be initiated in several ways, viz. thermally, photochemically, and catalytically. Photochemical initiation is usually by mercury vapor lamps which emit with maximum intensity for λ values in the range from 2500 to 3200 Å. The maximum molecular extinction coefficient, ϵ, for aldehydes corresponds to λ of about 2900 A [21] (Table 1). Photochemical initiation enables oxidations to be achieved at temperatures in the vicinity of $0°C$. In this zone, thermal initiation is practically negligible, which is an advantage in kinetic investigations. Nevertheless, since the light absorption effectiveness depends on various parameters, the amount of light actually absorbed is only known approxi-

TABLE 1
Extinction coefficient, ϵ, for various aldehydes

Aldehyde	λ_{max} (Å)	ϵ_{max} (l mole^{-1} cm^{-1})	Solvent
Acetaldehyde	2934	11.8	n-Hexane
Propionaldehyde	2895	18.2	
n-Butyraldehyde	2900	17.8	
n-Heptaldehyde	2923	24.4	

mately. For catalytic initiation, catalysts with a base of Cu, Mn, Co, etc. are usually used in the form of salts (naphthenates, stearates, etc.) or chelates.

For basic oxidations using silver oxide, addition of copper or iron oxide (Cu_2O, Fe_2O_3) enhances oxidation [22] but, under these conditions, the reaction occurs by a catalytic mechanism without radical chain propagation.

The kinetic examination of the liquid phase oxidation of aldehydes is usually done using initial rates for small conversion ratios. This enables fortuitous autoinhibition and catalysis phenomena to be eliminated. Experimentation must be carried out on a single aldehyde batch, since considerable differences may be observed when going from one bath to another. This is a serious handicap with regard to the absolute values of the rate coefficients determined.

1.3 OXIDATION PRODUCTS

The primary product of the photochemical or catalytic autoxidation of an aldehyde (RCHO) in the liquid phase by dissolved molecular oxygen is the corresponding peracid, RCO_3H, viz.

$$RCHO + O_2 = RCO_3H$$

The secondary reactions of peracid in the reaction medium have given rise to much controversy. Peracid is responsible for the formation of the acid, which is the major final product of the oxidation, but also (at least partly) of inhibitors which are produced in too small a quantity to be identified. The main controversy is concerned with the intermediate formation of an "X peroxide" by reaction of the peracid with the aldehyde, with the structure of this peroxide, and its mode of decomposition. Many attempts have been made to isolate and identify X peroxide, especially in the case of acetaldehyde.

The autoxidation of acetaldehyde has been the subject of detailed analytical research. In 1916, an unstable peroxide other than peracetic acid was discovered and isolated by Galitzenstein and Mugdan [7]. A peroxide compound stable at $-30°C$ was isolated by Kagan and Lubarsky [8] by causing peracetic acid to react on acetaldehyde. This X peroxide was shown to be present in a peracid solution by taking advantage of the difference in reactivity of these two compounds with potassium iodide [8,9]. Wieland [10] was the first to suggest that X peroxide had a structure corresponding to hydroxy-1-ethyl peroxyacetate, viz.

$$CH_3CHO + CH_3\overset{O}{\overset{\|}{C}}-OOH \rightarrow CH_3\overset{OH}{\underset{H}{\overset{|}{C}}}-O-O-\overset{O}{\overset{\|}{C}}CH_3$$

(I)

References pp. 122—124

The same structure was proposed by Kagan and Lubarsky [8].

In 1941, Losch [11] isolated an X peroxide (freezing point 20—22°C) with a molecular mass of 105 and an active oxygen content of 14—15%. At ambient temperature, this peroxide is transformed into acetic acid without loss of weight. Under the influence of a catalyst made of cobalt and copper salts, the peroxide produces 60% acetic anhydride. The formation of this peroxide has been interpreted according to the general mechanism reaction of peracids with carbonyl-containing derivatives, viz.

$$CH_3-\overset{+}{\underset{H}{C}}\overset{O^-}{\lVert} \quad + \quad \overset{+}{H}\cdots O-O-\overset{O}{\underset{}{\overset{\lVert}{C}}}-CH_3 \;\rightarrow\; CH_3-\underset{H}{\overset{OH}{C}}-O-O-\overset{O}{\overset{\diagdown}{C}}CH_3$$

(I)

However, since it is not easy to explain the formation of acetic anhydride from I, Bawn and Williamson [9] proposed a hydroperoxidic hydroperoxy-1-ethyl acetate structure whose formation involves the anion HO_2^-, viz.

$$CH_3-\overset{+}{\underset{H}{C}}\overset{O^-}{\lVert} \quad + \quad \underset{HOO^-}{\overset{O}{\overset{\lVert}{C}}-CH_3} \;\rightarrow\; CH_3-\underset{H}{\overset{OOH}{C}}-O-\overset{O}{\underset{CH_3}{C}}$$

(II)

The change to an anhydride in this case does not imply an intramolecular rearrangement. A third structure, which is that of an "isoozonide", has also been suggested by Wittig and Pieper [12].

In an infrared spectrophotometric investigation, Vasilyev and Emanuel [13] attribute the 847 cm^{-1} band to the O—O bond of a hydroperoxidic structure and favor structure II for the X peroxide. The authors base their argument on the fact that, in peracetic acid, this band is 856 cm^{-1} and, generally speaking, it is 835—855 cm^{-1} in hydroperoxides, Nevertheless, the band corresponding to the O—O bond is 840—842 cm^{-1} in acyl peroxides [14], whereas, in diacyl peroxides [15], it is 890—904 cm^{-1} and hence the argument put forward is not decisive.

As the result of a later infrared investigation, Niclause [16] suggests that the most plausible structure is the one proposed by Wieland. From these observations, made as part of the photochemical oxidation studies of several aldehydes, the authors conclude that it is actually compounds of type I that exist and may be in solution in three forms in equilibrium, viz.

$$RC\diagdown_{O-O}^{\overset{O\cdots HO}{\diagup}}\overset{\diagup}{\diagdown}CH-R \rightleftharpoons RC\diagdown_{O-O}^{\overset{O}{\diagup}}\diagdown_{CH-R}^{\diagdown}\underset{HO}{\diagup} \overset{+RCHO}{\rightleftharpoons} RC\diagdown_{O-O}^{\overset{O}{\diagup}}\diagdown_{CH-R}^{\diagdown}\underset{R-C\diagdown H}{\overset{O\cdots HO}{\diagup}}$$

In the case of acetaldehyde, X peroxide, when pyrolized in solution in a solvent such as acetone or butyl acetate under reduced pressure at about 80°C, reforms peracid and acetaldehyde [17], viz.

$$CH_3-C\diagdown_{O-O}^{\overset{O\cdots HO}{\diagup}}\overset{\diagup}{\diagdown}C-CH_3 \rightarrow CH_3-\overset{O}{\overset{\|}{C}}-OOH + CH_3CHO$$

This reaction scheme leads to the experimental stoichiometric law, based on the analysis of both the aldehyde and the overall peracids present as well as the oxygen consumed

$$\Delta[CH_3CHO] + \Delta[\text{peroxide}] = 2\Delta[O_2]$$

The checking of this law is an easy way of determining whether or not a parasite reaction exists in the system.

To sum up, the overall scheme for the oxidation of an aldehyde may be written as

$$RCHO + O_2 = R\overset{O}{\overset{\|}{C}}-OOH \xrightarrow[\text{catalysis}]{\text{homogeneous}} R\overset{O}{\overset{\|}{C}}-OH + \tfrac{1}{2}O_2\uparrow$$

$$+ RCHO$$

$$RC\diagdown_{O-O}^{\overset{O\cdots HO}{\diagup}}\overset{\diagup}{\diagdown}CHR \overset{\Delta}{\rightarrow} 2\,R\overset{O}{\overset{\|}{C}}-OH$$

$$\downarrow$$

$$(RC=O)_2O + H_2O \rightleftharpoons 2\,R\overset{O}{\overset{\|}{C}}-OH$$

However, in the presence of a different anhydride $(R_1C=O)_2O$, the peracid reacts with this anhydride to produce a mixed diacyl peroxide [18], viz.

$$RC\!-\!OOH + \begin{array}{c} R_1C \overset{\displaystyle O}{\underset{\displaystyle O}{\diagup}} \\ O \\ R_1C \diagdown_O \end{array} \rightarrow \begin{array}{c} RC \overset{\displaystyle O}{\diagdown} \\ O \\ R_1C \diagup^O \\ R_1C \diagdown_O \end{array} + R_1C\!-\!OH$$

As a reminder, in the case of liquid phase heterogeneous catalytic oxidation with a controlled pH in the presence of silver oxide, no intermediate peroxidic species is isolated. The non-radical reaction leads directly to the acid which is obtained in the form of a salt [19,20]. This highly special oxidation method will be dealt with in a separate section.

2. Oxidation of normal saturated aldehydes

Among paraffinic aldehydes, oxidation has been studied only for some straight chain derivatives, e.g. acetaldehyde, heptanal, decanal. We will deal with these cases one by one and attempt to describe the kinetic conclusions observed during thermal, photochemical, and catalytic oxidation. We will begin with decanal because the most quantitative thermal and photochemical oxidation results were obtained with this aldehyde.

2.1 OXIDATION OF n-DECANAL

The oxidation of n-decanal was examined in great detail by Cooper and Melville in 1951 [23]. Kinetic investigations were made under the conditions: at 350—700 torr and 0—30°C using n-decane as solvent. The results lead to the following conlusions with regard to the rate of thermal oxidation, V_{th}, and the rate of purely photochemical oxidation, V_{ph}.

(a) For thermal oxidation, the rate of oxygen absorption is proportional to the square root of the partial oxygen pressure, P_{O_2}, and to the 3/2 power of the aldehyde concentration, viz.

$$V_{th} = k_{th}[RCHO]^{3/2}[O_2]^{1/2}$$

(b) The rate of photochemical oxidation is proportional to the square root of light intensity, I, and to the aldehyde concentration, viz.

$$V_{ph} = k_{ph}[RCHO][O_2]^0 I^{1/2}$$

(c) The overall rate (V_g) of oxidation observed at all temperatures in the presence of light is equal to $V_{ph} + V_{th}$ and is expressed by

$$V_g = V_{ph} + V_{th} = \{k_{ph}I^{1/2} + k_{th}[RCHO]^{1/2}[O_2]^{1/2}\}[RCHO]$$

If it is assumed that thermal initiation is brought about by the reaction

RCHO + O$_2$ → radicals

then the rate of thermal oxidation can be written as

$$V_{th} = k_{th}[RCHO][O_2]^0[RCHO]^{1/2}[O_2]^{1/2}$$

Thus, if the terms relating to initiation are disregarded, both rate expressions are the same.

The mechanism suggested is

Initiation

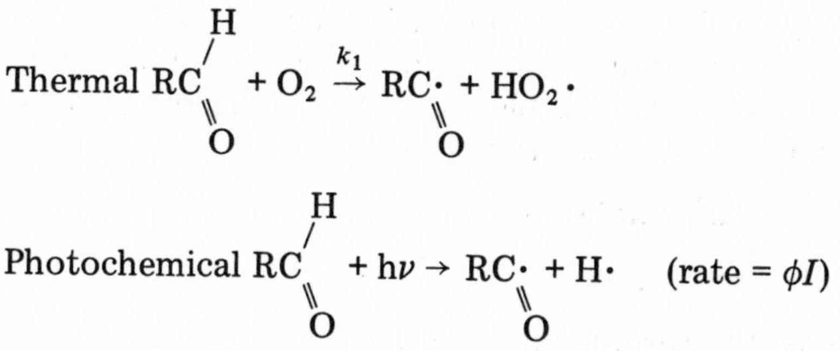

Thermal: RCHO + O$_2$ $\xrightarrow{k_1}$ RC·(O) + HO$_2$·

Photochemical: RCHO + hν → RC·(O) + H·   (rate = ϕI)

Propagation

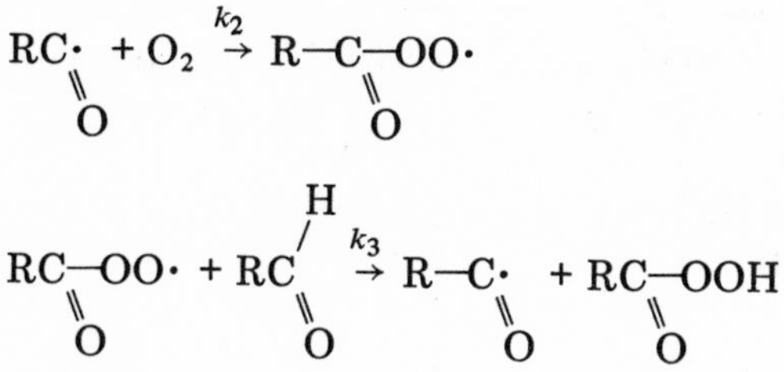

RC·(O) + O$_2$ $\xrightarrow{k_2}$ R—C(O)—OO·

RC(O)—OO· + RCHO $\xrightarrow{k_3}$ R—C·(O) + RC(O)—OOH

Terminations

2 RC·(O) $\xrightarrow{k_4}$

RC·(O) + RC(O)—OO· $\xrightarrow{k_5}$ } products

2 RC(O)—OO· $\xrightarrow{k_6}$

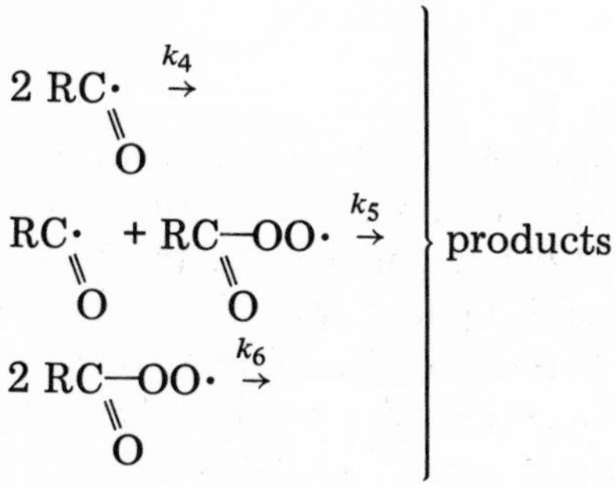

Assuming stationary concentrations of active species and that the chains are long, the expression obtained for the rate of oxygen absorption (when the pressure of oxygen is not too low) is

$$\frac{d[O_2]}{dt} = k_3 k_6^{-1/2}[RCHO][I + k_1[RCHO][O_2]]^{1/2}$$

Thus

$$V_{th} = k_3 k_6^{-1/2} k_1^{1/2}[RCHO]^{3/2}[O_2]^{1/2}$$

and

$$V_{ph} = k_3 k_6^{-1/2}[RCHO]I^{1/2}$$

which are similar to the empirical expressions.

By experimentally determining the rate of photochemical initiation, from an examination of the rate of inhibited n-decanal oxidation, and by determining the concentration of peroxidic species (obtained by the rotating sector method), it is possible to calculate the propagation (k_3) and rupture (k_6) coefficients which, at 5°C, are

$$k_3 = (7.2 \pm 0.2) \times 10^2 \text{ l mole}^{-1} \text{ s}^{-1}$$

$$k_6 = (7.5 \pm 1.2) \times 10^6 \text{ l mole}^{-1} \text{ s}^{-1}$$

From a knowledge of the various rate coefficients, it is possible to determine the rate of thermal initiation, the chain length, and the average lifetime of peroxidic radicals.

	Thermal oxidation	Photochemical oxidation
Rate of initiation (mole l^{-1} s^{-1})	4.2×10^{-9}	1.1×10^{-7}
Chain length	24,000	4000
Lifetime (s)	5.6	1.1
$[RO_2 \cdot]$ (mole l^{-1})	2.4×10^{-8}	1.2×10^{-7}

2.2 OXIDATION OF ACETALDEHYDE

Bäckström [24] found that the photochemical oxidation of acetaldehyde in the liquid phase led to the formation of peracetic acid as the primary product. Bowen and Tietz [25] examined the photochemical oxidation of acetaldehyde in the gas and liquid phases; the primary product was always peracetic acid. The kinetic results of this work are the same as those obtained in 1951 by Niclause and coworkers [26] in their later kinetic investigation of the oxidation in the liquid phase using a wider range of conditions.

Photochemical initiation is brought about by a UV lamp with a maximum intensity for λ between 2967 and 3200 Å. The temperature range chosen (-90 to $40°C$) enables thermal oxidation and gas phase oxidation to be made negligible. Since the vapor pressure of acetaldehyde is low under such conditions, the glass equipment can be used for partial oxygen pressures ranging from 0.2 to 1 atm.

The kinetic characteristics of this oxidation are similar to those obtained with decanal and are

(a) high quantum yield of about 10^3;

(b) rate of oxidation proportional to $I^{1/2}$;

(c) if the oxygen pressure is sufficiently high, the rate of oxidation is directly proportional to the aldehyde concentration and is independent of P_{O_2}. At low oxygen pressures, the rate increases with the pressure up to a limiting value V_∞ which corresponds to sufficient oxygen pressure (about 400 torr), i.e.

$$V_\infty = k V_1^{1/2} [RCHO]$$

$$V = F V_\infty$$

F depends to an increasing extent on P_{O_2} up to a maximum value of unity. It also depends to a decreasing extent on temperature. The activation energy for V_∞ is equal to 3.5 kcal mole^{-1}.

Bawn and Williamson [9] examined the catalyzed oxidation of acetaldehyde in solution in acetic acid at $25°C$. Whereas uncatalyzed oxidation has mediocre reproducibility, catalyzed oxidations are reproducible within 2%. The catalyst was cobalt acetate in solution in the cobaltous form. The partial oxygen pressure varied from 550 to 950 torr. Under such conditions, as in the case of photochemical oxidations, the stoichiometry of the reaction follows the overall equation

$$\frac{\text{Acetaldehyde consumed}}{\Delta[O_2]} = 2$$

In the case of the catalyzed oxidation, 2% of the oxygen can be found in the form of CO and CO_2.

The rate of oxidation is independent of shaking rate as soon as this is greater than 10 beats per second. The reaction rate is also independent of the oxygen pressure in the pressure range examined (>500 torr). The initial rate is proportional to the concentrations of catalyst and aldehyde, viz.

$$V = k[RCHO][Cat]$$

Adding water to the reaction medium brings about a decrease in the rate of oxidation. In this type of reaction, the catalyst probably affects the rate of initiation by causing an accelerated decomposition of the peracid according to

$$Co^{2+} + CH_3CO_3H \rightarrow Co^{3+} + CH_3CO_2\cdot + OH^-$$

$$Co^{3+} + CH_3CO_3H \rightarrow Co^{2+} + CH_3CO_3\cdot + H^+$$

$$CH_3CO_2\cdot + CH_3CHO \rightarrow CH_3CO_2H + CH_3\overset{O}{\overset{\|}{C}}\cdot$$

According to the authors, cobalt acetate catalyzes the decomposition of peracid and the related reaction rate is more or less proportional to the concentration of $CoAc_2$. This appears to imply a chain termination reaction other than one between peroxidic radicals.

In reality, the phenomena are complex. Thermal decomposition is superimposed on catalytic decomposition and, at the same time, both homolytic and heterolytic mechanisms occur. The fast equilibrium that is established between peracid and X peroxide also adds to the complexity of the system, viz.

$$CH_3\overset{O}{\overset{\|}{C}}-OOH + CH_3CHO \overset{K}{\rightleftharpoons} X \qquad (K_{25°C} = 0.27 \text{ mole } l^{-1})$$

At the same time, the catalyst reacts with X peroxide as well as with peracid. However, from the standpoint of peroxide transformation, the catalytic decomposition of peracid may be ignored. In the absence of a catalyst, only the X peroxide is decomposed at an appreciable rate, the kinetics being first-order, viz.

$$-\frac{d[X]}{dt} = k_{II}[X]$$

In the presence of catalyst ($[CoAc_2] = 8 \times 10^{-5}$ mole l^{-1}), the rate coefficient k'_{II} is equal to 0.50×10^{-3} s^{-1} and the value of the rate coefficient for the decomposition of the peracid itself is 0.58×10^{-6} s^{-1}.

In the absence of a catalyst, the formation of acetic acid from peracetic acid and acetaldehyde may occur [27] according to either

$$CH_3CHO + CH_3CO_3H \underset{k_{-I}}{\overset{k_I}{\rightleftharpoons}} X \underset{k_{II}}{\rightarrow} 2\,CH_3CO_2H \qquad (I)$$

or

$$CH_3CHO + CH_3CO_3H \underset{k_{-I}}{\overset{k_I}{\rightleftharpoons}} X$$

$$CH_3CHO + CH_3CO_3H \underset{k_{II}}{\rightarrow} 2\,CH_3CO_2H \qquad (II)$$

Just because peroxidic products were observed to disappear does not necessarily mean that one or the other of these mechanisms is valid. If a large excess of aldehyde is present, thus enabling the differential system corresponding to scheme (I + II) to be reduced to an integratable system, Vasilyev and Emanuel [27] have shown, from the appropriate experi-

Table 2
Values of the rate coefficients for the interaction of acetaldehyde and peracetic acid at different temperatures [27]

Temperature ($^\circ$C)	$\dfrac{k_I[CH_3CHO]}{k_{-I}}$	$\dfrac{k_{II}}{k_{-I}}$	$k_{-I} \times 10^4$ (s^{-1})
17.9	12.0	0.6	0.99
23.4	9.8	0.8	1.1
33.7	7.1	0.9	2.6
39.5	5.8	1.1	3.8

ments comparing the calculated values and experimental results, that scheme I conforms more closely to reality. According to these authors, the values of the rate coefficients are as shown in Table 2. The reproducibility of the experiments is not very satisfactory and the values of the pre-exponential terms are uncertain, but the three coefficients can be given as

$$k_I \quad = (10^2 - 10^5) \exp(-7000/RT) \text{ l mole}^{-1} \text{ s}^{-1}$$

$$k_{-I} = (10^5 - 10^8) \exp(-13,000/RT) \text{ s}^{-1}$$

$$k_{II} \quad = 10^8 \exp(-17,000/RT) \text{ s}^{-1}$$

The equilibrium constant $K = k_I/k_{-I}$ has the values shown in Table 3. The values of K are to be compared with that obtained by McNesby and Davis [28], i.e. $K_{25^\circ C} = 0.27$ mole l^{-1}. From Table 2, k_{II} at 23.4°C is 0.8 $\times$ 1.1 $\times$ 10^{-4} s^{-1}, although this value is rather low; the above authors actually found that $k_{II} = 2.5 \times 10^{-4}$ s^{-1} at 25°C. A comparison between the heat of reaction (ΔH) calculated from the activation energies and from the mean

TABLE 3
Values of the equilibrium constant $K = k_I/k_{-I}$ with toluene as solvent [27]
Toluene: $K = 1.6 \times 10^3 \exp(-5500/RT)$ mole l^{-1}.

Temperature ($^\circ$C)	K (mole l^{-1})
-20	0.25
0	0.52
19.3	1.13
20.1	1.22
30.4	1.87

values of the bond energies for the different possible X peroxides, viz.

$$CH_3-\underset{\underset{OH}{|}}{\overset{\overset{H}{|}}{C}}-O-O-\overset{\overset{O}{\|}}{C}-CH_3 \qquad \Delta H = -10 \text{ kcal mole}^{-1}$$

$$CH_3-\underset{\underset{OOH}{|}}{\overset{\overset{H}{|}}{C}}-O-\overset{\overset{O}{\|}}{C}-CH_3 \qquad \Delta H = -10 \text{ kcal mole}^{-1}$$

$$CH_3-\underset{\underset{OH}{|}}{C}\overset{O-O}{\underset{O}{<\hspace{-4pt}>}}C-CH_3 \qquad \Delta H = -20 \text{ kcal mole}^{-1}$$

does not enable a choice to be made between hydroxyethyl peroxyacetate and hydroperoxyethyl acetate. However, it appears that the isoozonide structure can be eliminated.

The rate at which X peroxide is formed is influenced by the nature of the solvent [27]. Figure 1 shows that the rate at which peracetic acid dis-

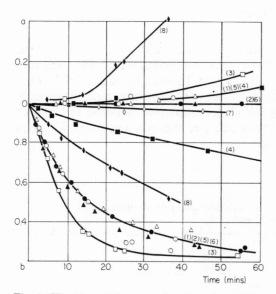

Fig. 1. Kinetics of the consumption of peroxyacetic acid and formation of acetic acid in various solvents. Temperature = 24.2°C; $[CH_3CO_3H]_0 = 0.273$ mole l^{-1}; $[CH_3CHO]_0 = 0.21$ mole l^{-1}. (a) Peroxyacetic acid; (b) acetic acid. 1, Carbon tetrachloride; 2, chloroform; 3, toluene; 4, benzene; 5, nitrobenzene; 6, methanol; 7, nitromethane; 8, acetone.

appears is more or less the same in carbon tetrachloride, chloroform, toluene, and benzene in which 70—78% of the peracid is transformed into X. The rate at which X peroxide is formed is considerably less in methanol, nitromethane, and acetone, which are capable of forming hydrogen bonds as shown by the infrared spectrophotometry of peracid in these solvents.

More recently, the kinetics and the effect of solvents on the oxidation catalysed by metal polyphthalocyanines and porphyrins (M—P) has been studied [71]. It was suggested that the initiation process is

$$M\text{—}P + O_2 \rightleftharpoons P\text{—}M^+\text{—}O_2^- \xrightarrow{CH_3CHO} P\text{—}M\text{—}OOH + CH_3CO$$

2.3 OXIDATION OF HEPTALDEHYDE

The photochemical oxidation of heptaldehyde has been studied in two extensive kinetic investigations [28,29] under slightly different conditions from the standpoint of temperature and solvents. (McNesby and Davis [28]: solvent, cyclohexane; oxygen pressure, 100—600 torr; temperature, 20—35°C. Lemaire [29]: solvent, decane; oxygen pressure, 50—720 torr; temperature, 0—17°C.)

Although, in both cases, thermal initiation is negligible, somewhat contradictory results have been obtained with regard to the influence of oxygen pressure on the rate of oxidation which, in both studies was found to be proportional to the square root of the light intensity. According to McNesby and Davis, the order with respect to oxygen is probably

$P_{O_2} < 200$ torr, first order

200 torr $< P_{O_2} < 370$ torr, zero order

$P_{O_2} > 370$ torr, first order

Considering the low degree of experimental reproducibility obtained by these authors, it is more probable that the effect of oxygen is closer to that determined by Niclause and his coworkers [31] who show that, at very low pressures ($\leqslant$10 torr), the rate of oxidation depends on P_{O_2} and is independent of P_{O_2} for pressures greater than 160 torr. In these pressure ranges, the expression for the rate of oxidation has the forms

$$V_{ph} = k_{ph} I^{1/2} [RCHO] \qquad P_{O_2} > 160 \text{ torr } (E = 4 \text{ kcal mole}^{-1})$$

$$V_{ph} = k'_{ph} I^{1/2} [O_2] \qquad P_{O_2} < 10 \text{ torr } (E = -0.5 \pm 1.5 \text{ kcal mole}^{-1})$$

The X peroxides produced by the reaction of perheptanoic acid and heptanal would have the structures [30]

TABLE 4
Infrared spectra of X_1, X_2 and X_3 peroxides from perheptanoic acid and heptanal [30] in CCl_4 solution
Cell thickness, 1 mm.

Frequency (cm^{-1})	Assignment
3618	Stretching vibration, OH (X_2)
3592	Stretching vibration, OH (X_1)
3465 ± 10	Stretching vibration, OH (X_3)
1778	Stretching vibration, C=O (X_2), (X_3)
1762	Stretching vibration, C=O (X_1)
1145, 1080	Not assigned
863	Stretching vibration, —O—O—

According to Niclause and his coworkers [30], the assignment of IR absorption bands for the peroxide is probably as given in Table 4.

2.4 OXIDATION OF n-BUTANAL

The photochemical oxidation of n-butanal [31,32] in the temperature range 0—30°C results in similar kinetics. At sufficiently high oxygen pressures, the activation energy is 1.8 kcal mole^{-1}.

Oxidation of n-butanal with cobalt and copper salts as catalyst gives peracid and acid. The values of the rate coefficient obtained with various cobalt and copper acetate concentrations confirms the absence of synergism between the catalytic action of cobalt and copper salts in the oxidation of aldehydes [65]. The catalytic oxidation of n-butanal enables butyric anhydride to be produced with high yields of about 50—60%. The best yields are obtained when the solvent consists of 20—30% anhydride, with the catalysts a mixture of cobalt and copper butyrates. In the presence of acetic anhydride, mixed anhydride and acetic acid are formed, viz.

$$
CH_3(CH_2)_2C\!-\!OOH + \begin{array}{c} CH_3C \\ \diagdown \\ O \\ \diagup \\ CH_3C \\ \diagdown \\ O \end{array} \rightleftharpoons \begin{array}{c} CH_3C \\ \diagdown \\ O \\ | \\ O \\ | \\ CH_3(CH_2)_2C \end{array} + CH_3CO_2H
$$

$$
\underset{\downarrow}{\overset{\text{(cat.)}}{}}
$$

$$
CH_3(CH_2)_2C\!-\!O\!-\!C\!-\!CH_3 + \tfrac{1}{2}O_2 \uparrow
$$

The transanhydrization leading to mixed diacyl peroxide is an easy reaction.

Oxidation of iso-butanal, by-product of n-butanal production by the hydroformylation of propene, also gives peroxyisobutyric acid. The kinetic diagram with ozone as initiator is classical [66]. Kinetic parameters were determined in a temperature range from 10 to 30°C [67].

3. Oxidation of unsaturated aldehydes

In the case of the oxidation of unsaturated aldehydes, the investigation is complicated by the fact that the aldehyde and the acid resulting from the transformation of peracid are liable to become polymerized. The double bond in a position α to the carbonyl group is not very reactive with regard to peracid, and so there is no epoxidation.

In this category of aldehydes, the only ones examined have been croton-aldehyde, acrolein and methacrolein.

3.1 OXIDATION OF 2-BUTENAL

The kinetics of the photochemical oxidation of 2-butenal in a pure state or in n-decane solution have been studied by Niclause and coworkers [34]. Oxidation taking place between 0 and 16°C under UV irradiation and a partial oxygen pressure of 18—700 torr produces kinetics similar to those of the photochemical oxidation of saturated aldehydes, viz.

(a) at low oxygen pressures, the rate is proportional to $I^{1/2}$ and P_{O_2}

$$V_{ph} = k'I^{1/2}[RCHO]_0[O_2] \qquad E \simeq \pm 0.5 \text{ kcal mole}^{-1}$$

(b) at high oxygen pressures, the rate no longer depends on the oxygen pressure

$$V_{ph} = k'I^{1/2}[RCHO][O_2]^0$$

By assuming a reaction mechanism similar to that for the oxidation of n-decanal, the "oxidability", $k_3 k_6^{-1/2}$, at $1.5°C$ has an approximate value of $8 \times 10^{-3} \, l^{1/2} \, \text{mole}^{-1/2} \, s^{-1/2}$, and $k_5 (k_4 k_6)^{-1/2}$ a value of about 0.5.

Maaraui et al. [72] have investigated the thermal oxidation in n-decane. The chain length was high, 1000—2000. They used inhibitors to investigate the mechanism of initiation and proposed the reaction

$$2 \text{ RCHO} + O_2 \rightarrow 2 \text{ RCO} + H_2O_2$$

with a rate coefficient, $k = 4.4 \times 10^{13} \exp(-25,000/RT) \, l^2 \, \text{mole}^{-2} \, s^{-1}$.

3.2 OXIDATION OF ACROLEIN AND METHACROLEIN

The liquid phase catalytic oxidation of acrolein and methacrolein has been the subject of various investigations [35—38]. Considering the natural tendency of reactants and products to become polymerized as well as the sensitivity of chain oxidations to inhibitors, it is not surprising that the findings of the different investigations are sometimes rather conflicting. Nevertheless, the primary oxidation products are exactly analogous to those of saturated aldehyde oxidation. Only the acid yields are affected, mainly as the result of the lack of acid stability. The overall scheme for the oxidation of acrolein can be written as

$$\text{acrolein} \rightarrow \text{peracrylic acid} \overset{+ \text{ acrolein}}{\rightleftharpoons} \text{X peroxide} \rightarrow \text{acrylic acid} \rightarrow \text{polymers}$$

The presence of the intermediate X has been shown [37] and its decomposition into acrylic acid has been examined.

In the case of acrolein oxidation in the presence of cobalt acetylacetonate, $Co(Acac)_3$, Table 5 gives the results obtained with different solvents [37]. The influence of solvents on both rate and selectivity may occur in a complex manner. Free acid selectivity depends in particular on the stability of this acid, because the oxidation of acrolein primarily produces acid almost quantitatively. Consequently, in a benzene—nitrobenzene mixture, acid is obtained with an 80% selectivity with conversions of 40% [39,40].

The influence of a catalyst on the rate of oxidation may occur in two ways, either by direct interaction with aldehyde or by the catalytic decomposition of peracid or X peroxide. Initiation by aldehyde—catalyst

TABLE 5
Oxidation of acrolein. Solvent effect [37]
Temperature, $30^{\circ}C$; $[\text{acrolein}]_0 = 3$ mole l^{-1}; $[\text{Co(Acac)}_3] = 1.5 \times 10^{-3}$ mole l^{-1}.

Solvent	$(d[O_2]/dt)_{max}$ [a]	Conversion (%)	Selectivity (%)
Formic acid	1.8	20.2	22.8
Acetic acid	6.36	29.4	36.1
Propionic acid	7.63	31.3	
Butyric acid	8.72	40.9	70.2
Valeric acid	8.12	39.7	36.6
Caproic acid	6.90	33.1	32.9
Capric acid	6.53	35.3	28.9
Benzene	5.82	39.6	32.3
Toluene	6.36	40.4	28.4
Carbon tetrachloride	4.45	39.2	30.9
Cyclohexane	3.09	32.2	23.7
n-Hexane	4.36	22.0	14.6
Nitrobenzene	2.18	34.7	7.6
Dimethyl sulfoxide	2.06	32.1	2.3
Water	0.26		
	9.08	80.3	5.6

[a] Calculated as $10^3 \, \Delta[O_2][\text{acrolein}] \, l^{-1} \, min^{-1}$.

interaction can be symbolized by

$$CH_2=CH-CHO + Co^{3+} \rightarrow CH_2=CH-\overset{\displaystyle O}{\overset{\|}{C}}{\cdot} + H^+ + Co^{2+}$$

It is generally accepted that there is the intermediate formation of a coordination complex. Cooper and Waters [41] propose, for initiation by divalent cobalt, the steps

$$RCHO + Co^{3+}(OH)(H_2O)_5 \rightarrow RC\overset{\displaystyle O\sim}{\underset{\displaystyle H---OH}{\overset{\|}{\diagup}}}Co^{3+}(H_2O)_5 \rightarrow R\overset{\cdot}{C}O + Co^{2+}$$

In the case of cobalt acetylacetonate, $Co(Acac)_2$, peroxidation is probably helped by the coordination of an acrolein molecule resulting in a complex capable of having two trans and cis configurations [37], viz.

trans   cis

It is certain that some solvents may compete with acrolein in such a coordination and hence may considerably disturb the rates.

Part of the initiation also comes from the catalytic decomposition of the peroxides present in the reaction medium. The kinetic examination of the decomposition of peracrylic acid in the presence of $Co(Acac)_3$, for example, leads to the empirical rate equation

$$-\frac{d[RCO_3H]}{dt} = k[RCO_3H]^{0.58}[Cat]^{0.56}$$

Ohkatsu et al. [37] propose a mechanism involving radical species and suggest, in particular, that cobalt is mainly in the +3 oxidation state. The above rate law clearly express the results of decomposition in the absence of oxygen. Since the reaction proceeds via radical species, the rate is obviously influenced by oxygen pressure and during the oxidation of acrolein, the existence of peroxide and its catalytic decomposition must also be taken into consideration.

With regard to the overall rate of the reaction, investigations show that, at low pressure, the rate law is

$$V = k[catalyst]^{0.5}[O_2]^{1.04}$$

illustrating how oxygen is fixed in a limited manner by the acyl radical.

The oxidation of methacrolein catalyzed by cobalt, added in the form of cobaltic acetate, usually produces methacrylic acid, but the reaction may be catalyzed by various transition metals [35] (Table 6). The kinetic study of the oxidation catalyzed by cobalt at concentrations of between 5×10^{-3} and 40×10^{-3} mole l^{-1} with aldehyde concentrations of $0.5 <$ [RCHO] < 4 mole l^{-1} shows that the rate of oxidation is independent of

TABLE 6

Activities of various catalysts and apparent yields of methacrylic acid in the oxidation of methacrolein [35]

Temperature, $40°C$; [methacrolein]$_0$ = 2.0 mole l^{-1}; oxygen off gas rate, 18 ml min^{-1}.

Catalyst [a]	Rate [b] $(\times 10^5)$	Aldehyde reacted [c] (%)	Yields [d] (%)	
			Gas Ch [e]	Tit.
$Co(OAc)_3$	26.5	66	35	79
Co-acetylacetonate	25.8	70	13	67
$Co(OAc)_2$	17.9	76	5	61
$Mn(OAc)_2$	11.7	41	28	
Cu-methacrylate	10.0	10 [f]	7	13
V_2O_5	11.8	82	9	55
HVO_3	11.0	54	10	81
SeO_2	2.4	43 [f]	13	

[a] 0.04 M catalyst. [b] Oxidation rate in mole O_2 l^{-1} s^{-1}. [c] Reaction time, four hours. [d] On aldehyde reacted. [e] By % peak area. [f] Reaction time, five hours.

TABLE 7

Cobaltic acetate catalyzed oxidation of methacrolein at 20°C
[methacrolein]$_0$ = 2.0 mole l^{-1}.

(a) Influence of catalyst concentration on rate [a] and products [b]

[Cat.] (mole l^{-1} × 10^{-3})	Rate [c] (×10^5)	Acetic acid (%)	Acrolein (%)	Methacrylic acid (%)
5	12.1	25	8	17
10	13.9	37	9	13
20	15.0	40	11	13
40	18.7	54	5	12

(b) Effect of O$_2$ pressure on rate [d]

O$_2$ pressure (torr)	Rate (×10^5)
695	19.1
765	21.0
791	24.6
864	29.4

[a] Experiments with off gas.
[b] % based on aldehyde consumed at approximately 60% reacted.
[c] Mole O$_2$ l^{-1} s^{-1}.
[d] 4 × 10^{-2} M catalyst.

the aldehyde concentration. The empirical expression for the rate would be

$$V = -\frac{d[O_2]}{dt} = k[cat]^{0.26}[O_2]$$

from the experimental results given in Table 7, and similar kinetic results are obtained with manganese. The apparent activation energy, measured for the manganese acetate catalyzed reaction, is 8.5 kcal mole^{-1}. These findings agree fairly satisfactorily with those obtained by Koshel et al. [36,42]. After an induction period that varies according to the catalyst, the oxidation of methacrolein is zero order. However, the anion bound to cobalt has great influence on the pre-exponential term in k_o and on the activation energy. For example, for cobalt oleate the pre-exponential term is approximately 10^7 and E is 12.5 kcal mole^{-1}, while for stearate, di-t-butylbenzoate, and phthalocyanine A is 10^{10} and E 17.5 kcal mole^{-1}.

To conclude this summary of findings concerning unsaturated aldehydes, a comparison can be made between the rates obtained under the same conditions for crotonaldehyde, acrolein, and methacrolein [35] (Table 8); it can be seen that the rates of oxidation are fairly similar. Under the same conditions, butyraldehyde proves to be much more oxidizable.

TABLE 8

Oxidation of various aldehydes in acetic acid. Influence of aldehyde on rate of oxidation

$[RCHO]_0 = 2$ mole l^{-1} ; catalyst, manganese acetate; concentration 4×10^{-2} mole l^{-1}.

Aldehyde	$-\dfrac{d[O_2]}{dt}$ (mole l^{-1} s^{-1}) ($\times 10^{-5}$)	Conversion (%)	Yield (%)	
			Acid	Peroxide
Butyraldehyde	31.0	47	61	5
Crotonaldehyde	5.3	23	25	9
Acrolein	1.0	13	29	33
Methacrolein	0.8	5	47	13

At this point, to complete the subject of the radical oxidation of substituted carboxaldehydes, mention should be made of the findings concerning glyceric aldehyde [43]. In the aqueous phase, the oxidation of this compound is a chain reaction. The accumulation of intermediate products such as glyceride and glycolic acids at the same time as acetic and formic acids and CO_2 indicates that this aldehyde has two reactive oxidation sites, i.e. the carbonyl group and the carbon α to this group.

4. Oxidation of benzaldehyde

Since Liebig [1], in 1835, observed that benzaldehyde was transformed into benzoic acid when left in the presence of air, a great deal of research has been done on the oxidation of this aldehyde. Kinetic investigations have been made on two types of system, one consisting of the oxidation of benzaldehyde alone and the other of its co-oxidation with other reactants.

4.1 AUTOXIDATION OF BENZALDEHYDE ALONE

The autoxidation of benzaldehyde produces perbenzoic acid as the primary product. This peracid reacts very rapidly with anhydrides to produce a mixed peroxide [44] and, in this case, an oxygen molecule is absorbed by benzaldehyde to produce two molecules of acid. According to Wittig and Pieper [12], the transformation probably takes place via the X peroxide intermediate which, however, has never been isolated.

4.1.1 Photochemical oxidation

A study of the initial stages of the photochemical oxidation enables the clearest results to be obtained. It has been found that benzaldehyde is oxidized according to the standard radical oxidation scheme, the initial rate being given by

$$V = kI^{1/2}[\text{RCHO}][\text{O}_2]^0 \simeq k_3 k_6^{-1/2} I^{1/2}[\text{RCHO}]$$

At a temperature of 5°C, the rate coefficients obtained with decane as the solvent and a benzaldehyde concentration of 1.98 mole l^{-1} are [23]

$$k_3 = 1.91 \times 10^3 \text{ l mole}^{-1} \text{ s}^{-1} \qquad E_3 = 1.8 \pm 0.5 \text{ kcal mole}^{-1}$$

$$k_6 = 2.1 \ \times 10^8 \text{ l mole}^{-1} \text{ s}^{-1} \qquad E_6 = 1 \text{ kcal mole}^{-1}$$

4.1.2 Thermal oxidation

Oxidation may be initiated by an agent that is able to decompose and produce free radicals. This is so with benzoyl peroxide (Bz_2O_2) and azobisisobutyronitrile (AIBN). Initiation by benzoyl peroxide produces initial rates which fit, in the experimental range defined by the boundary conditions $0.0446 \leqslant [\text{RCHO}]_0 \leqslant 0.224$ mole l^{-1}; $105 \leqslant P_{\text{O}_2} \leqslant 580$ torr; $20 \leqslant T \leqslant 35°C$; solvent, benzene. $[\text{Bz}_2\text{O}_2]_0 = 2.95 \times 10^{-3}$ mole per mole of solution, the equation

$$-\frac{d[\text{O}_2]}{dt} = k[\text{Bz}_2\text{O}_2]^{1/2}[\text{RCHO}][\text{O}_2]^0$$

At 25°C, $k = 0.85 \text{ l}^{1/2} \text{ mole}^{-1/2} \text{ s}^{-1/2}$ and the kinetics indicate that the usual mechanism applies ($k = k_3 k_6^{-1/2}$). Since the rate is proportional to the square root of the Bz_2O_2 concentration, the first-order decomposition of this peroxide is the only noteworthy source of radicals. The value of the product $k_3 k_6^{-1/2}$ is compatible with the photochemical results. The use of a less active initiator, AIBN, at normal temperatures gives kinetics which can only be explained by the intervention of a thermal initiation caused by the reaction of oxygen with benzaldehyde [45]. The results give the data, at 43°C.

rate of thermal initiation

$$k_3 k_6^{-1/2} = 0.288 \text{ l}^{1/2} \text{ mole}^{-1/2} \text{ s}^{-1/2}$$

$$V_1 \qquad = 2.11 \times 10^{-5} \text{ mole l}^{-1} \text{ s}^{-1}$$

Thermal autoxidation has been studied kinetically up to high degrees of conversion and the observations show the intervention of an appreciable inhibiting effect which is apparently the result of the formation of an inhibiting product at the stage of peracid reaction with aldehyde. The complete scheme thus becomes

$$\text{RCHO} + \text{O}_2 \xrightarrow{k_1} \text{RC}\overset{\text{O}}{\overset{\|}{\cdot}} + \text{HO}_2\cdot \tag{A}$$

$$\text{RC}\overset{\text{O}}{\overset{\|}{\cdot}} + \text{O}_2 \xrightarrow{k_2} \text{RC}\overset{\text{O}}{\overset{\|}{}}\!\!-\!\text{OO}\cdot$$

$$\text{RC}\overset{\text{O}}{\overset{\|}{}}\!\!-\!\text{OO}\cdot + \text{RCHO} \xrightarrow{k_3} \text{RC}\overset{\text{O}}{\overset{\|}{}}\!\!-\!\text{OOH} + \text{RC}\overset{\text{O}}{\overset{\|}{\cdot}}$$

$$\text{RC}\overset{\text{O}}{\overset{\|}{}}\!\!-\!\text{OOH} + \text{RCHO} \begin{cases} \nearrow 2\,\text{RCO}_2\text{H} \\ \searrow \text{inhibitor (AH)} + \text{RCO}_2\text{H} \end{cases}$$

$$\left.\begin{array}{l} 2\,\text{RC}\overset{\text{O}}{\overset{\|}{}}\!\!-\!\text{OO}\cdot \xrightarrow{k_6} \\[2mm] \text{RC}\overset{\text{O}}{\overset{\|}{}}\!\!-\!\text{OO}\cdot + \text{AH} \xrightarrow{k_7} \end{array}\right\} \text{products}$$

Numerical analysis enables the rate to be determined so that the experimental data can be accurately simulated as shown in Figure 2.

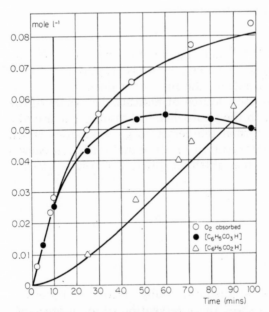

Fig. 2. Yields of the products in the oxidation of benzaldehyde. The lines represent the calculated results and the point are the experimental data. [Benzaldehyde]$_0$ = 0.640 mole l^{-1}; temperature, 43°C; solvent, o-dichlorobenzene; oxygen pressure, 1.25 atm. $\circ$, O$_2$ absorbed; $\bullet$, [C$_6$H$_5$CO$_3$H]; $\triangle$, C$_6$H$_5$CO$_2$H.

More recently, it has been reported [73] that the effect of phenolic inhibitors on the thermal oxidation of benzaldehyde and substituted benzaldehydes in acetic acid shows that initiation is due to reaction (A) and that there is no evidence for the third-order step, $2 \, RCHO + O_2$ (cf. Russian workers refs. 69, 70, and 72).

4.1.3 Catalytic oxidation

The oxidation of benzaldehyde catalyzed by manganese(II) and (III) acetate, cobalt(II) naphthenate, and cerium(IV) naphthenate has been studied by Kresge [46] with acetic acid as the solvent at a temperature of $50°C$ and an oxygen pressure of about 1 atm. In the case of oxidation in the presence of manganese with an aldehyde concentration of less than $0.5 \, \text{mole l}^{-1}$ and a manganese concentration of less than $10^{-5} \, \text{mole l}^{-1}$, the kinetics of the initial oxidation follow the empirical equation

$$-\frac{d[O_2]}{dt} = k[RCO_3H]^{1/2}[Mn]^{1/2}[RCHO] \tag{1}$$

which is compatible with the scheme

$$Mn^{n+} + RCO_3H \xrightarrow{k_1} RCO_3 \cdot \text{ or } RCO_2 \cdot$$

$$RCO_3 \cdot + RCHO \xrightarrow{k_3} RCO_3H + RCO \cdot$$

$$RCO \cdot + O_2 \xrightarrow{k_2} RCO_3 \cdot$$

$$RCO_3 \cdot + RCO_3 \cdot \xrightarrow{k_6} \text{products}$$

This mechanism gives the rate equation

$$-\frac{d[O_2]}{dt} = k_1^{1/2} k_3 k_6^{-1/2} [RCO_3H]^{1/2}[Mn]^{1/2}[RCHO]$$

which is similar to eqn. (1).

With a high manganese concentration, the catalyst appears to inhibit the reaction. If the concentration is higher than $10^{-3} \, \text{mole l}^{-1}$, the metal probably intervenes in the termination. With a high enough aldehyde concentration, the rate of oxidation becomes independent of RCHO concentration.

In the case of cobalt, the kinetics follow the empirical equation

$$-\frac{d[O_2]}{dt} = k[Co]^{1/2}[RCHO]^{3/2}$$

This equation is compatible with a standard chain mechanism involving an initiation of the type

$$RCHO + Co^{3+} \to Co^{2+} + RCO\cdot + H^+$$

However, it is certain that purely thermal autoxidation affects the author's results to some extent. Hendricks et al. [73] suggest that Co(II) and Co(III) are involved in both initiation and termination.

4.2 CO-OXIDATION OF SUBSTITUTED BENZALDEHYDES

The competitive oxidation of substituted benzaldehydes in acetic anhydride solution, in which the peracids are transformed into acetyl benzoyl peroxides thus eliminating peracid—aldehyde interaction, enabled the relative reactivities of the benzoylperoxy radicals to be determined with respect to a series of aldehydes [47]. In a mixture of two aldehydes, ther peroxidic radicals react according to the two pairs of competitive reactions

$$
R_1C\overset{O}{\underset{OO\cdot}{\Big\langle}}
\begin{cases}
+\,R_1CHO \overset{k_{11}}{\to} R_1C\overset{O}{\underset{OOH}{\Big\langle}} + R_1C\cdot\overset{O}{} \\
+\,R_2CHO \overset{k_{12}}{\to} R_2C\overset{O}{\underset{OOH}{\Big\langle}} + R_2C\cdot\overset{O}{}
\end{cases}
$$

$$
R_2C\overset{O}{\underset{OO\cdot}{\Big\langle}}
\begin{cases}
+\,R_1CHO \overset{k_{21}}{\longrightarrow} R_2C\overset{O}{\underset{OOH}{\Big\langle}} + R_1C\cdot\overset{O}{} \\
+\,R_2CHO \overset{k_{22}}{\longrightarrow} R_2C\overset{O}{\underset{OOH}{\Big\langle}} + R_2C\cdot\overset{O}{}
\end{cases}
$$

For long-chain radical oxidation, this leads to the expression

$$\frac{d[R_1CHO]}{d[R_2CHO]} = \frac{[R_1CHO]}{[R_2CHO]}\left(\frac{r_1[R_1CHO] + [R_2CHO]}{[R_1CHO] + r_2[R_2CHO]}\right)$$

for the consumption of aldehydes [48,49] with $r_1 = k_{11}/k_{12}$ and $r_2 = k_{22}/k_{21}$.

When the amounts of consumed and active oxygen and the acid distribution after peroxide reduction are known, it is possible to obtain the relative reactivities, which are summed up in Table 9.

A classification of aldehydes by increasing order of the Hammett con-

TABLE 9
Reactivity ratios in competitive oxidations of aldehydes at $30°C$

Aldehyde (1)	Aldehyde (2)	r_1 [a]	r_2 [a]
p-Chlorobenzaldehyde	Anisaldehyde	0.146	2.36
p-Chlorobenzaldehyde	p-Tolualdehyde	0.236	2.15
p-Chlorobenzaldehyde	m-Tolualdehyde	0.335	2.20
p-Chlorobenzaldehyde	p-Isopropylbenzaldehyde	0.285	1.87
p-Chlorobenzaldehyde	p-Cyanobenzaldehyde	1.15	0.46
p-Chlorobenzaldehyde	Benzaldehyde	0.540	1.62
m-Chlorobenzaldehyde	Benzaldehyde	0.378	1.95
Anisaldehyde	Benzaldehyde	2.22	1.00
p-Chlorobenzaldehyde	n-Butyraldehyde	0.278	10.73

[a] See text.

stant shows that the relative rates rise with the electron-donating power of the substituent. Since the reactivity of peroxidic radicals is independent of the hydrocarbon structure of the radical, the data are interpreted in terms of polar interaction in the transition state between the radical and the aldehyde, viz.

Generally speaking, the peroxidic radicals can be seen preferentially to attack the C—H bonds having a high electron density. The donor substituents increase the charge on CHO and thus cause the electrophile radical, RO_2, to come nearer. The stable forms in the transition state are the benzoate and the carbonium ion, thus facilitating the rupture and increasing the reactivity. The high reactivity of butyraldehyde, in particular, appears to stem from an interaction of this type, because, if the reactivity depended on the facility of obtaining the radical by the intervention of resonance stabilization, the order of the reactivities should be reversed since the presence of the phenyl group means that the benzoyl radical is the most resonance stabilized.

4.3 BENZALDEHYDE—OLEFIN CO-OXIDATION IN BENZENE SOLUTION

The autoxidation of benzaldehyde in the presence of olefins such as cyclohexene and α-methylstyrene with or without a catalyst enables an epoxide to be isolated as the principal olefinic product. This process has selectivities of 30—60% with mean olefin conversions of 40—70% [50—52]. The epoxidation is caused either by the reaction of peracid with the olefin or by the direct interaction of benzoylperoxy radicals with the double bond as is the case for the RO_2 products in the direct epoxidation of olefins by autoxidation [53—56].

The simplified reaction scheme can be symbolised as

$$
C_6H_5\overset{O}{\underset{\|}{C}}\!-\!OO\cdot
\begin{cases}
+ \text{ olefin} \xrightarrow{k_3'} \text{epoxide} + C_6H_5\overset{O}{\underset{\|}{C}}\!-\!O\cdot \\[2ex]
+ C_6H_5CHO \xrightarrow{k_3} C_6H_5\overset{O}{\underset{\|}{C}}\!-\!OOH + C_6H_5\overset{O}{\underset{\|}{C}}\cdot
\end{cases}
$$

$$
C_6H_5\overset{O}{\underset{\|}{C}}\!-\!OOH
\begin{cases}
+ \text{ olefin} \xrightarrow{k_E} \text{epoxide} + C_6H_5CO_2H \\[2ex]
+ C_6H_5CHO \xrightarrow{k_{II}} 2\,C_6H_5CO_2H
\end{cases}
$$

The values of k_3' have been determined (Table 10) for an aldehyde concentration of 3.57 mole l^{-1} and a temperature of 20°C, where k_3' is of the same order of magnitude as k_3 which is equal to 2.2×10^3 l mole^{-1} s^{-1}. In contrast, k_E is greater than k_{II}. Actually, $k_E^{(20°C)} = 1.27 \times 10^{-2}$ l mole^{-1} s^{-1} for cyclohexene [57] and $k_{II}^{(43°C)} = 0.018 \times 10^{-2}$ l mole^{-1} s^{-1}.

These data, coupled with the fact that the epoxide selectivity is not greatly influenced by the presence of cobalt, appear to indicate that the reactivity of the benzoylperoxy radical with regard to double bonds is very high. In the case of α-methylstyrene, which by itself in the presence of oxygen forms a polyperoxide [58], $[k_3 \,(30°C) = 7$ l mole^{-1} s^{-1} (polyperoxidation [59])] mainly epoxide and acetophenone are produced.

It should be mentioned that the rapidity of the intermediate radical

TABLE 10
Autooxidation of benzaldehyde in solution with olefins
$[C_6H_5CHO]_0 = 3.57$ mole l^{-1}; temperature, 20°C.

Olefin	[Olefin] (mole l^{-1})	k_3' (l mole^{-1} s^{-1})
2-Hexene	0.114	1.72×10^3
1-Hexene	0.160	0.138×10^3
Cyclopentene	0.120	3.66×10^3
1-Octene	0.051	0.236×10^3
Cyclohexene	0.079	1.57×10^3

rearrangement

$$C_6H_5-\overset{\overset{\text{O}}{\|}}{C}-O-O-\overset{\overset{\text{CH}_3}{|}}{\underset{\underset{\text{C}_6\text{H}_5}{|}}{\overset{\bullet}{C}}}-\overset{\bullet}{C}H_2 \rightarrow C_6H_5\overset{\overset{\text{O}}{\|}}{C}-O\cdot + CH_3-\overset{\overset{\text{O}}{\|}}{\underset{\underset{\text{C}_6\text{H}_5}{|}}{C}}\overset{\diagdown}{CH_2}$$

is particularly great; this must be attributed to the benzoylperoxy radical itself.

4.4 CO-OXIDATION OF BENZALDEHYDE WITH CYCLOHEXANONE

During the oxidation of benzaldehyde in cyclohexanone solution, the oxidation of cyclohexanone can be considered to be negligible [60]. On the other hand, there is competitive reaction of perbenzoic acid according to

$$C_6H_5\overset{\overset{\text{O}}{\|}}{C}-OOH \begin{cases} + \text{cyclohexanone} \overset{k'_{\text{II}}}{\rightarrow} \epsilon\text{-caprolactone} + C_6H_5CO_2H \\ + \text{RCHO} \overset{k_{\text{II}}}{\rightarrow} 2\ RCO_2H \end{cases}$$

The reactivity of cyclohexanone is greater than that of aldehyde

$$k_{\text{II}} = 2.17 \times 10^3 \exp(-10,000/RT)\ \text{l mole}^{-1}\ \text{s}^{-1}$$

$$k'_{\text{II}} = 17.8 \times 10^3 \exp(-10,000/RT)\ \text{l mole}^{-1}\ \text{s}^{-1}$$

In view of the rate coefficient ratio $k'_{\text{II}}/k_{\text{II}}$, by co-oxidation it is possible to produce caprolactone with good selectivities compared with aldehyde and ketone. Since the overall kinetic scheme for the reaction is known and all the kinetic coefficients have been estimated, it has been possible to make a computer simulation [60].

4.5 DIRECT DETERMINATION OF KINETIC COEFFICIENTS

The co-oxidation or competitive oxidation of aldehydes, either among themselves or with hydrocarbons, provides interesting information on the reactivity of peracyl radicals [61].

In this project, co-oxidation with 1,4-cyclohexadiene was chosen since its oxidation products are easy to analyze, because they involve formation of benzene from the $HO_2\cdot$ radicals which propagate the chain, the overall stoichiometry being

$$\text{(cyclohexadiene)} + O_2 \rightleftharpoons \text{(benzene)} + H_2O_2$$

Therefore, it has been possible to determine the relative reactivities, r_1 (1

TABLE 11
Rate coefficient ratios for cooxidations

RCHO	R'H	r_1 [a]	r_2 [b]	$r_1 \times r_2$
Heptanal	Cyclohexadiene	0.64	1.65	1.06
Octanal	Cyclohexadiene	0.64	1.65	1.06
Cyclohexanecarboxaldehyde	Cyclohexadiene	0.95	0.44	0.42
Pivaldehyde	Cyclohexadiene	2.15	0.36	0.77
Benzaldehyde	Cyclohexadiene	0.17	4.75	0.81
Heptanal	Benzaldehyde	~10	~0.1	~1.0

[a] $RCO_3 \cdot + RCHO \xrightarrow{k_{1,1}} RCO_3H + R\dot{C}_3$

$RCO_3 \cdot + R'H \xrightarrow{k_{1,2}} RCO_3H + R' \cdot$ $\quad r_1 = k_{1,1}/k_{1,2}$

[b] $HO_2 \cdot + R'H \xrightarrow{k_{2,2}} H_2O_2 + R' \cdot$

$HO_2 \cdot + RCHO \xrightarrow{k_{2,1}} H_2O_2 + R\dot{C}O$ $\quad r_2 = k_{2,2}/k_{2,1}$

for RCO_3) and r_2 (2 for HO_2), which are given in Table 11.

The values of r_1 and r_2 indicate that the reactivity of cyclohexadiene and aldehydes is approximately the same with regard to peracyl and hydroperoxy radicals. Nevertheless, in co-oxidation with cyclohexadiene or heptanal, benzaldehyde does not appear to be very reactive. This observation has also been made with the system butyraldehyde—benzaldehyde [47].

The results obtained for benzaldehyde—cyclohexene co-oxidation [51] appear to indicate that the reaction occurs with addition to the double bond, viz.

$$C_6H_5\overset{O}{\overset{\|}{C}}{-}OO\cdot \; + \; \overset{}{\underset{}{C}}{=}\overset{}{\underset{}{C} } \; \rightarrow \; \overset{O}{\overset{\diagup\diagdown}{\underset{}{C}}}{-}\overset{}{\underset{}{C}} \; + \; C_6H_5\overset{O}{\overset{\|}{-}C}{-}O\cdot$$

It is quite surprising that co-oxidation with cyclohexadiene does not produce the same phenomenon, at least partially. The existence of an inhibition phenomenon may also have to be considered.

The data on the coefficients $k_{1,2} = k_{1,1}/r_1$ and $k_{2,1} = k_{2,2}/r_2$ (Table 12) enables a comparison to be made between the reactivity of the $HO_2 \cdot$ radical and that of peroxyacyl radicals with regard to the same substrate. It is quite surprising to note that $k_{1,1}$ is very different from $k_{2,1}$ and $k_{1,2}$ from $k_{2,2}$. In general, for hydrocarbons, ethers, and secondary alcohols [62], reactivity depends on the oxidized molecule and is independent of the active species. However, except that $HO_2 \cdot$ is one radical compared with various hydrocarbon peroxidic radicals, it is possible that the cyclohexadiene oxidation mechanism is more complex than has been suggested and the interpretation of the results is uncertain.

The results of the co-oxidation of benzaldehyde with, successively,

TABLE 12

Rate coefficients [a] for reactions of HO$_2$ and peroxyacyl radicals at 0°C

Aldehyde	$k_{1,1}$ (l mole^{-1} s^{-1})	$k_{1,2}$ (l mole^{-1} s^{-1})	$k_{2,1}$ (l mole^{-1} s^{-1})	$k_{1,1}/k_{2,1}$	$k_{2,2}/k_{1,2}$
Heptanal	3.1×10^3	4.8×10^3	0.05×10^3	62	0.017
Octanal	2.9×10^3	6.1×10^3	0.05×10^3	78	0.013
Cyclohexane—carboxalde-hyde	1.1×10^3	1.2×10^3	0.181×10^3	6.0	0.068
Pivaldehyde	2.5×10^3	1.2×10^3	0.228	11	0.068
Benzaldehyde	12×10^3	70.6×10^3	0.017	700	0.0016

[a] See Table 11 for definitions of $k_{1,1}$ $k_{1,2}$ and $k_{2,1}$. $k_{2,2}$ refers to the reaction HO$_2$· + cyclohexadiene and has the value 0.082×10^3 l mole^{-1} s^{-1} at 0°C.

tetralin and cumene suggest that benzoylperoxy radicals have a very much higher degree of reactivity with regard to benzaldehyde than have other peroxy radicals, viz.

$k_3/k_3' \neq 920$; $k_3/k_3'' \neq 520$.

These differences in reactivity are, depending on the workers concerned, usually attributed to polar effects. Nevertheless, it seems apparent that interpretation is made uncertain by the extreme sensitivity of benzaldehyde oxidation to fortuitous inhibition phenomena. In other words, it is not certain that these rather surprising results cannot be attributed to the quantitative treatment of data from experiments involving considerable difficulties.

References pp. 122—124

5. Catalytic aldehyde oxidation without a chain mechanism

The oxidation of unsaturated α-aldehydes in an aqueous solution with a controlled basic pH ($12 \leqslant pH \leqslant 13$), catalyzed by silver, leads to the formation of the corresponding acids, which are stabilized in the form of salts, viz.

$$RCH=CH-CHO \xrightarrow[12 \leqslant pH \leqslant 13]{Ag,\, O_2} RCH=CH-\overset{\displaystyle O}{\overset{\|}{C}}-OH$$

Such oxidation does not involve radical species and is not inhibited by ordinary antioxidants. At the same time, it produces acids with high selectivities. Methacrolein is oxidized to methacrylic acid [35,42] and acrolein to acrylic acid at 4°C with yields of approximately 95%. Since the medium is a basic one, aldehyde is added gradually so as to prevent condensations.

Not very much research has been done on the kinetics of these reactions. The reaction scheme proposed by various authors [35,42] is

$$Ag\cdot Ag + O_2 \rightarrow Ag\cdot Ag \overset{O_2}{\underset{|}{}} \rightleftharpoons \overset{\overset{\textstyle O^-}{|}}{\underset{Ag \quad Ag}{\overset{O^+}{/\ \backslash}}}$$

$$\overset{\overset{\textstyle O^-}{|}}{\underset{Ag \quad Ag}{\overset{O^+}{/\ \backslash}}} + CH_2=CH-CHO \rightarrow \left[\begin{array}{c} CH_2=CH-\overset{O}{\overset{\|}{C}}-H \\ \overset{O}{\underset{Ag \quad Ag}{\overset{|}{\underset{/\ \backslash}{O}}}} \end{array} \right] \rightarrow$$

$$CH_2=CH-\overset{\displaystyle O}{\overset{\|}{C}}-OH + \overset{O}{\underset{Ag \quad Ag}{\overset{/\ \backslash}{}}}$$

Furfural [22] and 3,4-dihydro-2-pyran carboxaldehyde [63] can also be oxidized to the corresponding acids.

6. Conclusions

The liquid phase autoxidation of aldehydes by molecular oxygen is almost always a homogeneous reaction which is brought about by the intervention of active radical species. The initial kinetics for the different cases studied may be deduced from the general scheme of long kinetic chain radical oxidation already worked out for the oxidation of hydrocarbons by Bolland, Gee, and Semenov, in particular, (see Chap. 1), viz.

Initiation $\rightarrow$ radicals(r·) (rate = V_i)

$$r· + O_2 \rightarrow rO_2·$$

$$rO_2· + RCHO \rightarrow rO_2H + R\overset{O}{\underset{\|}{C}}·$$

Propagation $R\overset{O}{\underset{\|}{C}}· + O_2 \overset{k_2}{\rightarrow} R\overset{O}{\underset{\|}{C}}{-}OO·$

$$R\overset{O}{\underset{\|}{C}}{-}OO· + RCHO \overset{k_3}{\rightarrow} R\overset{O}{\underset{\|}{C}}{-}OOH + R\overset{O}{\underset{\|}{C}}·$$

Termination

$$2\ R\overset{O}{\underset{\|}{C}}· \overset{k_4}{\rightarrow}$$

$$R\overset{O}{\underset{\|}{C}}· + R\overset{O}{\underset{\|}{C}}{-}OO· \overset{k_5}{\rightarrow} \Big\}\ \text{inactive products}$$

$$2\ R\overset{O}{\underset{\|}{C}}{-}OO· \overset{k_6}{\rightarrow}$$

Assuming a quasi-steady state for active species, when chain lengths are large (>>100), the rate of oxygen absorption, which must not be confused with the rate of propagation, is written, no matter what the partial oxygen pressure may be, as

$$-\frac{d[O_2]}{dt} \sim Vp = k_3 [R\overset{O}{\underset{\|}{C}}{-}OO·][RCHO]$$

$$= k_3 k_6^{-1/2} V_i^{1/2} [RCHO] k_2 k_6^{1/2} [O_2]$$

$$\times \{k_3 k_4^{1/2} [RCHO] + k_2 k_6^{1/2} [O_2] + k_4^{1/2} k_3^{1/2} V_i^{1/2}\}^{-1}$$

assuming that $k_5 = k_4 k_6$.

This expression may have two limiting forms depending on whether the oxygen pressure, P_{O_2}, is sufficiently low or high. The oxygen concentration in the liquid phase, assuming there is sufficient agitation for saturation to be attained at all times, is proportional to P_{O_2}; $[O_2] = \chi P_{O_2}$.

(a) *Low oxygen pressures*
In this case,

$$[R\overset{O}{\underset{\|}{C}}{-}OO·] << [R\overset{O}{\underset{\|}{C}}·]$$

and the termination reactions are reduced to

$$2 \; RC\overset{\displaystyle O}{\underset{\displaystyle \|}{\cdot}} \; \overset{k_4}{\to} \; \text{inactive products}$$

$$-\frac{d[O_2]}{dt} = k_2 [RC\overset{\displaystyle O}{\underset{\displaystyle \|}{\cdot}}][O_2]$$

$$= k_2 k_4^{-\frac{1}{2}} V_i^{\frac{1}{2}} (\chi P_{O_2}) = V_\epsilon$$

The rate of oxidation is independent of the aldehyde concentration but is proportional to the oxygen concentration and the square root of the rate of initiation.

(b) High oxygen pressure

In this case, only peroxidic radicals exist in appreciable concentration in the reaction medium, and thus

$$-\frac{d[O_2]}{dt} = V_\infty = k_3 [RC\overset{\displaystyle O}{\underset{\displaystyle \|}{-}}OO\cdot][RCHO] = k_3 k_6^{-1/2} V^{1/2} [RCHO]$$

The products of the termination reaction were determined by several methods including oxygen labelling. As many as ten carbon dioxide molecules could be evolved per termination step in the oxidation of acetaldehyde and other aliphatic aldehydes, but no cage products were detected [68].

The apparent activation energies in the two extreme cases are, respectively

$$E_\epsilon = E_2 - \tfrac{1}{2}E_4 + \Delta H$$

$$E_\infty = E_3 - \tfrac{1}{2}E_6$$

where ΔH = the vaporization enthalpy of oxygen from the solution.

The rates of initiation depend on the type of activation chosen. In photochemical initiation, $V_i = 2 \phi I$, where I is the absorbed light intensity and ϕ = the efficiency coefficient. With an average intensity, rates of initiation of approximately 10^{-7} mole l^{-1} s^{-1} are attained. In thermal activation, autoinitiation by interaction between oxygen and aldehyde gives low V_i values of approximately 10^{-9} mole l^{-1} s^{-1} under standard laboratory conditions. When azonitrile is used, since the thermal decomposition rate of this product is approximately first order [62], V_i is given by

$$V_i = 2\alpha V_d$$

where $V_d = k_d [\text{azonitrile}]$ and α is about 0.5—0.7.

TABLE 13

Summary of the rate coefficients of elementary steps in aldehyde oxidations

Aldehyde	Initiation	$k_3 \times 10^{-3}$ (l mole^{-1} s^{-1})	$2k_6 \times 10^{-8}$ (l mole^{-1} s^{-1})	$k_3 k_6^{-1/2}$ (l$^{1/2}$ mole$^{-1/2}$ s$^{-1/2}$)	Temp. (°C)	Ref.
Benzaldehyde	$h\kappa$	1.2	17.6		0	61
Benzaldehyde	$h\nu$	1.91	2.1		5	23
Benzaldehyde	$h\nu$			0.16	25	23
Benzaldehyde	Bz_2O_2			0.85		64
Benzaldehyde	Thermal	4.17		0.288	43	45,60
Decanal (pure)	$h\nu$	0.72	0.075	0.27	5	23
Decanal (30% in n-decane)	$h\nu$	2.7	0.34	0.45	5	23
Octanal	$h\nu$	3.9	0.69	0.47	0	61
Heptanal	$h\nu$	3.1	0.54	0.39	0	61
Heptanal	$h\nu$			0.10	3	24,25,28
Acetaldehyde	$h\nu$	2.7	1.04	0.265	0	61
Cyclohexane—carboxaldehyde	$h\nu$	1.1	0.068	0.44	0	61
Pivaldehyde	$h\nu$	2.5	0.066	0.96	0	61
Crotonaldehyde	$h\nu$		0.008		1.5	61

Crotonaldehyde: $k_5(k_4k_6)^{-1/2} \simeq 0.500$ $k_2 k_4^{-1/2} = 13$ l$^{1/2}$ mole$^{-1/2}$ s$^{-1/2}$ [26,29,30].

Heptanal: $k_5(k_4k_6)^{-1/2} \simeq 0.3$ $k_2 k_4^{-1/2} = 400$ l$^{1/2}$ mole$^{-1/2}$ s$^{-1/2}$ [26,29,30].

References pp. 122—124

122

With regard to oxidations catalyzed by transition metals, the kinetics observed are not consistent. This can probably be attributed to the great reactivity of aldehydes, thus making experimentation especially delicate. Aldehydes have propagation rate coefficients, k_3, of approximately 10^3 l mole^{-1} s^{-1}, whereas for hydrocarbons they are approximately 1 l mole^{-1} s^{-1}. The most significant kinetic coefficients are summed up in Table 13.

It should also be emphasized that very little experimental kinetic research has been done on oxidations with high aldehyde conversion rates because such oxidations are made complex by the reaction of the peracid formed on the aldehyde present, by the complementary initiation caused by the peracid, and by the inhibition reactions. Likewise, very little kinetic data have been published on the oxidation of aldehydes on an industrial scale, particularly concerning the oxidation of acetaldehyde in acetic acid, the oxidation of aldehydes in peracetic acid, or the oxidation of acetaldehyde in acetic anhydride.

References

1 J. Liebig, Ann. Chem., 14 (1835) 139.
2 L. Bach, Compt. Redn., 124 (1897) 951.
3 A. Baeyer and V. Villiger, Chem. Ber., 33 (1900) 1569.
4 H.L.J. Bäckström, J. Am. Chem. Soc., 49 (1927) 1460.
5 J.R. McNesby and C.A. Heller, Chem. Rev., 54 (1954) 325.
6 M. Niclause, Sel. Chim., 15 (1956) 57.
7 E. Galitzenstein and M. Mugdan, U.S. Pat. 1, 179,421, 1916.
8 M.J. Kagan and G.O. Lubarsky, J. Phys. Chem., 39 (1935) 837.
9 C.E.H. Bawn and J.B. Williamson, Trans. Faraday Soc., 47 (1951) 735.
10 H. Wieland, Chem. Ber., 45 (1922) 2606; 54 (1921) 2358.
11 H. Losch, Arbeitsgemeinschaft Methoden der Organischen Chemie, Taqund, Berlin, 1941, p. 23.
12 G. Wittig and G. Pieper, Ann. Chem., 142 (1941) 546.
13 R.F. Vasilyev and N.M. Emanuel, Bull. Acad. Sci. USSR, Div. Chem. Sci., (4) (1956) 375.
14 Yu.A. Oldekop, A.N. Sevchenko, I.P. Zyatkov, G.S. Tylina and A.P. Elnitskii, Bull. Acad. Sci. USSR, Div. Chem. Sci., 128 (1949) 1211.
15 S.S. Ivanchev, A.I. Yurzencho and Yu.N. Anisimov, Russ. J. Phys. Chem., 39 (1965) 1009.
16 M. Niclause, Rev. Inst. Fr. Pet. Ann. Combust. Liq., XXIII (1968) 219.
17 B. Philipps, F.C. Frostick and P.S. Starcher, J. Am. Chem. Soc., 79 (1957) 598.
18 W.R. Jorissen, Z. Phys. Chem., 22 (1897) 34.
19 M.I. Farberov and G.N. Koshel, Kinet. Catal. (USSR), 6 (1965) 666.
20 M.I. Farberov and G.N. Koshel, J. Appl. Chem. USSR, 39 (1966) 2101.
21 A.E. Gillam and E.S. Stern, Introduction to Electronic Absorption Spectroscopy in Organic Chemistry, Arnold, London, 1957.
22 R.J. Harrison and M. Hoyle, Org. Synth., 36 (1956) 37.
23 H.R. Cooper and H.W. Melville, J. Chem. Soc., (1951) 1984, 1994; Proc. R. Soc. London, Ser. A, 216 (1952) 175.
24 H.L.J. Bäckström, Z. Phys. Chem., 5 (1934) 99.
25 E.J. Bowen and E.L. Tietz, J. Chem. Soc., (1930) 234.

26 P. Fille, M. Niclause and M. Letort, Bull. Soc. Chim. Fr., (1952) 436; Compt. Rend., 236 (1953) 1489; J. Chim. Phys., 53 (1956) 8.
27 R.F. Vasilyev and N.M. Emanuel, J. Phys. Chem. USSR, (1956) 375.
28 J.R. McNesby and T.W. Davis, J. Am. Chem. Soc., 76 (1954) 148.
29 J. Lemaire, Thesis, Nancy, 1962.
30 J. Lemaire, M. Niclause and N. Parant, Rev. Inst. Fr. Pet. Ann. Combust. Liq., XX (1965) 1703.
31 K. Shimomura, Nippon Kagaku Zasshi, 78 (1957) 1326.
32 K. Shimomura, Nippon Kagaku Zasshi, 82 (1961) 1314.
33 N.G. Kostyuk, S.V. Lvov, V.B. Falkovskii, A.V. Starkov and N.M. Levina, J. Appl Appl. Chem. USSR, 35 (1962) 2021.
34 X. Deglise, J. Lemaire and M. Niclause, Rev. Inst. Fr. Pet. Ann. Combust. Liq., XXIII (1968) 793.
35 W.F. Brill and F. Lister, J. Am. Chem. Soc., 26 (1960) 565.
36 G.N. Koshel, M.I. Farberov and Yu.A. Moskvichev, J. Appl. Chem. USSR, 37 (1964) 2287.
37 Y. Ohkatsu, M. Takeda, T. Hara, T. Osa and A. Misono, J. Chem. Soc. Jpn. Ind. Chem. Sect., 69 (1966) 2130; Bull. Chem. Soc. Jpn., 40 (1967) 1413, 1893, 2111.
38 V.A. Shushunov, V.A. Redoshkin and Yu Golubev, J. Appl. Chem. USSR, 35 (1962) 832.
39 F. Lanos and G. Clément, Fr. Pat. PV 980,234, 1964.
40 F. Lanos and G. Clément, Fr. Pat. PV 960,060, 1964.
41 T.A. Cooper and W.A. Waters, J. Chem. Soc., (1964) 1538.
42 M.I. Farberov and G.N. Koshel, Kinet. Catal. (USSR), 6 (1965) 666.
43 L.M. Andronov and Z.K. Maizus, Bull. Acad. Sci. USSR, Div. Chem. Sci., 3 (1967) 519.
44 W.R. Jorissen and P.A.A. Van den Beek, Rec. Trav. Chim., 46 (1927) 42; 47 (1928) 301.
45 J.P. Franck, I. Sérée de Roch and L. Sajus, Bull. Soc. Chim. Fr., (1969) 1947, 1957.
46 E.N. Kresge, Thesis, Floride, 1961 (in English).
47 C. Walling and E.A. McElhill, J. Am. Chem. Soc., 73 (1951) 2927.
48 F.R. Mayo and C. Walling, Chem. Rev., 46 (1950) 191.
49 J. Alagy, G. Clément and J.C. Balaceanu, Bull. Soc. Chim. Fr., (1959) 1325; (1960) 1495; (1961) 1792.
50 T. Ikawa, T. Fukushima, M. Muto and T. Yanagihara, Can. J. Chem., 44 (1966) 18.
51 T. Ikawa, H. Tomizawa and T. Yanagihara, Can. J. Chem., 45 (1967) 1900.
52 E. Niki and Y. Kamiya, Bull. Chem. Soc. Jpn., 40 (1967) 583.
53 W.F. Brill, J. Am. Chem. Soc., 85 (1963) 141.
54 W.F. Brill and B.J. Barone, J. Org. Chem., 29 (1964) 140.
55 I. Sérée de Roch and J.C. Balaceanu, Bull. Soc. Chim. Fr., (1964) 1393.
56 F.R. Mayo, J. Am. Chem. Soc., 78 (1956) 967.
57 B.M. Lynch and K.H. Pausacker, J. Chem. Soc., (1955) 1525.
58 F.R. Mayo, J. Am. Chem. Soc., 79 (1957) 2497.
59 J.A. Howard and K.U. Ingold, Can. J. Chem., 44 (1960) 1113.
60 J.P. Franck, Thesis, Paris, 1968.
61 G.E. Zaikov, J.A. Howard and K.U. Ingold, Can. J. Chem., 47 (1969) 3017.
62 L. Sajus, International Symposium on Oxidation, San Francisco, 1967, p. 45.
63 C.G. Overberger, H. Biletch, A.B. Finestone, J. Lilker and J. Herbert, J. Am. Chem. Soc., 75 (1952) 2078.
64 M.F.R. Mulcahy and I.C. Watt, Proc. R. Soc. London, Ser. A, 216 (1953) 10; J. Chem. Soc., (1954) 2971.
65 N.N. Lebedev and M.N. Manakov, Kinet. Katal., 15 (1974) 703.

66 S. Miyajima, T. Inukai, H. Harada, R. Yosuizawa, T. Hirai, K. Matsunaga and M. Harada, Bull. Chem. Soc. Jpn., 47 (1974) 2051.
67 B.I. Chernyak and L.A. Andrianova, Neftikhimiya, 14 (1974) 97.
68 N.A. Clinton, T.G. Trailor and R.A. Kenley, 170th National Meeting of the American Chemical Society, Chicago, Aug. 24—29 1975, Abstr. N Org.55.
69 S.A. Maslar and E.A. Blyumberg, Russ. Chem. Res., 45 (1976) 155.
70 E.T. Denisov, Russ. J. Phys. Chem., 52 (1978) 919.
71 Y. Ohkatsu, T. Hara and T. Osa, Bull. Chem. Soc. Jpn., 50 (1977) 696; Y. Ohkatsu, O. Sekiguchi and T. Osa, Bull. Chem. Soc. Jpn., 50 (1977) 701.
72 M.A. Maaraui, N.V. Nikipanchuk and B.I. Chernyak, Kinet. Katal., 19 (1978) 396.
73 C.F. Hendricks, H.C.A. van Beek and P.M. Heertjes, Ind. Eng. Chem. Prod. Res. Dev., 16 (1977) 270; 17 (1978) 260.

Chapter 3

The Oxidation of Alcohols, Ketones, Ethers, Esters and Acids in Solution

E.T. DENISOV

1. Introduction

Alcohols, ketones, and acids are formed as intermediates in the liquid phase oxidation of hydrocarbons [1] and are subject to further conversions. Therefore, investigation of the mechanisms of such conversions is necessary for the correct understanding of hydrocarbon oxidation. Moreover, the oxidation of alcohols and ketones is of scientific interest proper. The role of polar media and hydrogen bonding in chain oxidation is studied, particularly for alcohols and ketones. Alcohols are very convenient for the investigation of ionic oxidation reactions. The oxidation of certain alcohols is of interest for technology. For example, acetic acid and ethyl acetate may be produced by the oxidation of ethanol, and acetone and hydrogen peroxide by the oxidation of 2-propanol.

2. Oxidation of alcohols

2.1 THE KINETICS AND PRODUCTS OF ALCOHOL OXIDATION

2.1.1 Primary alcohols

Oxidation of methanol in the liquid phase is slow. At 81—145°C with azodi-isobutyronitrile and t-butyl peroxide as initiators, the oxidation products are formaldehyde, formic acid, hydrogen peroxide, and methyl formate [2].

The oxidation of ethanol was studied in detail by Zaikov et al. [3—7]. The main products of the oxidation in a steel autoclave under a pressure of 50—95 atm at 145—230°C are acetic acid and ethyl acetate with hydrogen peroxide and acetaldehyde as intermediates. Formic acid and methyl formate are produced in small amounts. The oxidation of ethanol proceeds with autocatalysis. Acetaldehyde is oxidized not only to acetic acid but also to ethyl acetate by disproportionation [8]

$$2 \text{ CH}_3\text{CHO} = \text{CH}_3\text{COOCH}_2\text{CH}_3$$

Addition of acetaldehyde to ethanol at the start of the oxidation accelerates the reaction, but, at a high degree of oxidation, it is retarded by the inhibiting action of resins formed from acetaldehyde. The sequence of ethanol oxidation conversions is

$$CH_3CH_2OH \rightarrow \begin{cases} H_2O_2 \rightarrow \text{decomposition} \\ CH_3CHO \rightarrow CH_3COOH \underset{H_2O}{\overset{EtOH}{\rightleftharpoons}} CH_3COOEt \\ \qquad\quad \rightarrow CH_3COOEt \end{cases}$$

Ethanol can be oxidized selectively to ethyl acetate in the presence of acetaldehyde and phosphoric acid [9]. The ethyl acetate yield is 89% at 145°C in the presence of 0.3 mole % of H_3PO_4 and 7 mole % of CH_3CHO (extent of conversion 97 mole %).

The oxidation of ethanol was studied in the gas [3,5] as well as in the liquid phase [4,10] without a solvent and in benzene solution under comparable temperature conditions in a steel reactor. The composition of the oxidation products in benzene and in the gas phase was found to be very similar. Decomposition of peroxy radicals with scission of C—C bonds, possibly on the metal wall of the reactor, was rather extensive both in the gas phase and in benzene solution. The polar medium of ethanol favoured the reaction $RO_2 \cdot + RH$ (Sect. 3.2.1). Ethanol in acetic acid medium was oxidized to the latter in the presence of Co, Mn, or Ce acetates as catalysts [11—15]. Acetaldehyde was added to accelerate the reaction.

Octadecanol oxidizes with the formation of acids, esters and carbonyl compounds [16]. Stearic acid predominates among the acids thus indicating primary oxidation of the alcohol group.

Hydroperoxide formaldehyde and formic acid are produced by the oxidation of ethylene glycol (35—95°C in chlorobenzene) [17]. Oxidation proceeds with autocatalysis. The rate of oxidation (W) measured by the consumption of oxygen is

$$W \sim [O_2]^{3/2}[RH]^{3/2}$$

In the absence of oxygen, the peroxide formed decomposes by a first-order reaction with a rate coefficient $k = 9 \times 10^{-5}\ s^{-1}$ at 75°C and $E = 18.7$ kcal mole^{-1}.

2.1.2 Secondary alcohols

Low-temperature photochemical oxidation of secondary alcohols produces oxyhydroperoxides [18—20]. Oxidation at 90—130°C proceeds autocatalytically to form the appropriate ketone and hydrogen peroxide [21—25]. Decay of H_2O_2 is observed at a high degree of oxidation, its concentration going through a maximum. For 2-propanol [25], $[H_2O_2]_{max} \approx$ 4 mole l^{-1}, whereas for cyclohexanol [23] $[H_2O_2]_{max} \approx 1$ mole l^{-1}. Such a difference is primarily connected with the disappearance of H_2O_2 by reaction with ketone. Conversions of H_2O_2 in cyclohexanol oxidation [26] were studied in detail. Hydrogen peroxide disappears in cyclohexanol at a rate of 2.6×10^{-4} $[H_2O_2]^{0.85}$ mole l^{-1} s^{-1} (120°C, $[H_2O_2]_0 = 0.1-$ 1.5 mole l^{-1}) and $E_a = 27$ kcal mole^{-1}. It is decomposed partly on the glass reactor wall (38% at 120°C in a Pyrex reactor) and partly under the action of oxycyclohexyl radicals. Hydrogen peroxide also reacts with cyclohexanone to form acids and lactone. Acids formed by the oxidation of cyclohexanol accelerate the decay of H_2O_2.

Ketones react with H_2O_2 forming peroxides. Cyclohexanone [22,27—30] produces the compounds

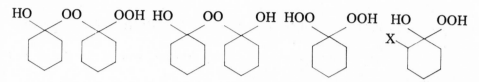

$$(X = Cl, Br)$$

Antonovsky and Terent'ev [31] have studied the interconversions of various cyclohexanone peroxides by IR spectroscopy and obtained

$$H_2O_2 + \quad \underset{k_-}{\overset{k_+}{\rightleftharpoons}}$$

$k_+ = 11 \exp(-5,500/RT)$ l mole^{-1} s^{-1} \qquad (CCl$_4$, 20—40°C)

$k_- = 1.1 \times 10^{11} \exp(-21,000/RT)$ s^{-1}

$K = \dfrac{k_+}{k_-} = 1.0 \times 10^{-10} \exp(15,500/RT)$ l mole^{-1}

$$+ \quad \underset{k_-}{\overset{k_+}{\rightleftharpoons}}$$

$k_+ = 6.0 \times 10^6 \exp(-12,000/RT)$ l mole^{-1} s^{-1} \quad (CCl$_4$, 20—40°C)

$k_- = 1.4 \times 10^2 \exp(-8,500/RT)$ s^{-1}

$K \ = 4.3 \times 10^4 \exp(-3,500/RT)$ l mole^{-1}

$$\underset{k_-}{\overset{k_+}{\rightleftharpoons}} \quad +$$

$k_+ = 5.0 \times 10^4 \exp(-12,000/RT)$ s^{-1} \qquad (CCl$_4$, $C_{per} = 0.001$ mole l^{-1}, 20—45°C)

References pp. 195—203

$$k_- = 22 \exp(-4{,}000/RT) \, 1 \, \text{mole}^{-1} \, \text{s}^{-1}$$

$$K = 2.3 \times 10^3 \exp(-8{,}000/RT) \, \text{mole} \, 1^{-1}$$

The equilibrium constant in cyclohexanol for the reaction

calculated from kinetic data [32] is $K = 2.2 \times 10^{-5} \exp(6{,}700/RT) \, 1 \, \text{mole}^{-1}$ (110—130°C).

The peroxides obtained from methyl ethyl ketone and H_2O_2 in the presence of mineral acid [33] are

The results for diethyl ketone are similar [34]. The interaction between H_2O_2 and ketones plays a great part in alcohol oxidation.

2.2 CHAIN MECHANISM OF ALCOHOL OXIDATION

Alcohol oxidation proceeds by a chain mechanism, as does that of hydrocarbons. The structure of the primary oxidation products shows that the alcohol group is the first to be oxidized. It is the C—H bond near the O—H group that is attacked. This bond is more readily broken than that in the corresponding hydrocarbon because the odd electron of an oxyalkyl radical interacts with the p-electrons of oxygen. The dissociation energies of C—H bonds in methane, methanol, ethane, and ethanol [35]; viz. $D_{C-H} = 102$ (CH_4) and 92 ($H-CH_2OH$); 98 (C_2H_6) and 88 kcal mole^{-1} [$CH_3CH(OH)-H$], are in agreement with the above. The O—H bond in alcohols is strong: $D_{O-H} = 102$ kcal mole^{-1}. Methyl radicals attack the $>C(OH)-D$ bond and do not attack the $>CHO-D$ bond in the molecule of deuterated 2-propanol [36]. The oxygen molecule is converted into hydrogen peroxide in the course of alcohol oxidation, as has been shown in experiments [37,38] with ^{18}O, viz.

$$>CHOH + {}^{18}O={}^{18}O \rightarrow >C=O + H-{}^{18}O-{}^{18}O-H$$

The kinetics of the initiated oxidation of alcohols is the same as for hydrocarbon oxidation [3,39—42], the rate equation being

$$W = const. \times [RH][O_2]^0[I]^{1/2}$$

where I = initiator, at oxygen pressures above 100 torr and $k_i[I] << W$ (k_i = rate coefficient for decomposition of I). This is consistent with the scheme for alcohol oxidation

$$I \xrightarrow{k_i} r\cdot \xrightarrow{RH} R\cdot$$

$$R\cdot + O_2 \rightarrow RO_2\cdot$$

$$RO_2\cdot + RH \xrightarrow{k_p} ROOH + R\cdot$$

$$RO_2\cdot + RO_2\cdot \xrightarrow{k_t} molecular\ products$$

Assuming a stationary state

$$W - W_i = \frac{k_p}{\sqrt{2k_t}} [RH][O_2]^0 \sqrt{W_i}$$

where $W_i = k_i$ [1]. This scheme is valid when P_{O_2} is sufficiently high so that $[R\cdot] << [RO_2\cdot]$, and reactions $R\cdot + R\cdot$ and $R\cdot + RO_2\cdot$ are negligible in comparison with $RO_2\cdot + RO_2\cdot$; the rate of free radical formation from H_2O_2 produced by oxidation is very low compared with $k_i[I]$. When the chains are sufficiently long

$$W = \frac{k_p}{\sqrt{2k_t}} [RH]\sqrt{W_i}$$

This expression is valid for the calculation of the ratio of rate coefficients. The general scheme given above may be considered as two alternnative mechanisms, viz.

$$\left.\begin{array}{l} >\overset{\cdot}{C}OH + O_2 \rightarrow >C(OH)OO\cdot \\ >C(OH)OO\cdot + >CH(OH) \rightarrow >C(OH)OOH + >\overset{\cdot}{C}OH \\ >C(OH)OOH \rightleftharpoons >C=O + H_2O_2 \end{array}\right\} \quad (1)$$

$$\left.\begin{array}{l} >C(OH)OO\cdot \rightarrow >C=O + HO_2\cdot \\ >C(OH)H + HO_2\cdot \rightarrow H_2O_2 + >\overset{\cdot}{C}OH \\ H_2O_2 + >C=O \rightleftharpoons >C(OH)OOH \end{array}\right\} \quad (2)$$

The hydroxyperoxy radical propagates the chain in (1) and the hydroperoxy radical in (2). As both mechanisms lead to the same products, they can be distinguished on the ground of kinetic data only. If (1) is true, the addition of H_2O_2 to alcohol must result in some replacement of hydroxyperoxy radicals by hydroperoxy radicals by the reaction

$$>C(OH)OO\cdot + H_2O_2 \rightarrow >C(OH)OOH + HO_2\cdot$$

since hydroperoxides undergo a similar reaction. The substitution of $HO_2\cdot$ for $>C(OH)OO\cdot$ will be indicated by a change in the kinetics. The kinetics of initiated cyclohexanol oxidation alter after addition of H_2O_2 [44] (see Sect. 2.4). However, no such difference, with and without added H_2O_2, is observed in the presence of a base (Na_2HPO_4) [45] (see Sect. 2.5). Therefore it might be suggested that mechanism (1) is valid for the oxidation of alcohols at low temperatures ($<130-150°C$) in the absence of H_2O, bases, and acids. Mechanism (2) seems to be valid for the oxidation of alcohols in the presence of H_2O, bases, and acids, as well as at higher temperatures.

The ratio of the rate coefficients, i.e. $k_p/\sqrt{2k_t}$ is a measure of the reactivity of the organic substance towards oxidation. It depends first of all on the dissociation energy of the C—H bond. The ratio increases with decreasing D_{C-H} (Table 1). In passing from one alcohol to another, the change in $k_p/\sqrt{2k_t}$ is essentially due to changes in k_p since k_t seems to change only slightly.

Absolute rate coefficients, k_p and k_t, measured for cyclohexanol only [48,49] by the rotating sector technique at $50-70°C$ are $k_p = 1.8 \times 10^8 \exp(-12,600/RT)$ l mole^{-1} s^{-1}, $2k_t = 7.5 \times 10^8 \exp(-3,600/RT)$ l mole^{-1} s^{-1}. The initiation rate coefficient for azodi-isobutyronitrile decomposition in cyclohexanol is [49] $k_i = 5.0 \times 10^{17} \exp(-35,000/RT)$ s^{-1}.

The reaction between two hydroxyperoxyradicals seems to be that of disproportionation

$$>C(OH)OO\cdot + >C(OH)OO\cdot \rightarrow >C=O + O_2 + >C(OH)OOH \qquad (3)$$

The rate coefficient measured by the pulse radiolysis technique [51] is $2k_t = 1.8 \times 10^7$ l mole^{-1} s^{-1}, i.e. one order of magnitude higher than that measured by the sector technique [49]. The oxygen dissolved in cyclohexanol seems to be rapidly consumed on irradiation. Free hydroxyalkyl radicals $R\cdot$ disappear partly by bimolecular interaction ($R\cdot + R\cdot$), the rate coefficient of which is high [51], 3.4×10^8 l mole^{-1} s^{-1}. This is in agreement with the radiolysis yields obtained, viz. $G_{ketone} = 6.4$ and $G_{ROOH} = 0.34$, whereas they should be the same if only reaction (3) occurs.

Peroxy radicals of cyclohexene react with alcohols at $60°C$ with the following rate coefficients [270] (l mole^{-1} s^{-1}): 5.6 ($C_6H_5CH_2OH$), 2.5 (cyclohexanol), 2.0 (i-PrOH), 1.9 (EtOH), 1.2 (n-BuOH) and 0.3 (MeOH).

TABLE 1
The ratio $k_p/\sqrt{2k_t}$ for alcohols

Alcohol	Temp. (°C)	O_2 pressure (Torr)	$k_p/\sqrt{2k_t}$	Ref.
CH_3OH	81—145	3×10^3—6×10^3	$6.0 \times 10^4\ \exp(-13,700/RT)$	2
CH_3OH	145	3×10^3—6×10^3	2.7×10^{-3}	2
CH_3OH	75	760	2.3×10^{-5}	46
CH_3CH_2OH	75	760	3.3×10^{-5}	39
CH_3CH_2OH	145	7600	3.8×10^{-3}	2
$(CH_3)_2CHOH$	80	100—760	1.2×10^{-3}	39
$(CH_3)_2CHOH$	86—138	1.5×10^3—2.2×10^3	$3 \times 10^4\ \exp(-12,000/RT)$	40
$(CH_3)_2CHOH$	20	760	7.9×10^{-5}	47
$(CH_3)_2CHOH$	145	7600	1.8×10^{-2}	2
$CH_3CH(OH)C_2H_5$	65—75	760	$1.0 \times 10^5\ \exp(-13,000/RT)$	42
$CH_3CH(OH)C(CH_3)_3$	84—104	760	$5.0 \times 10^4\ \exp(-12,600/RT)$	42
cyclohexanol (H, OH)	80	100—760	1.6×10^{-3}	39
cyclohexanol (H, OH)	50—75	200—800	$7.6 \times 10^3\ \exp(-10,800/RT)$	48,49
cyclohexanol (H, OH)	50—75	760	$4.0 \times 10^3\ \exp(-10,500/RT)$	268
cyclohexanol (H, OH)	80—100	760	$2.1 \times 10^4\ \exp(-12,000/RT)$	269
cyclohexanol (H, OH)	65—111	300—950	$7.5 \times 10^3\ \exp(-11,000/RT)$	41
cyclohexanol (H, OH)	90—120	760	$5.1 \times 10^3\ \exp(-10,300/RT)$	50
CH_3 cyclohexanol (H, OH)	50—100	760	$2.1 \times 10^8\ \exp(\pm18,600/RT)$	268
CH_3 cyclohexanol (H, OH)	80—100	760	$1.8 \times 10^4\ \exp(-12,000/RT)$	269
CH_3 cyclohexanol (H, OH)	50—75	760	$2.3 \times 10^5\ \exp(-13,000/RT)$	268
CH_3 cyclohexanol (H, OH)	80—100	760	$9.3 \times 10^4\ \exp(-12,800/RT)$	269
CH_3—cyclohexanol (H, OH)	50—75	760	$1.3 \times 10^4\ \exp(-11,000/RT)$	268
CH_3—cyclohexanol (H, OH)	80—100	760	$2.7 \times 10^4\ \exp(-12,200/RT)$	269
cyclohexanol (CH_3, OH)	80—100	760	$3.7 \times 10^4\ \exp(-13,400/RT)$	269
cyclohexenol (H, OH)	55—75	760	$7.6 \times 10^3\ \exp(-9,000/RT)$	42
$CH_3CH(OH)C_6H_5$	65—95	760	$1.15 \times 10^4\ \exp(-10,500/RT)$	42

References pp. 195—203

2.3 THE FORMATION OF FREE RADICALS IN ALCOHOL OXIDATION

2.3.1 Generation of chains

In the absence of initiators, free radicals are formed in the oxidation by reactions of alcohol molecules (RH) with oxygen [52]

$$RH + O_2 \longrightarrow R\cdot + HO_2\cdot - q_1$$

$$2\,RH + O_2 \longrightarrow R\cdot + H_2O_2 + R\cdot - q_2$$

The heat of $HO_2\cdot$ solvation is +8, of O_2 +4, and of H_2O_2 +13 kcal mole^{-1} [53]. Then, assuming that the heats of solvation of RH and R$\cdot$ are equal, $q_1 = 31$ and $q_2 = 19$ kcal mole^{-1} if $D_{R-H} = 82$ kcal mole^{-1}, i.e. the termolecular reaction is energetically more favoured than the bimolecular one.

The mechanism of free radical formation in cyclohexanol was studied by the inhibitor technique [54]. The termolecular reaction was found to be predominant with a rate

$$W = k[RH]^2[O_2]$$

$$k = 8.3 \exp(-16,000/RT)\,l^2\,\text{mole}^{-2}\,s^{-1}\,.$$

The low pre-exponential factor (<10 compared with 10^5 for tetralin [55]) can be accounted for by the higher degree of solvation of the activated complex compared with that of the reactant molecules. The activation energy is approximately equal to the reaction endothermicity assuming that $D_{C-H} = 80$ kcal mole^{-1}.

2.3.2 Decomposition of hydrogen peroxide to free radicals

In the absence of initiators, alcohol oxidation is self-accelerating (see Sect. 2.1) due to the formation of peroxides, as in the case of hydrocarbon oxidation. The mechanism of free radical formation from peroxides was studied for 2-propanol [25] and cyclohexanol [32,56]. The kinetics of hydrogen peroxide formation obeys the equation [25]

$$\frac{[H_2O_2]^{1/2}}{[RH]} = \frac{[H_2O_2]_0^{1/2}}{[RH]_o} + Kt$$

where K is a constant down to $[H_2O_2] = 0.7$ mole l^{-1}. This equation is consistent with the chain mechanism of alcohol oxidation provided H_2O_2 is the single source of free radicals and the initiation rate $w_i = k_i[H_2O_2]$. In this case, $K = k_p\sqrt{k_i}/2\sqrt{2k_t}$. The values of K were determined from experimental results and k_i was calculated to be $1.1 \times 10^7 \exp(-23,000/RT)$ s^{-1} in the range 99–130°C. The rate coefficient for free radical formation from H_2O_2 in cyclohexanol measured by the inhibitor technique

[56] at 120—140°C, is $k_i = 0.9 \times 10^7 \exp(-23,500/RT)$ s^{-1}.

The following reactions may be suggested for first-order decay of H_2O_2 to free radicals in alcohol

$$H_2O_2 \rightarrow 2\ HO\cdot + q_1 \qquad q_1 = -50 \text{ kcal mole}^{-1} \tag{3}$$

$$H_2O_2 + R—H \rightarrow HO\cdot + H_2O + R\cdot + q_2$$

$$q_2 = 118 - 50 - 85 = -17 \text{ kcal mole}^{-1} \tag{4}$$

$$H_2O_2 + R'OH \rightarrow HO\cdot + H_2O + R'O\cdot + q_3$$

$$q_3 = 118 - 50 - 102 = -35 \text{ kcal mole}^{-1} \tag{5}$$

$$H_2O_2 + 2\ R'OH \rightarrow 2\ H_2O + 2\ R'O\cdot - q_4$$

$$q_4 = (2 \times 118) - (2 \times 102) - 50 = -18 \text{ kcal mole}^{-1} \tag{6}$$

Reactions (4) and (6) are consistent with the experimental value of $E = 23$ kcal mole^{-1}, since $E \gg |{-}q|$ for an endothermic reaction. The k_i value remains unchanged after dilution of cyclohexanol with both chlorobenzene and decane. Therefore the mechanism seems to be

$$\overset{\overset{\textstyle H}{\textstyle |}}{2\ R'OH + H_2O_2 \rightleftharpoons R'OH...O—O....HOR'} \rightarrow R'O\cdot + 2\ H_2O + R'O\cdot \tag{6}$$
$$\underset{\textstyle H}{|}$$

When $[H_2O_2] > 1$ mole l^{-1}, the bimolecular reaction becomes important [56]

$$W_i = k_{i1}[H_2O_2] + k_{i2}[H_2O_2]^2$$

where $k_{i2} = 6.8 \times 10^9 \exp(-29,100/RT)$ l mole^{-1} s^{-1} (120—140°). The most probable mechanism is

$$HOOH + R'OH \rightleftharpoons HOOH.....O\overset{\diagup H}{\underset{\diagdown R'}{}}$$

$$HOOH + HOOH \rightleftharpoons HOO\overset{\diagup H}{....}HOOH \rightarrow HO\cdot + H_2O + HO_2\cdot$$

$$q_2 = -50 - 90 + 118 = -22 \text{ kcal mole}^{-1}$$

Ketones play an important part in the decomposition of peroxides to free radicals. With H_2O_2, they form hydroxyhydroperoxides (see Sect. 2.1)

$$\mathord{>}C{=}O + H_2O_2 \rightleftharpoons \mathord{>}C(OH)OOH$$

Hydroxyhydroperoxide decomposition to free radicals is more rapid than that of H_2O_2. The contribution from peroxide in the form of hydroxyhydroperoxide increases with increasing ketone concentration. The rate of free radical formation rises accordingly, as confirmed by use of the inhibitor technique for the system cyclohexanol—cyclohexanone—H_2O_2 [32], the steps being

$$>C=O + H_2O_2 \underset{K}{\rightleftharpoons} >C\overset{OH}{\underset{OOH}{\Big\langle}} \overset{k_D}{\rightarrow} >C\overset{OH}{\underset{O\cdot}{\Big\langle}} + HO\cdot$$

with $k_D = 2.2 \times 10^4 \exp(-16{,}200/RT)$ s^{-1} and $K = 0.12$ l mole^{-1} (120°C), 0.11 (110°C) and 0.10 (100°C). The rate of H_2O_2 decay to free radicals also increases with increasing acetone concentration in 2-propanol [25], being given by

$$W_i = k_{i1}[H_2O_2] + k'[H_2O_2][\text{acetone}]$$

Obviously, H_2O_2 and acetone also form a hydroxyhydroperoxide which rapidly decomposes to free radicals, with a rate coefficient at 118°C of $k' = k_D K = 1.2 \times 10^{-6}$ l mole^{-1} s^{-1}.

2.3.3 Mechanism of cyclohexanol oxidation

The mechanism of cyclohexanol oxidation has been studied in detail and is rather complex [23,26,32,48—50,57,58]. Various reactions involving H_2O_2 decomposition and cyclohexanone oxidation play the main part in the later stages of the process.

Hydrogen peroxide decay in cyclohexanol oxidation occurs by several routes. (a) By reaction with hydroxycyclohexyl radicals [57]

$$>\overset{\cdot}{C}OH + H_2O_2 \rightarrow >C=O + H_2O + HO\cdot$$

The concentration of dissolved oxygen decreases with increasing reaction rate. Consequently, the concentration of hydroxycyclohexyl radicals increases because $[>\overset{\cdot}{C}OH] \propto [RH][RO_2\cdot]/[O_2]$ and decomposition of H_2O is accelerated. (b) By heterogeneous hydrogen peroxide decomposition at the reactor wall. (c) By cyclohexanone oxidation by H_2O_2 [26] and subsequent increase in the rate of H_2O_2 disappearance as cyclohexanone accumulates. (d) By cyclohexanone oxidation to acids which accelerate the decomposition of H_2O_2 [26]. Of interest are experiments on cyclohexanol oxidation in the presence of H_2O_2 at the start of the reaction [58]. The $[H_2O_2]$ value appears to increase to $[H_2O_2]_{max}$. The stationary concentration of H_2O_2 in cyclohexanol containing no oxidation products is much higher than $[H_2O_2]_{max}$ in experiments without addition of H_2O_2 (Fig. 1). The reason is that cyclohexanone and the products of its oxidation react with H_2O_2, so that decay of the latter becomes more and more extensive as cyclohexanone accumulates [58]. Similarly, cyclohexanone and its products hinder the accumulation of H_2O_2.

Quantitative calculation of the rates of product formation during cyclo-

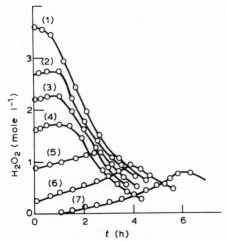

Fig. 1. Kinetic curves of H_2O_2 production in the oxidation of cyclohexanol at 120°C with initial concentrations of H_2O_2 of (1) 3.68, (2) 2.70, (3) 2.18, (4) 1.58, (5) 0.86, (6) 0.19 and (7) 0 mole l^{-1}.

hexanol oxidation at 120°C [59] made use of the equations

$$\frac{d[H_2O_2]}{dt} = 1.8 \times 10^{-2}[RH]\sqrt{W_i}$$

$$- 1.45 \times 10^{-4}[H_2O_2](1 + 1.25[R'{=}O]) + 1.7 \times 10^{-3}[H_2O_2][acid]$$

$$\frac{d[R'{=}O]}{dt} = \{1.8 \times 10^{-2}[RH] - 3.6 \times 10^{-3}[R'O]\}\sqrt{W_i}$$

$$+ 5.4 \times 10^{-5}[H_2O_2] - 1.2 \times 10^{-4}[H_2O_2][R'{=}O]$$

$$\frac{d[acid]}{dt} = 0.43 \times 10^{-4}[H_2O_2][R'{=}O] + 1.8 \times 10^{-3}[R'{=}O]\sqrt{W_i}$$

$$\frac{d[lactone]}{dt} = 0.3 \times 10^{-4}[H_2O_2][R'{=}O] + 1.8 \times 10^{-3}[R'{=}O]\sqrt{W_i}$$

$$W_i = 9 \times 10^{-7}[H_2O_2] + 6 \times 10^{-7}[H_2O_2]^2 + 3 \times 10^{-6}[H_2O_2][R'{=}O]$$

The results of the calculations are in good agreement with the experimental data.

2.4 PHOTO-OXIDATION AND RADIATION-INDUCED OXIDATION OF ALCOHOLS

2.4.1 Photo-oxidation

Photo-oxidation of alcohols yields the same products as are produced by thermal oxidation [20,60—62] (see Sect. 2.1).

The photo-oxidation at room temperature is a free radical non-chain reaction, as shown by the low quantum yield ($\phi \lesssim 1$ [60,61]) and the close-to-zero energy of activation. This conclusion is confirmed by the values of $k_p/\sqrt{2k_t}$ (Table 1). Non-chain oxidation occurs when

$$\frac{k_p}{\sqrt{2k_t}}\,[RH]\sqrt{W_i} < W_i$$

i.e.

$$\frac{k_p}{\sqrt{2k_t}} < \frac{\sqrt{W_i}}{[RH]}$$

Let us assume that $W_i = 10^{-6}$ mole l^{-1} s^{-1} and $[RH] = 10$ mole l^{-1}, then $k_p/\sqrt{2k_t} < 10^{-4}$ $l^{1/2}$ mole$^{-1/2}$ $s^{-1/2}$ and ethanol oxidation will occur by a chain mechanism only at $t \gtrsim 100°C$, whereas that of 2-propanol occurs at $t \gtrsim 37°C$. Oxidation of 2-propanol by UV irradiation in the presence of anthraquinone produces anthrahydroquinone radicals [62] (detected by the ESR technique). This proves the radical mechanism of photo-sensitized alcohol oxidation.

Certain features of alcohol photo-oxidation are connected with the presence of sensitizers. As found by Bäckström [60], the rate of photo-oxidation of 2-propanol (with benzophenone as sensitizer) is inversely proportional the oxygen pressure. Dependence of the 2-propanol oxidation rate on alcohol concentration and on light intensity, I, (with anthraquinone as sensitizer) is expressed as

$$W \sim \frac{I[RH]}{a + b[RH]}$$

The proposed initial stage of alcohol photo-oxidation, consistent with the above facts (A-sensitizer), is

Subsequent conversions may be represented by the steps

$$R\cdot + O_2 \rightarrow RO_2\cdot$$

$$\cdot AH + O_2 \rightarrow HO_2\cdot + A$$

$$RO_2\cdot + RO_2\cdot \rightarrow \text{ketone} + O_2 + ROOH$$

$$RO_2\cdot + HO_2\cdot \rightarrow ROOH + O_2$$

$$ROOH \rightleftharpoons \text{ketone} + H_2O_2$$

The photo-oxidation of primary alcohols, e.g. ethanol [61], is more complex. The primary oxidation products are H_2O_2, aldehyde, and acid, the acid yield increasing with increasing P_{O_2} and decreasing [A] [61]. The reaction mechanism suggested is [61]

$$2 \, CH_3CH(OH)OO \cdot \rightarrow 2 \, CH_3COOH + H_2O_2$$

$$CH_3CH(OH)OO \cdot + CH_3\dot{C}HOH \rightarrow CH_3CH(OH)OOH + CH_3CHO$$

$$CH_3CH(OH)OO \cdot + HO_2 \cdot \rightarrow CH_3CH(OH)OOH + O_2$$

$$CH_3CH(OH)OO \cdot + \cdot AH \rightarrow CH_3CH(OH)OOH + A$$

2.4.2 Radiation-induced oxidation

Radiation-induced oxidation of alcohols is a non-chain reaction when the temperature is sufficiently low and $k_p/\sqrt{2k_t} < \sqrt{W_i}/[RH]$. In this case, the reaction rate is proportional to the irradiation intensity (I) and is virtually independent of temperature [46,63—69].

Above a certain temperature, the oxidation becomes a chain reaction. The increase in radiolytic yield, G, with temperature and a dependence of G on I of the type [46]

$$G = a + bI^{-1/2}$$

are typical for radiation-induced chain oxidations.

The composition of alcohol radiolysis products is very different in the presence and absence of O_2, due to the reactions [65]

$$CH_3OH \rightsquigarrow CH_3 \cdot, \, \cdot H, \, \cdot CH_2OH$$

In the absence of oxygen

$$\cdot CH_3 + CH_3OH \rightarrow CH_4 + \dot{C}H_2OH$$

$$H \cdot + CH_3OH \rightarrow H_2 + \cdot CH_2OH$$

$$2 \cdot CH_2OH \rightarrow HOCH_2CH_2OH$$

$$2 \cdot CH_2OH \rightarrow CH_2O + CH_3OH$$

In the presence of oxygen

$$\cdot CH_3 + O_2 \rightarrow CH_3OO \cdot$$

$$H \cdot + O_2 \rightarrow HO_2 \cdot$$

$$\cdot CH_2OH + O_2 \rightarrow HOCH_2OO \cdot$$

$$2 \, HOCH_2OO \cdot \rightarrow HCOOH + CH_2O + H_2O + O_2$$

$$\searrow HCOOOCH_3 + H_2O + O_2$$

$$HOCH_2OO \cdot + HO_2 \cdot \rightarrow CH_2O + H_2O_2 + O_2$$

$$CH_3OO \cdot + HOCH_2OO \cdot \rightarrow CH_3OOH + CH_2O + O_2$$

TABLE 2
Values of G for methanol at different O_2 concentrationes

Product	G (mole l^{-1}) at 20°C in CH_3OH			
	$[O_2] = 0$ (ref. 65)	$[O_2] = 10^{-3}$ (ref. 65)	$[O_2] = 10^{-3}$ (ref. 68)	$[O_2] = 10^{-3}$ (ref. 46)
H_2	4.98	1.9	1.28	3.0
CH_4	0.43	0.18		
Peroxide	0	3.1	2.69	4.4
CH_2O	2.2	8.7	3.78	4.0
$(CH_2OH)_2$	3.2	0.1		
HCO_2H	0	1.5		
$HCOOOCH_3$				2.8
$\Delta[CH_3OH]$	9.3	11.2		
$\Delta[O_2]$				7.7

The values of G for methanol and other primary alcohols are listed in Tables 2 and 3. Radiation-induced oxidation of n-butanol was studied by Komarov et al. [66,67] over a wide temperature range.

Hughes and Makada [47] have studied the radiation-induced oxidation of 2-propanol in acid and alkaline aqueous solutions at 20°C. Alcohol oxidation in acid solution at $[RH] > 0.1$ mole l^{-1} proceeds by a chain mechanism

$$G = a + b\ I^{-1/2}$$

and

$$G = G_0 + C[RH]$$

No chain reactions occur in alkaline solutions: $G = G_0$ and does not depend on I and $[RH]$. This can be explained by the equilibrium

$$HO_2\cdot\ \rightleftharpoons\ H^+ + O_2^-\cdot$$

and the low reactivity of $O_2^-\cdot$.

TABLE 3
Values of G for the radiolytic oxidation of primary alcohols

RCH_2OH	Temp. (°C)	Solvent, concentration of alcohol (mole l^{-1})	G_{H_2}	G_{RCHO}	$G_{H_2O_2}$	G_{RCCOH}	F
CH_3CH_2OH	20	H_2O, 8.9×10^{-3}		2.4	3.2		6
CH_3CH_2OH	25	H_2O, 3.4×10^{-2}		2.6	4.15		6
$C_2H_5CH_2OH$	20	n-PrOH	1.4	9.5	7.6	4.4	6
n-$C_3H_7CH_2OH$	20	H_2O, 0.76		1.5	1.3	1.9	6
n-$C_3H_7CH_2OH$	104	H_2O, 0.76		4.6	1.1	2.2	6

TABLE 4
Rate parameters for the reaction of ozone with alcohols in CCl_4 solution

Alcohol	k at 25°C (1 mole^{-1} s^{-1})	$\log(A/1$ mole^{-1} s^{-1})	E (kcal mole^{-1})	Ref.
C_2H_5OH	0.35	6.7	9.8	271
$(CH_3)_2CHOH$	0.89	7.3	10.0	271
n-C_4H_9OH	0.54	7.3	10.3	271
sec-C_4H_9OH	1.6	7.1	9.4	271
t-C_4H_9OH	9.8×10^{-3}	4.6	9.0	271
(cyclohexanol)	2.0	6.7	8.7	272

2.4.3 Oxidation with ozone

Oxidation of cyclohexanol with a mixture of O_2 and O_3 at 80—100°C proceeds by a chain mechanism [70]. The rate of free radical formation is 1000 times lower than that of ozone consumption and the activation energy for chain initiation by ozone is 11 kcal mole^{-1}. The cyclohexanone formed is oxidized by ozone without formation of free radicals.

Ozone reacts with alcohols by a bimolecular process with rate coefficients which are much higher than those of the reaction $RO_2\cdot$ + alcohol (Table 4). For example, in the case of cyclohexanol, $k(RO_2\cdot + RH) = 4.3 \times 10^{-2}$ mole^{-1} s^{-1} and $k(O_3 + RH) = 2.0$ l mole^{-1} s^{-1} (25°C), i.e. ozone reacts 50 times faster than the peroxy radical. This difference in rate coefficients stems from different activation energies: $E = 12.0$ kcal mole^{-1} for the reaction $RO_2\cdot$ + RH and only 8.7 kcal mole^{-1} for the reaction O_3 + RH. The following mechanism is proposed [272].

$$>\text{CHOH} + O_3 \rightarrow >\dot{\text{C}}\text{OH} + HO_3\cdot \begin{cases} >\text{C}=\text{O} + H_2O_2 + \frac{1}{2}O_2 \\ >\dot{\text{C}}\text{OHO}_2 + \text{HO}\cdot \end{cases}$$

Acetone was found in the reaction products of ozone with t-butanol [271]. Thus t-butoxy radicals are formed and ozone reacts with the OH group of the alcohol. The low A-factors and activation energies are probably connected with the association of alcohol molecules through hydrogen bonding.

2.5 INHIBITORS OF ALCOHOL OXIDATION

2.5.1 Mechanism of reaction of alcohol peroxy radicals with phenols and aromatic amines

Phenols and aromatic amines inhibit the oxidation of alcohols, breaking the chains by reactions with peroxy radicals as they do in the case of hydrocarbon oxidation [1]. At the same time, the mechanism of this reaction for alcohols differs from that for hydrocarbons. While in hydrocarbon oxidation [1] the rate is given by

$$W(\text{InH} + \text{RO}_2 \cdot) = k_{\text{InH}}[\text{InH}][\text{RO}_2 \cdot]$$

for alcohols [71,72]

$$W(\text{InH} + \text{RO}_2 \cdot) = k_{\text{InH}}[\text{InH}][\text{RO}_2 \cdot] + k'_{\text{InH}}[\text{InH}][\text{RO}_2 \cdot]^2$$

Such a relationship is consistent with the mechanism

$$\text{RO}_2 \cdot + \text{InH} \xrightarrow{k_{\text{InH}}} \text{ROOH} + \text{In}$$

$$\text{RO}_2 \cdot + \text{InH} \overset{K}{\rightleftharpoons} \text{RO}_2 \cdot ...\text{InH}$$

$$\text{RO}_2 \cdot + \text{HIn} ... \text{RO}_2 \cdot \overset{k}{\rightarrow} \text{ROOH} + \text{InOOR} \rightarrow \text{products}$$

with $k'_{\text{InH}} = Kk$. In the case of α-naphthol in cyclohexanol at $120°C$ [71], $k_{\text{InH}} = 1.1 \times 10^3$ l mole^{-1} s^{-1} and $k'_{\text{InH}} = 4.8 \times 10^9$ l^2 mole^{-2} s^{-1}; in a mixture of chlorobenzene (70%) with cyclohexanol (30%) [72] at $120°C$, $k_{\text{InH}} = 2.2 \times 10^3$ l mole^{-1} s^{-1} and $k'_{\text{InH}} = 7 \times 10^8$ l^2 mole^{-2} s^{-1}. It is interesting to compare these values with those for hydrocarbon: in cyclohexane [73] ($75°C$), $k_{\text{InH}} = 3.9 \times 10^5$ l mole^{-1} s^{-1}, which is two orders of magnitude higher than for cyclohexanol. Such a large difference is accounted for by the fact that most inhibitor molecules are linked by hydrogen bonds to cyclohexanol and are not attacked by peroxy radicals (see Sect. 6.2).

The reaction of free radicals with aromatic amines and phenols may proceed by two routes, abstraction of H from the O—H or N—H bond and addition of the radical to an inhibitor. Free radicals and atoms, $H\cdot$, $HO\cdot$, $CH_3\cdot$ and $C_6H_5\cdot$, are known to add to aromatic compounds. Both abstraction and addition are observed simultaneously in alcohol oxidation (the Boozer and Hammond mechanism [74]), while hydrocarbon peroxy radicals only abstract H-atoms from inhibitors *.

Let us discuss the conditions under which the reversible addition of $RO_2\cdot$ to the inhibitor can be observed. Inasmuch as termination of chains in oxidation occurs in general by three reactions, the rate being

$$W_{\text{I}} = 2k_t[\text{RO}_2 \cdot]^2 + 2k'_{\text{InH}}[\text{InH}][\text{RO}_2 \cdot]^2 + 2k_{\text{InH}}[\text{InH}][\text{RO}_2 \cdot]$$

the reversible addition could be observed under the conditions

$$2k'_{\text{InH}}[\text{InH}][\text{RO}_2 \cdot]^2 > 2k_t[\text{RO}_2 \cdot]^2 \tag{I}$$

---

* The dependence of cumene oxidation rate on [PhOH] is of the type $\overset{\cdot}{W} \sim \sqrt{W_i}/[\text{PhOH}]$[74] and is not due to the addition of $RO_2\cdot$ to phenol, but is a consequence of the exchange reaction

$$\text{PhO}\cdot + \text{ROOH} \rightarrow \text{PhOH} + \text{RO}_2 \cdot$$

as found by Thomas [75].

$$2k'_{\text{InH}}[\text{InH}][\text{RO}_2\cdot]^2 > 2k_{\text{InH}}[\text{InH}][\text{RO}_2\cdot] \tag{II}$$

Condition (I) reduces to $[\text{InH}] > k_t/k'_{\text{InH}}$. When this condition is fulfilled

$$[\text{RO}_2\cdot] \cong \frac{k_{\text{InH}}}{2k'_{\text{InH}}}\left\{\sqrt{1 + \frac{2W_I k'_{\text{InH}}}{k^2_{\text{InH}}[\text{InH}]^2}} - 1\right\}$$

and condition (II) becomes $[\text{InH}] < W_i k'_{\text{InH}}/4k^2_{\text{InH}}$.

Consequently, the mechanism of reversible addition would become evident only when the inequalities

$$k_t/k'_{\text{InH}} < [\text{InH}] < W_I k'_{\text{InH}}/4k^2_{\text{InH}}$$

are fulfilled, i.e. only when

$$4k_t k^2_{\text{InH}} < W_I k'^2_{\text{InH}}$$

or

$$k_{\text{InH}} k'_{\text{InH}} < \tfrac{1}{2}\sqrt{W_I/k_t}$$

In the case of cyclohexanol at $120°C$, $\sqrt{k_t} = 5.5 \times 10^2\ l^{1/2}\ \text{mole}^{-1/2}\ s^{-1/2}$, $k_{\text{InH}} = 1.1 \times 10^3\ l\ \text{mole}^{-1}\ s^{-1}$ and $k'_{\text{InH}} = 4.8 \times 10^9\ l^2\ \text{mole}^{-2}\ s^{-1}$ (α-naphthol) and thus the above inequality is fulfilled when $W_I > 6 \times 10^{-8}\ \text{mole}\ l^{-1}\ s^{-1}$. For hydrocarbons $k_{\text{InH}} \approx 10^5\ l\ \text{mole}^{-1}\ s^{-1}$ and for the same values of k'_{InH} and k_t the inequality is fulfilled when $W_I > 6 \times 10^{-4}$, whereas usually $W_I = 10^{-8}-10^{-6}\ \text{mole}\ l^{-1}\ s^{-1}$. For this reason, the mechanism of reversible addition is not observed for hydrocarbons, but is observed for alcohols where abstraction of H from the inhibitor is inhibited by hydrogen bonding.

2.5.2 Regeneration of aromatic amines in alcohol oxidation

With hydrocarbons, two chains are normally terminated by one molecule of the inhibitor (of phenol or aromatic amine), i.e. the stoichiometric coefficient of the inhibitor is $f = 2$ [1]. With cyclohexanol, multiple termination of chains takes place with the same molecule of the inhibitor, as found for α-naphthylamine [76]. This was observed for many aromatic amines in primary as well as in secondary alcohols [77,78]. To the first approximation, the mechanism of inhibitor regeneration consists of reaction of the alcohol hydroxyperoxy radical with the radical In·, reducing to InH with simultaneous termination of the chain [79], viz.

$$>\text{C(OH)OO}\cdot + \text{InH} \rightarrow >\text{C(OH)OOH} + \text{In}\cdot$$

$$>\text{C(OO}\cdot)\text{OH} + \text{In}\cdot \rightarrow \text{InH} + >\text{C}=\text{O} + \text{O}_2$$

Reduction of In· by the hydroxyperoxy radical was established in the following way [79]. Tetraphenylhydrazine added to cyclohexanol in the presence of an initiator decomposed to diphenylnitrogen radicals which were reduced to diphenylamine. No reduction was observed in the absence of initiator. The mechanism of regeneration of inhibitors was studied in detail for α-naphthylamine in cyclohexanol [44]. Three kinds of radical, hydroxyalkyl, hydroxyperoxy, and hydroperoxy, were found to be present in the alcohol. The ratio of their concentrations depends on $[O_2]$, $[>CH(OH)]$, and H_2O_2, i.e. they are produced and disappear by the reactions

$$>\!\overset{\cdot}{C}OH + O_2 \rightarrow >\!C(OH)OO\cdot$$

$$>\!C(OH)OO\cdot + H_2O_2 \rightarrow >\!C(OH)OOH + HO_2\cdot$$

$$>\!C(OH)OOH \rightarrow >\!C\!=\!O + H_2O_2$$

$$HO_2\cdot + >\!CHOH \rightarrow H_2O_2 + >\!\overset{\cdot}{C}OH$$

Each of these radicals react with In· by two routes, reduction of In· to InH and addition to In· (resulting in the disappearance of InH). For example

$$In\cdot + HO_2\cdot \xrightarrow{k_{In}} InH + O_2$$

$$In\cdot + HO_2\cdot \xrightarrow{k'_{In}} InOOH$$

The stoichiometric coefficient $f = 2(1 + k_{In}/k'_{InH})$ and naturally depends on the kind of radical with which In· reacts. As found in experiments on the oxidation of cyclohexanol with α-naphthylamine ($120°C$, $W_I = 1.86 \times 10^{-6}$ mole l^{-1} s^{-1}), f increases with oxygen pressure from 16.6 ($P_{O_2} = 0$) to 48 ($P_{O_2} = 760$ torr). It follows that the degree of amine regeneration by reaction with hydroxyperoxy radicals is higher than that for hydroxyalkyl

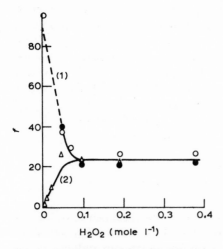

Fig. 2. Stoichiometric coefficient, f, as a function of $[H_2O_2]$ for α-naphthylamine in alcohols oxidation at $75°C$. (1) In cyclohexanol (○) and 2-propanol (●); (2) in t-butanol (△).

radicals. Experiments on the addition of H_2O_2 to cyclohexanol (Fig. 2) have shown that hydroxyperoxy radicals are responsible for oxidation, and the degree of amine regeneration is higher for hydroxyperoxy radicals than for $HO_2 \cdot$. The ratio k_{In}/k'_{In} is 47 ± 5 for $>C(OH)OO \cdot$, 9 ± 2 for $HO_2 \cdot$ and 6 ± 2 for $>\dot{C}OH$ with α-naphthylamine in cyclohexanol at $75°C$ [44]. The role of hydroxyalkyl radicals in the regeneration of $In \cdot$ is small: at $P_{O_2} = 760$ torr and $120°C$, 95% of $In \cdot$ invert to InH by reaction with peroxy radicals [44].

2.5.3 Inhibition of alcohol oxidation by quinones

Hydroxyperoxy radicals reduce $In \cdot$ to InH due to the exothermicity of the reaction

$$>C(OH)OO \cdot + In \cdot \rightarrow >C=O + O_2 + InH + q$$

If $D_{>C(OO \cdot)O-H} = 75$ kcal mole^{-1}, then $q = 30$ kcal mole^{-1}. Therefore, hyroxyperoxy radicals, in contrast to alkylperoxy radicals, display a dual reactivity. They can take part both in oxidation and in reduction reactions and they would be expected to react not only with radicals but with molecules of the oxidizing agent, with quinones for example. The kinetics of 2-propanol oxidation in the presence of benzoquinone has been studied [80]. Quinones are known to terminate chains in hydrocarbon oxidation only by reactions with alkyl radicals [1]. In alcohol oxidation, quinone terminates chains by reaction with hydroxyalkyl as well as with hydroxyperoxy radicals [80]. At $71°C$ and $P_{O_2} = 760$ torr, 86% of chain termination is due to the reaction $>C)OH)OO \cdot +$ quinone. The rate coefficient is $k_Q(>C(OH)OO \cdot +$ quinone$) = 3.2 \times 10^3$ l mole^{-1} s^{-1} and $k_Q/k_p = 1.0 \times 10^4$. Just as in the case of aromatic amines, $f > 2$; $f = 23$ for quinone, i.e. quinone is regenerated in the reactions

$$Q + >C(OH)OO \cdot \rightarrow \cdot QH + O_2 + >C=O$$

$$HQ \cdot + >C(OH)OO \cdot \rightarrow Q + >C(OH)OOH$$

2.5.4 Chemiluminescence in oxidations inhibited by aromatic amines in the presence of alcohols

Chemiluminescence in liquid phase oxidation arises from the recombination of two peroxy radicals [81], excited molecules of ketone formed by the reaction

$$RO_2 \cdot + RO_2 \cdot \rightarrow alcohol + O_2 + ketone^*$$

being the source of the radiation

$$ketone^* \rightarrow ketone + h\nu$$

Addition of an inhibitor (an aromatic amine, for example) decreases the

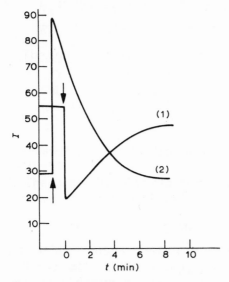

Fig. 3. Chemiluminescence intensity as a function of time after addition of α-naphthyl-amine (10^{-4} mole l^{-1}) at 75°C and $W_i = 5.4 \times 10^{-7}$ mole l^{-1} s^{-1}. (1) In cumene oxida-tion; (2) in chlorobenzene in the presence of 5 vol.% t-butanol.

concentration of peroxy radicals and the rate of the reaction $RO_2 \cdot + RO_2 \cdot$. The intensity of chemiluminescence consequently drops. Direct propor-tionality between the rate ratio W/W_0 and $\sqrt{I/I_0}$ is observed (I and I_0 are the intensities of chemiluminescence in the presence and absence of inhibitor, respectively). The situation is different when aromatic amines are added in the presence of alcohol. These amines hinder oxidation by reacting with peroxy radicals. However, addition of aromatic amines to oxidizing alcohol does not lower but increases the intensity of chemi-luminescence [82] (Fig. 3). This is connected with the formation of another source producing more intense chemiluminescence than the reac-tion of two peroxy radicals. The following features are characteristic of chemiluminescence in the presence of aromatic amines.

(a) Chemiluminescence arises only in the presence of O_2 and of an initiator, i.e. when peroxy radicals are formed.

(b) Chemiluminescence is detected only in the presence of an activator (e.g. 9,10-dibromanthracene).

(c) Chemiluminescence becomes more intense when aromatic amine is added to the system in the presence of an alcohol or of another polar compound (e.g. dioxan, water).

(d) The intensity of chemiluminescence is proportional to the rate of initiation.

(e) The intensity of chemiluminescence increases with amine concentra-tion, tending to some limiting value. All these facts are in agreement with the scheme

$$RO_2\cdot + AmH \rightarrow ROOH + Am\cdot$$

$$RO_2\cdot + Am\cdot \rightarrow {}^1\Pi^*$$

$${}^1\Pi^* + \text{polar molecule} \rightarrow {}^3\Pi^*$$

$${}^3\Pi^* + A \rightarrow \Pi + A^*$$

$$A^* \xrightarrow[({}^1\Pi^*)]{} A + h\nu$$

An excited singlet molecule ($^1\Pi^*$) is produced by the reaction between peroxy and amine radicals. This molecule is rapidly deactivated in a non-polar medium by the radiationless transition mechanism. However, in the presence of polar molecules, e.g. alcohol, a fraction of the excited molecules pass to the triplet excited state ($^3\Pi^*$). The triplet state lifetime is sufficiently long for this molecule to meet an activator molecule (A) and impart its energy to it. The excited activator molecule then emits light. Thus, chemiluminescence in oxidation can arise, not only by the reaction $RO_2\cdot + RO_2\cdot$, but also by reaction of $RO_2\cdot$ with other radicals. An important part in this process is played by polar molecules. It might be that the hydrogen bonding between alcohol and a product of the reaction $RO_2\cdot + Am\cdot$, is responsible for the transition of excited molecules from the singlet to the triplet state.

2.6 NEGATIVE CATALYSIS IN ALCOHOL OXIDATION

2.6.1 Negative redox catalysis

Hydroxyperoxy radicals can induce both oxidation and reduction. If the inhibitor is present in two states, oxidized and reduced, and each state reacts with hydroxyperoxy radicals only, terminating the chains, then negative catalysis will take place, each inhibitor molecule terminating chains an infinite number of times. This is the case on addition of $CuSO_4$ to cyclohexanol [83]. Cupric ions in a concentration of 10^{-5} mole l^{-1} virtually stop the initiated oxidation of cyclohexanol. The mechanism of the retarding action of cupric ions is

$$>C(OH)OO\cdot + Cu^{2+} \xrightarrow{k} Cu^+ + H^+ + O_2 + >C=O$$

$$>C(OH)OO\cdot + Cu^+ \xrightarrow{H^+} Cu^{2+} + >C(OH)OOH$$

The first stage is suggested to be the rate-limiting one. Dependence of the oxidation rate on $[Cu^{2+}]$ is expressed as

$$\frac{W_o}{W} - \frac{W}{W_o} = 2\frac{k}{\sqrt{2k_t}}\frac{[Cu^{2+}]}{\sqrt{W_i}}$$

TABLE 5

Rate coefficients for the reactions of hydroxyperoxy radicals with compounds of transition metals in alcohol

Compound [a]	$k/\sqrt{2k_t}$	k $(1\ mole^{-1}\ s^{-1})$	$k'/2k_t$ $(1\ mole^{-1})$
Cyclohexanol [83,273] 75°C			
$Cu^{2+}aq\ (9\%\ H_2O)$	5.1×10^3	1.0×10^7	
$CuSt_2$	1.7×10^3	3.4×10^6	
$MnSt_2$	3.9×10^3	7.8×10^6	
$CoSt_2$	1.6×10^2	3.2×10^5	
$CeSt_3$	64	1.3×10^5	
$FeSt_3$	7.6	1.5×10^4	
2-Propanol [274] 71°C			
$Cu(DH)$	410		9.5×10^5
$Cu(SalH)_2$	100		4.4×10^4
$Cu(DfH)_2$	49		3.0×10^5
$Cu(DH)_2NH_3I$	30		1.3×10^5
$Co(DH)_3$	0.24		6.5×10^4
$Co(DH)_2NH_3I$	0.25		1.1×10^5
$Co(DH)_2PyI$	57		9.5×10^4
$Co(DH)_2NH_3Cl$	4.1		5.2×10^5
$Co(DH)_2Py_2$	10		2.5×10^4
$Fe(DH)_2Py_2$	0.8	6.0	6.0×10^3
$[Co(DCH_3)_2]ClO_4$	1.0		2.0×10^8

[a] $(DH)_2$ = dimethylglyoxime; $(DfH)_2$ = diphenylglyoxime; SalH = salicylaldoxime; Py = pyridine; StH = stearic acid.

The rate coefficient, k, is $3.2 \times 10^6\ 1\ mole^{-1}\ s^{-1}$ (75°C, cyclohexanol + 9% of H_2O).

Similar results were obtained when transition metal stearates were added to cyclohexanol (Table 5). The dioxymine complexes of Co, Cu and Fe retard oxidation of 2-propanol [274] by termination of chains. The rate of termination obeys the equation

$$W_t = 2k[Me][RO_2 \cdot] + 2k'[Me][RO_2 \cdot]^2 + 2k_t[RO_2 \cdot]^2$$

The values of $k/\sqrt{2k_t}$ and $k'/2k_t$ are given in Table 5. The scheme suggested is [274]

$$Co^{3+}I_n + RO_2 \cdot \rightarrow X \cdot \rightarrow Co^{2+}I_n$$
$$X \cdot + RO_2 \cdot \rightarrow Co^{3+}I_n$$
$$Co^{2+}I_n + RO_2 \cdot \rightarrow Co^{3+}I_n$$
$$X \cdot + RH \rightarrow Co^{2+}I_n + R \cdot$$

$\Big\}$ + molecular products

Not only transition metals but also I^- and Br^- ions inhibit the oxidation of alcohol [84]. Iodine, both in the form of I$^-$ and of I$_2$, inhibits the initiated oxidation. Cyclohexanone, the product of cyclohexanol oxidation, hinders this inhibition, reacting in the form of enol with I$_2$. This leads to critical phenomena in the oxidation. A sharp transition from non-inhibited to inhibited reaction is observed when the concentration of I$_2$, and W_i, are changed. Inhibition is caused by the reactions

$$>\overset{\cdot}{C}OH + I_2 \longrightarrow \; >C=O + HI + I\cdot$$

$$>C(OH)OO\cdot \; + I_2 \longrightarrow \; >C=O + O_2 + HI + I\cdot$$

$$>C(OH)OO\cdot \; + I\cdot \longrightarrow \; >C=O + O_2 + HI$$

$$>C(OH)OO\cdot \; + HI \longrightarrow \; >C(OH)OOH + I\cdot$$

Inhibition of initiated cyclohexanol oxidation by Br$^-$ is peculiar. It starts a certain time after the addition of Br$^-$ and the rate of the inhibited oxidation does not depend on the Br$^-$ concentration. Cyclohexanone has no effect. Obviously, the inhibiting action is not due to Br$^-$ ions but to bromine oxides and bromoxygen acids.

2.6.2 Inhibition by bases

Initiated oxidation of alcohols is inhibited by bases [45,47]. For example, 4×10^{-4} M NaOH lowers the rate of initiated oxidation of cyclo-hexanol by a factor of ten (75°C, $W_i = 5.6 \times 10^{-7}$ mole l^{-1} s^{-1}, 10% of H$_2$O), and 4×10^{-3} M NaOH completely stops the oxidation [45]. The

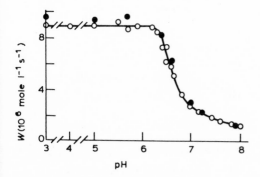

Fig. 4. Cyclohexanol oxidation rate as a function of pH (pH is given for aqueous solutions) at 75°C without (○) and with (●) the addition of 0.1 mole l^{-1} of H$_2$O$_2$. $W_i = 5.6 \times 10^{-7}$ mole l^{-1} s^{-1}, 10 vol.% H$_2$O.

References pp. 195—203

inhibiting action of bases is connected with the equilibrium

$$>C(OH)OO\cdot \xrightleftharpoons{OH^-} >C{=}O + H_2O + O_2^-\cdot$$

The ion-radicals $O_2^-\cdot$ are less reactive than peroxy radicals towards alcohols, but they enter into a fast reaction with each other, viz.

$$O_2^-\cdot + O_2^-\cdot \rightarrow O_2^- + O_2$$

In aqueous solution [85], $k(O_2^-\cdot + O_2^-\cdot) = 1.5 \times 10^7 \, l \, mole^{-1} \, s^{-1}$. Therefore, in the presence of a base, when all peroxy radicals readily convert to $O_2^-\cdot$, chain oxidation stops. The dependence of the oxidation rate on pH is shown in Fig. 4. Addition of H_2O_2 does not change the rate of oxidation. This may be explained by the decomposition of hydroxyalkyl radicals to ketone and $HO_2\cdot$ in the presence of bases (see Sect. 2.1).

Alcohol oxidation is more strongly inhibited by sodium bicarbonate than by alkali. This seems to be connected with additional inhibition by HCO_3^- ions by the reactions [86]

$$HCO_3^- + >C(OH)OO\cdot \rightarrow CO_3^-\cdot + >C(OH)OOH$$

$$CO_3^-\cdot + >C(OH)OO\cdot \rightarrow HCO_3^- + >C{=}O + O_2$$

2.7 CATALYSIS IN THE OXIDATION OF ALCOHOLS

2.7.1 Catalysis by transition metals

Transition metal ions inhibit alcohol oxidation by reacting with hydroxyperoxy radicals (see Sect. 2.6), viz.

$$Me^{n+} + >C(OH)OO\cdot \xrightarrow{k_1} Me^{n+1} + >C(OH)OO^-$$

$$Me^{n+1} + >C(OH)OO\cdot \xrightarrow{k_2} Me^{n+} + H^+ + O_2 + >C{=}O$$

As well as terminating chains, these ions may initiate chains by reactions with H_2O_2 and RH, viz.

$$Me^{n+} + H_2O_2 \xrightarrow{k_3} Me^{n+1} + \cdot OH + HO^-$$

$$Me^{n+1} + H_2O_2 \xrightarrow{k_4} Me^{n+} + H^+ + HO_2\cdot$$

$$Me^{n+1} + RH \xrightarrow{k_5} Me^{n+} + R\cdot + H^+$$

Acceleration of oxidation will be observed only when initiation is predominant over termination, i.e. when

$$k_3[Me^{n+}][H_2O_2] + k_4[Me^{n+1}][H_2O_2] + k_5[Me^{n+1}][RH]$$
$$> (k_1[Me^{n+}] + k_2[Me^{n+1}])[>C(OH)OO\cdot]$$

This is the case for 2-propanol oxidation in the presence of Co^{2+} and H_3PO_4 and Cr^{3+} and H_3PO_4 [87,88]. The rate of reaction is a maximum at $[H_3PO_4]/[Co^{2+}] = 1$, is independent of oxygen pressure at $P_{O_2} > 70$ torr, and is proportional to $[Co^{2+}]^{1/2}[RH]^2$.

2.7.2 Catalysis by bromide ions

Br^- ions inhibit the oxidation of alcohols if the latter do not contain H_2O_2. However, in the presence of H_2O_2, 2-propanol oxidation is accelerated by Br^- [89], as the latter induces decomposition of H_2O_2 leading to free radical formation. The rate of initiation by reaction of Br^- with H_2O_2 is [89]

$$W_i = k_A[Br^-][H_2O_2] + k_B[Br^-][H_2O_2]^2$$

At $70°C$, $k_A = 1.2 \times 10^{-4}$ l mole^{-1} s^{-1} and $k_B = 1.3 \times 10^{-3}$ l^2 mole^{-2} s^{-1}. Consumption of hydrogen peroxide by reaction with Br^- is much faster than its normal decomposition to free radicals, the rate of which is only 2% of that of H_2O_2 decay ($70°C$, $[KBr] = 5.8 \times 10^{-3}$ and $[H_2O_2] = 0.22$ mole l^{-1}, 90% of 2-propanol and 10% of H_2O). The mechanism suggested is [89]

$$Br^- + H_2O_2 \rightarrow HBrO + OH^-$$

$$RH + HOBr \rightarrow Br\cdot + H_2O + R\cdot$$

$$H_2O_2 + HOBr \rightarrow Br\cdot + H_2O + HO_2\cdot$$

$$>CHOH + HOBr \rightarrow >C=O + H_2O + HBr$$

Decomposition of hydroxyperoxides to free radicals is also accelerated by bromide ions [90]. The reaction of hydroperoxide with Br^- is second order, in contrast to that of hydrogen peroxide, i.e.

$$W_i = k[Br^-][ROOH]$$

where

$$k = 4.0 \times 10^8 \exp(-19,500/RT) \text{ l mole}^{-1} \text{ s}^{-1}$$

for t-butyl hydroperoxide in 2-propanol.

2.7.3 Catalysis by hydrogen ions

Hydrogen ions are known to catalyze various heterolytic reactions, in particular the heterolytic decay of peroxides. Recently, strong mineral acids have been found to accelerate the oxidation of 2-propanol by decomposing H_2O_2 to give free radicals [91]. It is only the hydrogen ions that cause decomposition to free radicals, since addition of salts (for example $KClO_4$ instead of $HClO_4$) does not cause decomposition of H_2O_2, and bases (pyridine, water) inhibit the decomposition of H_2O_2 to radicals in the presence of acid [92,93]. The rates of free radical formation from H_2O_2 under the action of acids are given by

$$W_i = k_i[H_2O_2]^2[HA]$$

for $[H_2O_2] \leqslant 1.5 \times 10^{-2}$ mole l^{-1} and $[HA] \leqslant 0.01$ mole l^{-1}. In 2-propanol at $70°C$, $10^2 k_i = 2.0 \ l^2$ mole^{-2} s^{-1} (H_2SO_4), 1.8 (HClO$_4$, and 1.9 (HCl), i.e. the rate coefficients are the same for different acids [92]. In the case of HClO$_4$, $k_i = 3.3 \times 10^{15}$ exp($-27,000/RT$) l^2 mole^{-2} s^{-1} in the temperature $60—75°C$. Hydrogen peroxide decomposes under the action of H$^+$ to molecular products, along with production of free radicals, and this is, in fact, the main route of H_2O_2 decay. The rate of H_2O_2 decomposition to molecular products is given by [92]

$$W = k[H_2O_2][HClO_4]$$

where $k = 5.9 \times 10^{13}$ exp($-24,700/RT$) l mole^{-1} s^{-1} (2-propanol, $60—75°C$). Therefore the kinetics of heterolytic H_2O_2 decay are different from those of the decomposition to radicals. The latter reaction represents 1% of the overall decomposition of H_2O_2 at $70°C$, $[HClO_4] = 0.02$ and $[H_2O_2] = 0.02$ mole l^{-1}. The rate coefficients of heterolytic decay of H_2O_2 in the presence of H_2SO_4 and HClO$_4$ are similar (3.9×10^{-2} and 3.2×10^{-2} l mole^{-1} s^{-1} at $70°C$, respectively), but that for HCl is different (5.9×10^{-2}). The proposed mechanism is

$$HA + ROH \rightleftharpoons ROH_2^+ + A^-$$

$$ROH_2^+ + H_2O_2 \overset{K}{\rightleftharpoons} H_3O_2^+ + ROH$$

$$H_3O_2^+ + H_2O_2 \overset{k_1}{\rightarrow} H_3O^+ + HO\cdot + HO_2\cdot \qquad (7)$$

$$H_3O_2^+ + ROH \overset{k_2}{\rightarrow} H_3O^+ + H_2O + {>}C{=}O$$

For this scheme

$$W = (k_1[H_2O_2] + k_2[ROH])[H_3O_2^+] \cong k_2[ROH] \times [H_3O_2^+]$$

$$= k_2 K[H_2O_2][ROH_2^+] = k_2 K[H_2O_2][HA]$$

if dissociation is complete, and

$$W_i = k_1[H_2O_2][H_3O_2^+] = k_1 K[H_2O_2]^2[HA]$$

Reaction (7) is the source of free radicals. It is similar to the reaction

$$H_2O_2 + H_2O_2 \rightarrow HO_2\cdot + H_2O + HO\cdot \qquad -22 \text{ kcal mole}^{-1}$$

but is more advantageous energetically ($q_1 = -18$ kcal mole^{-1}), as the reaction

$$H_3O_2^+ + H_2O \rightarrow H_2O_2 + H_3O^+$$

is exothermic to the extent of 4 kcal mole^{-1}. The above mechanism explains the increase in the rate of cyclohexanol oxidation in emulsions with decreasing pH [94]. Acids induce decomposition to free radicals not only of H_2O_2 but also of hydroperoxides [95]. In 2-propanol, $HClO_4$ decomposes t-butyl hydroperoxide at a rate given by

$$W_i = k_i[ROOH]^2[HClO_4]$$

$$[ROOH] \leqslant 3 \times 10^{-2}, [HClO_4] \leqslant 10^{-2} \text{ mole } l^{-1},$$

$$k_i = 4.0 \times 10^{15} \exp(-28,000/RT) \, l^2 \text{ mole}^{-2} \text{ s}^{-1} \, (60-75°C).$$

2.7.4 Oxidation of alcohols with palladium salts

$PdCl_2$ oxidises primary and secondary alcohols to aldehydes and ketones, respectively [96,97]. Catalytic oxidation of alcohols with oxygen may be performed in the presence of cupric ions, viz.

$$RCH_2OH + PdCl_2 \rightarrow RCHO + 2 HCl + Pd$$

$$Pd + 2 Cu^{2+} \rightarrow Pd^{2+} + 2 Cu^+$$

$$2 Cu^+ + \tfrac{1}{2} O_2 + 2 H^+ \rightarrow 2 Cu^{2+} + H_2O$$

2.7.5 Oxidation of alcohols in the presence of heterogeneous catalysts

Primary alcohols (glycerine [98], cinnamic alcohol [99], octanol [100], $C_6H_5OCH_2CH_2OH$ [101,102], ethylene glycol [103], n-butanol [104], n-propanol [105]) are oxidized to aldehydes in the presence of a platinum catalyst. The mechanism of oxidation of n-propanol in the presence of platinum at 40—85°C was investigated [105]. The reaction proceeded both on the surface of platinum (30%), and also in the bulk liquid by a chain mechanism. It was found in experiments using diphenylamine as inhibitor that chains were generated on the platinum surface, without participation of O_2, with an activation energy of 5.5 kcal mole^{-1}. The chain length appeared to be 300.

2.8 OXIDATION OF ALCOHOLS IN BASIC SOLVENTS

Addition of bases to alcohols inhibits the chain oxidation due to the equilibrium

$$HO_2 \cdot \rightleftharpoons H^+ + O_2^- \cdot$$

and the low reactivity of $O_2^-\cdot$ (see Sect 2.6). However, in the presence of strong bases in considerable concentration when a substantial part of the molecules of alcohol are in the form of alcoholate ions, oxidation is rapid. Le Berre and Etienne observed [106,107] fast the oxidation of alcoholates of primary and secondary alcohols at $20°C$ in benzene and tetrahydrofuran. Primary alcohols are oxidized to acids, secondary to ketones. The stoichiometry of the oxidation depends on the experimental conditions. One and 0.5 mole of oxygen are consumed per mole of alcoholate in non-polar and polar solvents, respectively. Thus

$$(C_6H_5)_2CHOH \xrightarrow[C_6H_6]{t\text{-BuOK}} (C_6H_5)_2CO + KOOH$$

$$(C_6H_5)_2CHOH \xrightarrow[(CH_2)_4O]{t\text{-BuOK}} \tfrac{1}{2}(C_6H_5)_2CO + \tfrac{1}{2}(C_6H_5)_2CHOH + KO_2$$

Oxidation of alcoholates yields ketyl radicals [108,109]. They are formed in alkaline medium by reaction of the alcoholate with ketone [110]

$$B + {>}C\overline{HO} + {>}C{=}O \rightleftharpoons BH^+ + 2 {>}\overset{\cdot}{C}{-}\overline{O}$$

Alkali metal benzhydrolate oxidizes in toluene [111] and benzene [112] as well as in t-butanol [113] with autocatalysis that is produced by the KO_2 formed in the oxidation [113]. The induction period disappears when KO_2 is added to a solution of potassium benzhydrolate. The kinetics of oxidation of sodium benzhydrolate was studied by Pereshein et al. [111]. The maximum oxidation rate appears to be approximately proportional to $[RONa][O_2]/[ROH]$, the activation energy being 12 kcal mole^{-1}. The inhibiting action of alcohol (benzhydrol and t-butanol) on the oxidation of metal benzhydrolates was noted by Russell et al. [110]. No deuterium exchange was observed during the oxidation of potassium benzhydrolate in t-butanol. Thus no dianions are produced from benzhydrolate ion by the equilibrium reaction

$$B^- + (C_6H_6)_2CHO^- \rightleftharpoons BH + (C_6H_5)_2\overline{C{-}O}$$

The most probable mechanism seems to be

$$B^- + (C_6H_5)_2CHO^- + O_2 \rightarrow (C_6H_5)_2CO + O_2^{2-} + HB$$

$$O_2^{2-} + O_2 \rightleftharpoons 2 O_2^-\cdot$$

$$O_2^-\cdot + (C_6H_5)_2CHO \rightarrow HO_2^- + (C_6H_5)_2CO^-\cdot$$

$$(C_6H_5)_2CO^-\cdot + O_2 \rightarrow (C_6H_5)_2CO + O_2^-\cdot$$

Oxidation of fluorenol and xanthenol alcoholates in t-butanol and mixtures with pyridine and dimethylsulfoxide is accelerated by nitrobenzene [113]. The mechanism suggested is

$$Ar_2C\overline{HO} + \overline{B} \rightleftharpoons Ar_2CO^{2-} + BH$$

$$Ar_2CO^{2-} + O_2 \rightleftharpoons Ar_2CO^- + O_2^- \rightleftharpoons Ar_2C(O^-)OO^- \rightleftharpoons Ar_2CO + O_2^{2-}$$

$$Ar_2CO^{2-} + ArNO_2 \rightarrow Ar\overline{NO}_2\cdot + Ar_2\overline{CO}\cdot$$

$$Ar\overline{NO}_2\cdot + O_2 \rightarrow ArNO_2 + O_2^-\cdot$$

$$Ar_2\overline{CO}\cdot + O_2 \rightarrow Ar_2CO + \overline{O}_2\cdot$$

Methanol is oxidized by O_2 in the presence of CH_3ONa and cupric phenanthroline complex [114] to form formaldehyde. The proposed mechanism is

$$Cu(phen)^+ + O_2 \rightarrow Cu(phen)^{2+} + O_2^-\cdot$$

$$O_2^-\cdot + CH_3OH \cdot Cu(phen)^{2+} \rightarrow HO_2\cdot + CH_2O + H^+ + Cu(phen)^+$$

$$HO_2^- + Cu(phen)^{2+} \rightarrow HO_2\cdot + Cu(phen)^+$$

$$HO_2\cdot \rightleftharpoons O_2^-\cdot + H^+$$

$$HO_2^- + Cu(phen)^+ \rightarrow Cu(phen)O^+ + OH^-$$

$$(phen)CuO^+ + CH_3OH \rightarrow (phen)CuOH^+ + \dot{C}H_2OH$$

The radical $(t\text{-Bu})_2NO\cdot$ accelerates this reaction [115] and it is suggested that this is due to the reaction

$$(t\text{-Bu})_2NO\cdot + CH_3OH\cdot Cu(phen)^{2+} \rightarrow (t\text{-Bu})_2NOH + CH_2O + H^+$$
$$+ Cu(phen)^+$$

2.9 CO-OXIDATION OF ALCOHOLS AND AROMATICS

Hydroperoxy radicals have been found to hydroxylate benzene [2]. After H_2O_2 is formed in alcohol oxidations

$$>C(OH)OOH \rightleftharpoons >C=O + HOOH$$

$HO_2\cdot$ radicals are produced by

$$>C(OH)OO\cdot + HOOH \rightarrow >C(OH)OOH + HO_2\cdot$$

When 2-propanol is oxidized in the presence of benzene, the latter is hydroxylated to form phenol [116]. Phenol is oxidized in the course of the reaction giving a resin displaying a strong inhibiting action. Hydroxylation of benzene is observed in 2-propanol oxidation at temperatures of 150—200°C. Phenol accumulates in concentrations up to 0.2 mole l^{-1}. The ratio of rate coefficients is

$$\frac{k(HO_2\cdot + C_6H_6 \rightarrow C_6H_5OH)}{2\,k^{1/2}(HO_2\cdot + HO_2\cdot)} = 4.2 \times 10^2 \exp(-11{,}700/RT)\ l^{1/2}$$

$$mole^{-1/2}\ s^{-1/2}$$

The ratio of $k(HO_2\cdot + (CH_3)_2CHOH)/k(HO_2\cdot + C_6H_6)$ is 57 at 137°C. Hydroxylation of toluene with formation of cresols, in parallel with oxidation at the methyl group, takes place on co-oxidation of toluene and 2-propanol at 160—210°C [117]. Cresols are formed in a concentration of 0.03 mole l^{-1} when 2-propanol and toluene, in a ratio of 7 : 3, are oxidized at 165°C for 80 min [117]. The ratio of cresol isomers is *ortho* : *meta* :

para = 2 : 1 : 1. When toluene is hydroxylated by Fenton's reagent (·OH as hydroxylating agent), the ratio of isomers is different [118]; *ortho* : *meta* : *para* = 55 : 15 : 30. The ratio of rate coefficients is

$$k(HO_2 \cdot + C_6H_5CH_3 \rightarrow cresol)/\sqrt{2k_t}$$

$$= 17.8 \exp(-9,100/RT) \, l^{1/2} \, mole^{-1/2} \, s^{-1/2}$$

and

$$k(HO_2 \cdot + C_6H_5CH_3 \rightarrow cresol)/k(HO_2 \cdot + (CH_3)_2CHOH) = 25 \, (110°C).$$

3. Oxidation of ketones

3.1 THE PRODUCTS OF KETONE OXIDATION

Ketones, as well as hydrocarbons, are oxidized by a chain mechanism with the participation of peroxy radicals as chain carriers. The weakest bond is the α-C—H bond (due to the carbonyl group). Therefore, oxidation proceeds chiefly at the α-C atoms. The oxidation of a ketone produces α-keto hydroperoxy as the primary intermediate product, with its subsequent conversion to different oxygen-containing products. The composition of the oxidation products depends on the chemical structure of the ketone and the experimental conditions.

3.1.1 Acetone

Photochemical oxidation of acetone at room temperature yields peroxide [119,120], acids [119—121] (acetic acid [121]), aldehydes [119—121] (in particular formaldehyde [121]), and CO_2 [120]. Methane and ethane are produced in small amounts [120]. Under pressure at 180—200°C, acetone is oxidized to peroxide (apparently CH_3COCH_2OOH), methylglyoxal, formaldehyde, acetic and formic acids, H_2O, and CO_2 [122]. The oxidation produces, after 400 min, 0.25 mole l^{-1} methylglyoxal, 6×10^{-3} mole l^{-1} formaldehyde, 1.05 mole l^{-1} acetic acid, and 0.14 mole l^{-1} formic acid at 190°C and a pressure of 40 atm. The conversion to oxidation products occurs by two parallel routes [122]

$$CH_3COCH_3 \xrightarrow{O_2} CH_3COCHO + H_2O \rightarrow CH_3COOH + CO_2 + H_2O(80—90\%)$$

$$CH_3COCH_3 \xrightarrow{O_2} CH_3COOH + CH_2O \rightarrow CH_3COOH + HCOOH(10—20\%)$$

3.1.2 Methyl ethyl ketone and other aliphatic ketones

Oxidation of methyl ethyl ketone at $100-145°C$ under pressure has been studied in detail [123]. The intermediate products of this reaction are hydroperoxide and diacetyl, and the main oxidation products are acetic acid and ethyl acetate. The sequence of processes is

$$CH_3COCH_2CH_3 \rightarrow CH_3COCHCH_3 \begin{cases} \nearrow CH_3COOH + CH_3CHO \rightarrow \tfrac{1}{2}CH_3COOEt \\ \searrow CH_3COCOCH_3 \rightarrow 2\ CH_3COOH \end{cases}$$
$$\underset{OOH}{|}$$

Ethanol, methanol, acetone, methyl acetate, CO, and CO_2 are formed in small amounts. In the presence of Co and Ni acetates, methyl ethyl ketone is oxidized selectively to diacetyl at $80°C$ [124].

Oxidation of di-n-propyl ketone ($110-120°C$) produces butyric and propionic acids in equimolar amounts [125,126]. The sequence of reactions is suggested to be

$$(CH_3CH_2CH_2)_2CO \rightarrow CH_3CH_2\underset{\underset{OOH}{|}}{C}HCOCH_2CH_2CH_3$$
$$\downarrow$$
$$CH_3CH_2COOH \leftarrow CH_3CH_2CHO + CH_3CH_2CH_2COOH$$

The products of methyl n-hexyl ketone oxidation are capronic and acetic acids, capronic aldehyde, and diketone ($130°C$) [126]. The hydroperoxide formed by the oxidation of di-iso-propyl ketone is relatively stable [127]. Decay of hydroperoxide produces acetone and isobutyric acid

$$(CH_3)_2CHCOCH(CH_3)_2 \overset{O_2}{\rightarrow} (CH_3)_2\underset{\underset{OOH}{|}}{C}COCH(CH_3)_2$$
$$\downarrow$$
$$(CH_3)_2CO + (CH_3)_2CHCOOH$$

Oxidation of iso-propyl methyl ketone at $70-130°C$ produces methanol, acetone, acetic acid and isopropyl acetate [128]. Koslenkova et al. [129] have studied the oxidation of higher aliphatic ketones ($C_{11}-C_{13}$).

3.1.3 Cyclohexanone

The primary product of cyclohexanone oxidation is α-ketohydroperoxide [130], subsequently converted to diketone, the semialdehyde of adipic acid [130,131], and adipic acid [130,131]. Valeric and caproic acids and caprolactam were detected among the oxidation products [130].

The mechanism of cyclohexanone oxidation has been found [275,276] to be more complicated than suggested by Pritzkow [130] and is shown in

156

the scheme

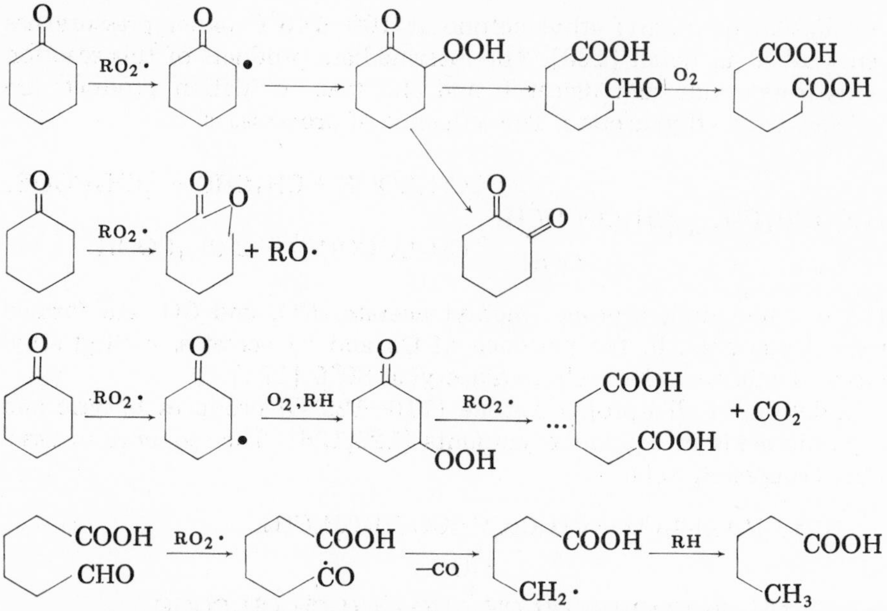

$$\text{(cyclohexanone)} \xrightarrow{RO_2\cdot} \text{(radical)} \longrightarrow \text{(2-OOH-cyclohexanone)} \longrightarrow \text{COOH, CHO} \xrightarrow{O_2} \text{COOH, COOH}$$

$$\text{(cyclohexanone)} \xrightarrow{RO_2\cdot} \text{(bicyclic peroxide)} + RO\cdot$$

$$\longrightarrow \text{(1,2-cyclohexanedione)}$$

$$\text{(cyclohexanone)} \xrightarrow{RO_2\cdot} \text{(radical)} \xrightarrow{O_2, RH} \text{(4-OOH-cyclohexanone)} \xrightarrow{RO_2\cdot} \cdots \text{COOH, COOH} + CO_2$$

$$\text{COOH, CHO} \xrightarrow{RO_2\cdot} \text{COOH, }\dot{C}O \xrightarrow{-CO} \text{COOH, }CH_2\cdot \xrightarrow{RH} \text{COOH, }CH_3$$

The kinetics of carbon monoxide and dioxide generation in the oxidation of cyclohexanone labelled with a ^{14}C carbonyl group has been investigated [279]. It was suggested that CO and CO_2 were formed by the decay of acyl and acyl peroxy radicals.

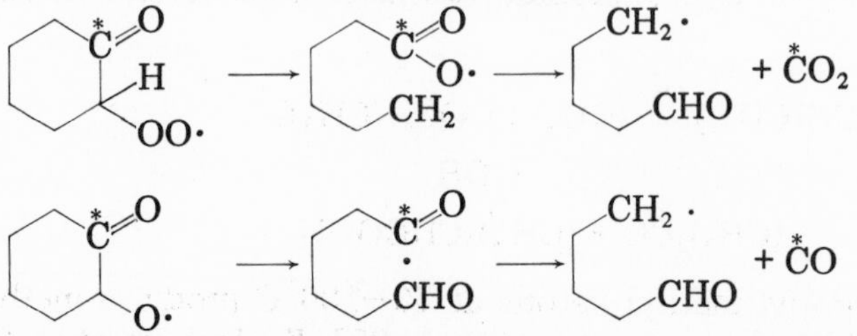

$$\overset{*}{C}(=O)\text{—H, }OO\cdot \longrightarrow \overset{*}{C}(=O)\text{—}O\cdot, CH_2 \longrightarrow CH_2\cdot, CHO + \overset{*}{C}O_2$$

$$\overset{*}{C}(=O), O\cdot \longrightarrow \overset{*}{C}(=O)\cdot, CHO \longrightarrow CH_2\cdot, CHO + \overset{*}{C}O$$

The following compounds are formed by UV-irradiated oxidation of cyclohexanone in methanol at $40°C$ [132] in the presence of sulfuric acid (yields in parentheses): $(CH_3O)_2CH(CH_2)_4CO_2CH_3(45\%)$, $HOCH_2(CH_2)_4$-$CO_2CH_3(15\%)$, $CH_3O_2C(CH_2)_4CO_2CH_3(15\%)$. In acetic anhydride, the products are $CHO(CH_2)_4CO_2COCH_3(65\%)$ (cyclohexanone)$-OCOCH_3$ (30%).

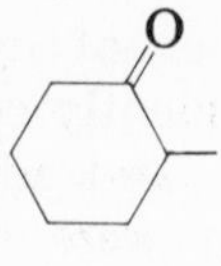

3.1.4 Other ketones

α-Tetralone oxidation at $70-100°C$ yields α-diketone, hydroperoxide and products I and II [133].

COOH ... CH₂OH structures

$$\text{(I)} \quad \begin{array}{c} \text{COOH} \\ \text{CH}_2\text{CH}_2\text{CHO} \end{array} \qquad \text{(II)} \quad \begin{array}{c} \text{CH}_2\text{OH} \\ \text{CH}_2\text{CH}_2\text{COOH} \end{array}$$

o-Chlorophenylbenzylketone is converted to *o*-chlorobenzoic acid and benzaldehyde during photochemical oxidation at 25°C [134].

Diacetyl oxidizes at 80°C with the formation of acetic acid, CO_2, methyl acetate, methylglyoxal, methanol, peroxide, and formaldehyde [135].

Unsaturated ketones oxidize with the formation of hydroperoxide in the α-position to the double bond [136] (cf. olefins), viz.

$$RCOCH=CHCHR_1R_2 \xrightarrow{O_2} RCOCH=CHC(OOH)R_1R_2$$

3.1.5 Ketohydroperoxide

Formation of α-ketohydroperoxide in the course of an oxidation is observed for cyclohexanone [130] and β,β'-dimesitylpropiophenone [137]. Hydrogenation of the hydroperoxides formed yields α-ketoalcohols. α-Ketohydroperoxides decomposes to acid and aldehyde according to Rieche [138], viz.

$$\underset{\underset{OOH}{|}}{R_1CHCOR_2} = R_1CHO + R_2COOH$$

The composition of the oxidation products of ketones of low-molecular weight is in agreement with this mechanism. α-Ketohydroperoxide is decomposed in a parallel reaction to diketone. Organic acids accelerate hydroperoxide decomposition in hydrocarbon solutions [139]. When acid accumulates in the course of the reaction, the decomposition proceeds autocatalytically. When it is added to the hydroperoxide solution, the rate of decomposition is approximately proportional to the acid concentration [139]. An ionic mechanism

$$\underset{\underset{OOH}{|}}{R_1CHCOR_2} \xrightarrow{AcOH} \underset{\underset{OOH_2^-}{|}}{R_1CHCOR_2} \xrightarrow[-H_2O]{} \underset{\underset{O^+}{|}}{R_1CHCOR_2} \rightarrow R_1C^+HOCOR_2$$

$$\xrightarrow[-H^+]{H_2O}$$

$$R_1CHO + R_2COOH$$

is suggested. The composition of the products of cyclohexanone oxidation in methanol and acetic anhydride in the presence of acid [132] is consistent with this mechanism.

3.1.6 The formation of acids by the oxidation of ketones

As shown above, the formation of acids by the oxidation of ketones is accompanied by scission of the α-C—C bond. Acids formed by paraffin

TABLE 6
Composition of acids in the oxidation of paraffin hydrocarbons

Hydrocarbon	Yields of acids (%)							Ref.
	C_1	C_2	C_3	C_4	C_5	C_6	C_7	
n-Hexane	4	55	19	22				142
n-Hexane	6	60	17	17				143
n-Heptane	6	44	25	19	6			144
n-Heptane	7	58	21	14				143
n-Decane	10	23	16	13	9	6	7	141

oxidation are known to be produced from ketones. Assuming that ketones are attacked at the α-CH_2 group and only the α-C—C bond of the ketone is broken, the mean number of carbon atoms in acids formed by the oxidation of paraffins (C_nH_{2n+2}) would be $n/2$, and the amount of acids with $n - m$ atoms would be equal to that of acids with m atoms. This is not in agreement with experiment. The mean number of carbon atoms in acids formed by the oxidation of n-heptane (140—150°C) is 2.76 instead of 3.5 [140] and in those obtained from n-decane (140°C) 2.80 instead of 5 [141]. The equality $[C_{n-m}$-acid] = $[C_m$-acid] is also not fulfilled, as seen from Table 6.

Lower acids are formed in larger amounts than higher acids. This disproportion cannot be explained by the oxidation of higher acids, since the latter is slow and has no effect on the composition of acid products as found in experiments with labelled acids [140,141].

It was found for the oxidation of heptane [140] that up to 90% of CO_2 is formed in parallel with the acids and only 10% by decarboxylation of acids. Oxy- and ketoacids (up to 18% of all acids) were found to be produced in parallel with fatty acids in the oxidation of n-decane [141]. All the above facts are inconsistent with the assumption that the α-C—C bond only is broken on oxidation of ketones. Undoubtedly, some ketones are oxidized with scission of two C—C bonds. This conclusion is confirmed by the prevailing amount of lower fatty acids and parallel formation of CO_2 and acids in the oxidation of paraffins. Obviously, not only the α-CH_2 group but also other CH_2 groups are attacked in the ketone molecule. This results in the formation of bifunctional compounds with subsequent oxidation to acids, oxyacids, and ketoacids. The competing attack by peroxy radicals at the α-CH_2 and other CH_2 groups will be discussed later.

3.2. ELEMENTARY STEPS OF KETONE OXIDATION

3.2.1 Chain propagation

The reaction of ketones with oxygen is a chain process in which the chain propagates by the alternating steps

$$R\cdot + O_2 \rightarrow RO_2\cdot$$

$$RO_2\cdot + RH \rightarrow ROOH + R\cdot$$

When the oxygen pressure is sufficiently high (>100 torr), chains are terminated by the interaction of peroxy radicals, and the rate-limiting propagation step is $RO_2\cdot + RH$. The rate of initiated oxidation of ketones is expressed as

$$W = \frac{k_p}{\sqrt{2k_t}}[RH]\sqrt{W_i} + W_i$$

The rate coefficients, k_p, were found from the values of $k_p/\sqrt{2k_t}$ (Table 7) and of k_t measured by the chemiluminescence technique [81]. In the case of acetone, the ratio $k_p/\sqrt{2k_t}$ was found to be [145] 7.5×10^5 $\exp(-15{,}500/RT)$ $l^{1/2}$ $mole^{-1/2}$ $s^{-1/2}$. Assuming that k_t for acetone oxidation is equal to that for methyl ethyl ketone [146], viz. $2k_t = 2 \times 10^7$ $\exp(-1{,}600/RT)$ l $mole^{-1}$ s^{-1}, k_p for acetone is 3.4×10^9 $\exp(-16{,}300/RT)$ l $mole^{-1}$ s^{-1}.

Values of k_p for some ketones are given in Table 8. It can be seen that oxidation of the methyl group of acetone is slow, that of the ketones with a CH_2 group is faster, and still more rapid is that of methyl i-propyl ketone with a tertiary C—H bond. However, partial k_ps for one attacked C—H bond of ketone must be calculated for correct comparison of different C—H bond reactivities. Values of $k_{p,C-H}$ are given in Table 9.

The relative reactivity of the α-C—H bond of ketones at 100°C varies

TABLE 7
Values of $k_p/\sqrt{2k_t}$ for some ketones

Ketone	Temp. (°C)	$k_p/\sqrt{2k_t}$ ($l^{1/2}$ $mole^{-1/2}$ $s^{-1/2}$)	Ref.
Acetone	95—120	7.5×10^5 $\exp(-15{,}500/RT)$	145
Methyl ethyl ketone	35— 75	27.5 $\exp(-7600/RT)$	146
Methyl n-propyl ketone	50— 90	1.5×10^2 $\exp(-9000/RT)$	146
Methyl i-propyl ketone	40— 80	2.8×10^2 $\exp(-7600/RT)$	146
Diethyl ketone	70	5.45×10^{-4}	277
Di-n-propyl ketone	70	4.52	277
Di-n-butyl ketone	70	6.63	277
Methyl octyl ketone	70	4.40	277
Ethyl n-heptyl ketone	60— 80	7.6×10^2 $\exp(-9800/RT)$	277
n-Butyl amyl ketone	70	4.16	277
Diamyl ketone	70	4.56	277
Di-n-heptyl ketone	70	4.00	277
Cyclohexanone	55— 80	4.6×10^3 $\exp(-11{,}200/RT)$	147
2-Methylcyclohexanone	60— 80	6.7×10^5 $\exp(-13{,}800/RT)$	278
3-Methylcyclohexanone	65— 85	6.5×10^4 $\exp(-12{,}900/RT)$	278
4-Methylcyclohexanone	65— 85	7.2×10^3 $\exp(-11{,}200/RT)$	278

TABLE 8

Rate coefficients, k_p, for ketones and corresponding Arrhenius parameters

Ketone	Temp. [a] (°C)	$\log\{A \cdot$ (l mole^{-1} s^{-1})$\}$	E_a (kcal mole^{-1})	k_p at 100°C (l mole^{-1} s^{-1})	Ref.
Acetone	95—120	9.53	16.3	0.93	145
Methyl ethyl ketone	35— 75	5.10	8.4	1.5	146
Methyl n-propyl ketone	50— 90	5.92	9.8	1.5	146
Methyl i-propyl ketone	40— 80	4.76	7.2	3.5	146
Cyclohexanone	55— 80	7.38	12.0	2.1	147

[a] The temperature range over which the ratio $k_p/\sqrt{2k_t}$ was determined.

as primary : secondary : tertiary = 1 : 4.5 : 22.6. It will be seen from a comparison of cyclohexane and cyclohexanone that the carbonyl group facilitiates abstraction of H from the CH_2 group by the peroxy radical; $k_{p,C-H}$(cyclohexanone) : $k_{p,C-H}$(cyclohexane) = 5.6. In the case of trimethylpentane and methyl i-propyl ketone, the ratio is 11.

The following relation between E_a and D_{C-H} of the attacked bond was established [150,151], viz.

$$\Delta E_a = 0.45 \Delta D$$

If this relation is true for ketones, then the dissociation energies of α-C—H bonds in ketones may be estimated. Let D_{C-H} in cyclohexane be 89 kcal mole^{-1}, then D_{C-H} for the cyclohexanone α-CH_2 group will be 89 $- \Delta D =$ 85 kcal mole^{-1}. since $\Delta D = 1.8/0.45 = 4$ kcal mole^{-1}. If $D_{C-H,t}$ in isopentane is taken as 85 kcal mole^{-1} (as in isobutane), then D_{C-H} for methyl i-propyl ketone will be 85 $- \Delta Q = 85 - 4 = 81$ kcal mole^{-1}. The decrease in strength of the α-C—H bond of ketones in comparison with

TABLE 9

Partial rate coefficients, $k_{p,C-H}$, for ketones and some hydrocarbons

Compound	Attacked group	$k_{p,C-H}$ at 100°C (l mole^{-1} s^{-1})	E_a (kcal mole^{-1})	Ref.
Acetone	CH_3	0.155	16.3	145
Methyl ethyl ketone	CH_2	0.75	8.4	146
Methyl n-propyl ketone	CH_2	0.75	9.8	146
Cyclohexanone	CH_2	0.52	12	147
Cyclohexane	CH_2	0.093	13.8	148
Methyl i-propyl ketone	C—H	3.5	7.2	146
2,3,4-Trimethylpentane	C—H	0.32	9.1	149

hydrocarbon is due to stabilization of the radical formed by interaction of the odd electron with π-electrons of the carbonyl group. It will be noted that conjugation of the free valence with the C=O group is low in comparison with that in radicals of the allyl type where $\Delta D = 18$ kcal mole^{-1} ($\Delta D = D_{\text{propane}} - D_{\text{propene}}$). This may be explained by displacement of π-electrons towards the oxygen atom, the low electron density on the C atom, and the relatively small overlapping of the orbits of the odd and the π-electrons.

The reaction of peroxy radicals with ketone is that between two dipolar particles in a polar medium. The role of the medium in methyl ethyl ketone oxidation has been studied in detail [152—157]. The rate coefficient, k_p, decreases with dilution of methyl ethyl ketone by a non-polar solvent (benzene, n-decane, etc.). The change of k_p is caused by the nonspecific solvation of reacting particles and activated complexes. The relationship between k_p and the dielectric constant, ϵ, is expressed by the Kirkwood equation

$$\log k_p = \log k_p^0 - \frac{1}{2.3kT} \frac{\epsilon - 1}{2\epsilon + 1} \sum \frac{\mu_i^2}{r_i^3}$$

where

$$\sum \frac{\mu_i^2}{r_i^3} = \frac{\mu_{\text{RH}}^2}{r_{\text{RH}}^3} + \frac{\mu_{\text{RO}_2\cdot}^2}{r_{\text{RO}_2\cdot}^3} - \frac{\mu_{\ddagger}^2}{r_{\ddagger}^3}$$

k being Boltzman's constant, μ the dipole moment, and r the particle radius. The following expressions were obtained for mixtures of methyl ethyl ketone with benzene at 35—75°C.

$$\log k_p = -1.55 + 2.1 \frac{\epsilon - 1}{2\epsilon + 1} \quad (50°C)$$

$$\log k_p^0 = 15.05 - \frac{24500}{4.57T}$$

$$E_p = 23.5 - 33 \cdot \frac{\epsilon - 1}{2\epsilon + 1} \; ; \qquad E_p = 7 - \frac{25}{\epsilon}$$

The dipole moment of the activated complex RO$_2$· ... HR calculated from the dependence of $\log k_p$ on $(\epsilon - 1)/(2\epsilon + 1)$ is 8.1×10^{-18} esu cm (8.1 Debye).

When methyl ethyl ketone is oxidized in chlorobenzene, no linear dependence of $\log k$ on $(\epsilon - 1)/(2\epsilon + 1)$ is observed [155]. This may be explained by specific interaction between reacting particles and the solvent.

An ester (probably lactone) and an acid (adipic) are produced in parallel with hydroperoxide in initiated cyclohexanone oxidation at 80—110°C [158]. The peroxy radical is assumed not only to abstract a H atom from the ketone but also to add to the carbonyl group with subsequent decay of the adduct, viz.

Parallel formation of acids and oxyacids occurs in undecanone-6 oxidation and is explained by isomerization of the peroxy radical [159], viz.

$$\underset{RCCHCH_2CH_2R'}{\overset{O\ \ OO\cdot}{\parallel\ /}} \rightarrow \underset{RC-CH-CH_2\dot{C}HR'}{\overset{O\ \ OOH}{\parallel\ |}} \xrightarrow{O_2,RH} \underset{RC-CH-CH_2CHR'}{\overset{O\ \ OOH\ \ \ \ OOH}{\parallel\ |\ \ \ \ \ \ |}}$$

$$\rightarrow RCOOH + HCOCH_2CH(OOH)R'$$

3.2.2 Chain termination

Ketone oxidation chains terminate when two peroxy radicals react with each other. This is the main reaction of chain termination if the ketone contains no inhibitor and the oxygen pressure is sufficiently high for fast conversion of $R\cdot$ to $RO_2\cdot$. The values of k_t measured by the chemiluminescence technique [81] are shown in Table 10.

TABLE 10
Rate coefficients and Arrhenius parameters for reaction between two peroxy radicals

Ketone	Temp. (°C)	$2k_t$ at 75°C (l mole^{-1} s^{-1})	A (l mole^{-1} s^{-1})	E (kcal mole^{-1})	Ref.
Methyl ethyl ketone	35—75	6.30	7.30	1.6	146
Methyl n-propyl ketone	50—90	6.41	7.48	1.7	146
Methyl i-propyl ketone	40—80	6.48	7.48	1.6	146
Cyclohexanone	75	6.43			147

The rate coefficients are seen to be very close for all the ketones studied. A reaction between two ketoperoxy radicals may be assumed to proceed in the same way as that between alkylperoxy radicals [160,161], viz.

$$2 >\text{CHOO}\cdot \rightarrow >\text{CHOO}\!-\!\text{OOCH}< \rightarrow >\text{CHOH} + O_2 + >\text{C}=O$$

k_t depends on the polarity of the medium [152—156]. When methyl ethyl ketone is diluted by a non-polar solvent (benzene, CCl_4 n-decane), k_t decreases and the Kirkwood equation

$$\log(2k_t) = 3.18 + 6.2 \frac{\epsilon - 1}{2\epsilon + 1} \qquad (50°C)$$

is obeyed. The activated complex dipole moment calculated from this equation appears to be 11.2×10^{-18} esu cm. When $\epsilon = 1$, E_a (termination) = 6.5 kcal mole^{-1}; when $\epsilon = \infty$, $E_a = 1.2$ kcal mole^{-1} [from the relation $E = E - (q/\epsilon)$].

3.2.3 Generation of chains

When the oxidation of methyl ethyl ketone is conducted in a steel reactor, free radicals are formed at a rate given by [162]

$$W_i = k_i[\text{RH}][O_2]$$

with

$$k_i = 3 \times 10^9 \exp(-30,000/RT) \, l \, \text{mole}^{-1} \, s^{-1}$$

As initiation is 10 times slower for a glass reactor, the free radicals seem to be produced on the metallic wall. The rate of chain generation in cyclohexanone is 2.8×10^{-8} mole l^{-1} s^{-1} at 120° and 4×10^{-8} at 130°C, $E_a = 17.5$ kcal mole^{-1} (glass reactor) [163].

The chains are generated in cyclohexanone by the bimolecular reaction

$$\text{RH} + O_2 \rightarrow \text{R}\cdot + \text{HO}_2\cdot$$

with a rate coefficient $k_i = 1.0 \times 10^9 \exp(-24,000/RT) \, l \, \text{mole}^{-1} \, s^{-1}$ [324].

3.2.4 Degenerate chain branching

In the later stages of ketone oxidation, free radicals are formed from α-ketohydroperoxide. In cyclohexanone, α-ketohydroperoxide decomposes by a first-order reaction [164] with a rate coefficient $k_i = 5.9 \times 10^7 \exp(-20,400/RT)$ s^{-1}. Ketone takes part in the formation of free radicals from hydroperoxide (see below).

Two peroxides are formed in the oxidation of methyl ethyl ketone [165], an α-ketohydroperoxide and a peroxide denoted as X. Both

TABLE 11

Rate coefficients of the forward and back reactions and $K(=k_f/k_b)$ for addition of hydroperoxides to cyclohexanone in CCl_4 [167]

Hydro-peroxide	k_f (20°C) (l mole^{-1} s^{-1})	E (kcal mole^{-1})	k_b (20°C) (s^{-1})	E (kcal mole^{-1})	K (20°C) (l mole^{-1})
t-Butyl	1.83×10^{-4}	7.5	3.5×10^{-4}	10	0.52
Cumyl	1.17×10^{-4}	7.5	4.5×10^{-4}	10.5	0.26
Pinyl	1.17×10^{-3}	7.0			
Tetrallyl	3.0×10^{-3}	7.0	1.0×10^{-3}	9.5	3.0

decompose to free radicals by first-order reaction with

$$k = 1.15 \times 10^{10} \exp(-23,000/RT) \text{ s}^{-1} \text{ for } \alpha\text{-ketohydroperoxide}$$

and

$$k = 6.9 \times 10^{9} \exp(-21,200/RT) \text{ s}^{-1} \text{ for peroxide X}$$

In methyl ethyl ketone oxidation, free radicals are formed not only from peroxides but also from diacetyl, another intermediate product, which decomposes to radicals by a unimolecular reaction with a rate coefficient [166] of

$$k = 2.9 \times 10^{13} \exp(-35,400/RT) \text{ s}^{-1}$$

3.3 FORMATION OF FREE RADICALS BY REACTIONS OF KETONES WITH HYDROPEROXIDES

Hydroperoxides add to the carbonyl group of ketones by an equilibrium reaction to form hydroxyperoxide.

$$>C=O + ROOH \underset{k_b}{\overset{k_f}{\rightleftharpoons}} >C\overset{\displaystyle OOR}{\underset{\displaystyle OH}{<}}$$

The kinetics of such reactions in CCl_4 were studied by Antonovskii and Terent'ev [167] by IR spectroscopy (Table 11).

The rate coefficients for hydroxyperoxide decomposition to free radicals is different from that for hydroperoxides. Therefore, addition of hydroperoxide to ketone changes the rate of free radical formation. This was first found for the system cyclohexanone—t-butyl hydroperoxide [168] with chlorobenzene as solvent. The rate of initiation increases with ketone concentration at a constant concentration of hydroperoxide. The initiation mechanism is

$$\text{ROOH} \xrightarrow{k_1} \text{RO}\cdot + \text{HO}\cdot$$

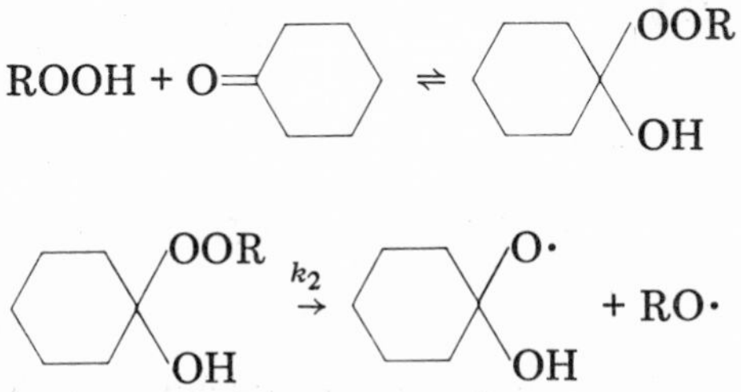

(X)

The rate of free radical formation in this system is

$$W_i = k_1[\text{ROOH}] + k_2[\text{X}]$$

where [X] = K[ketone] [ROOH]

$$W_i = k_1[\text{ROOH}] + k_2K[\text{ketone}][\text{ROOH}]$$

$$k_i = \frac{W_i}{[\text{ROOH}] + [\text{X}]} = \frac{k_1 + k_2K[\text{ketone}]}{1 + K[\text{ketone}]}$$

or

$$\frac{k_1}{k_i - k_1} = \frac{1}{(\alpha - 1)K}\,\frac{1}{[\text{ketone}]} + \frac{1}{\alpha - 1}$$

where $\alpha = k_2/k_1$. The rate coefficients for t-butyl hydroperoxide and cyclohexanone are [168]

$$k_1 = 3.6 \times 10^{12} \exp(-33{,}000/RT)\ \text{s}^{-1}$$

$$k_2 = 3.6 \times 10^{9} \exp(-26{,}000/RT)\ \text{s}^{-1}$$

$$K = 6.9 \times 10^{-7} \exp(11{,}000/RT)\ \text{l mole}^{-1}$$

The rate coefficients estimated for the system cyclohexyl hydroperoxide + cyclohexanone in chlorobenzene, are [169]

$$k_1 = 6.3 \times 10^{11} \exp(-32{,}000/RT)\ \text{s}^{-1}$$

$$k_2 = 4.0 \times 10^{4} \exp(-15{,}200/RT)\ \text{s}^{-1}$$

$$K = 1.4 \times 10^{-5} \exp(7900/RT)\ \text{l mole}^{-1}$$

When cyclohexyl hydroperoxide decomposes in the presence of cyclohexanone at 130°C with cyclohexane as solvent, ε-cyclohexyloxycaproic acid is formed [170], apparently by recombination of radicals in the solvent cage, viz.

Cyclohexanone accelerates the decomposition of cyclohexyl hydroperoxide [170].

The rate coefficients obtained for the system cumyl hydroperoxide—cyclohexanone at 120°C in chlorobenzene are [168] $k_1 = 2.1 \times 10^{-6} s^{-1}$, $k_2 = 5.7 \times 10^{-6} s^{-1}$, and $K = 1 \, l \, mole^{-1}$. In ketone medium, α-hydroperoxide exists in the form of hydroxyperoxide, as the equilibrium is shifted towards this peroxide. Thus, free radicals are formed in ketone from the hydroxyperoxide. The decay of α-ketohydroperoxide in cyclohexanone is 30 times faster than in chlorobenzene [164]. Experimental data yield

$$ROOH + 3 \, ketone \overset{K}{\rightleftharpoons} X \overset{k_2}{\to} free \, radicals$$

$$ROOH \overset{k_1}{\to} free \, radicals$$

$$k_1 = 1.2 \times 10^{-5} s^{-1} \quad (120°C)$$

$$k_2 = 3.0 \times 10^{-4} s^{-1} \quad (120°C)$$

$$k_2 = 5.9 \times 10^7 \, exp(20,400/RT) \, s^{-1}$$

$$K = 0.38 \, l^3 \, mole^{-3} \quad (120°C)$$

The situation is different for methyl ethyl ketone [165], where decomposition of free α-ketohydroperoxide is faster than that of the α-ketohydroperoxide adduct with the ketone, the rate coefficients being

$$k_1 = 3.4 \times 10^{13} \, exp(-27,400/RT) \, s^{-1} \quad (7.1 \times 10^{-4} s^{-1} \, at \, 70°C)$$

$$k_2 = 1.15 \times 10^{10} \, exp(-23,000/RT) \, s^{-1} \quad (2.0 \times 10^{-4} s^{-1} \, at \, 70°C)$$

$$K = 0.80 \, l \, mole^{-1} \quad (70°C)$$

The kinetics of methyl ethyl ketone oxidation shows that two peroxides are formed [165]. The rate coefficients for peroxide X (suggested to be $CH_3CH_2COCH_2OOH$) are

$$k_1 = 4.0 \times 10^{12} \, exp(-24,700/RT) \, s^{-1},$$

$$k_2 = 6.9 \times 10^9 \, exp(-21,200/RT) \, s^{-1}, \, and$$

$K = 1.4 \text{ l mole}^{-1}$ (70°C).

Thus free radicals are formed from peroxide produced by addition of α-ketohydroperoxides to the carbonyl group of ketones. In some cases, this accelerates, and in others hinders, the decomposition of peroxide to free radicals. It will be noted that α-ketohydroperoxides decompose to free radicals more rapidly than do the hydroperoxides of hydrocarbons.

3.4 OXIDATION OF KETONES IN THE PRESENCE OF ACIDS AND BASES

Enols are more reactive than ketones and acids accelerate enolization of ketones. Therefore, in the presence of acids and oxidizing ions, oxidation of ketone proceeds via its enolic form. Kooymen and coworkers [171,172] have found that, at 130°C in the presence of manganese acetate, aceto-phenone oxidizes in acetic and butyric acids at a rate equal to that of enolization. A linear dependence of log k on σ ($\rho = -0.7$) was observed for the oxidation of a number of substituted acetophenones. The proposed mechanism is

$$C_6H_5COCH_3 \xrightarrow{CH_3COOH} C_6H_5C(OH)=CH_2 \xrightarrow{Mn(III)}$$

$$C_6H_5COCH_2\cdot \xrightarrow{O_2} C_6H_5COCH_2OO\cdot \xrightarrow{Mn(II),H^+}$$

$$C_6H_5COCH_2OOH \rightarrow C_6H_5COOH + CH_2O$$

Oxidation of methyl ethyl ketone in H_2O in the presence of catalysts (Fe^{3+} and the complexes Cu^{2+}-pyridine, Fe^{3+}-phenanthraline, Mn^{2+}-phenanthraline) proceeds by a peculiar mechanism [173,174]. No chain reaction with propagation by the step $RO_2\cdot + RH$ takes place under these conditions. Hydrogen bonding, $RO_2\cdot ... HOH$, hinders this reaction. Oxidation starts with attacks on the enol form of the ketone by metal ions to form a radical $R\cdot$ followed by

$$R\cdot \xrightarrow{O_2} RO_2\cdot \xrightarrow{ion} ROOH$$

The hydroperoxide decomposes both to molecular products (acetic acid and acetaldehyde) and to free radicals (formation of $HO\cdot$ is suggested) which attack ketone molecules. Thus the overall rate of oxidation is much higher than that of enol oxidation by metal ions.

Ketones are rapidly oxidized with oxygen in the presence of strong bases (alkali, alcoholates) [175—185] to form acids (by C—C bond scission) and condensation products. The yield of acids may be increased by selection of the solvent [185]. Selective oxidation of cyclic ketones to dibasic acids was found for hexamethylphosphoramide in the presence of sodium methylate [185]. The mechanism suggested was (B = base)

$$R_1\overset{O}{\underset{\|}{C}}CH_2R_2 + B \rightarrow R_1\overset{\bar{O}}{\underset{|}{C}}=CHR_2 + BH^+$$

References pp. 195—203

$$\text{R}_1\overset{\overline{\text{O}}}{\underset{|}{\text{C}}}=\text{CHR}_2 + \text{O}_2 \rightarrow \text{R}_1\overset{\text{O}}{\underset{\|}{\text{C}}}-\overset{\cdot}{\text{C}}\text{HR}_2 + \text{O}_2^-\cdot$$

$$\text{R}_1\overset{\text{O}}{\underset{\|}{\text{C}}}-\overset{\cdot}{\text{C}}\text{HR}_2 + \text{O}_2 \rightarrow \text{R}_1\overset{\text{O}}{\underset{\|}{\text{C}}}-\overset{\text{OO}\cdot}{\underset{|}{\text{C}}}\text{HR}_2$$

$$\text{R}_1\overset{\text{O}}{\underset{\|}{\text{C}}}-\overset{\text{OO}\cdot}{\underset{|}{\text{C}}}\text{R}_2 + \text{R}_1\overset{\overline{\text{O}}}{\underset{|}{\text{C}}}=\text{CHR}_2 \rightarrow \text{R}_1\overset{\text{O}}{\underset{\|}{\text{C}}}-\overset{\text{O}\overline{\text{O}}}{\underset{|}{\text{C}}}\text{HR}_2 + \text{R}_1\overset{\text{O}}{\underset{\|}{\text{C}}}-\overset{\cdot}{\text{C}}\text{HR}_2$$

$$\text{R}_1\overset{\text{O}}{\underset{\|}{\text{C}}}-\overset{\text{O}\overline{\text{O}}}{\underset{|}{\text{C}}}\text{HR}_2 \rightarrow \text{R}_1\overset{\overline{\text{O}}}{\underset{\backslash}{\text{C}}}_{\diagdown\text{O}} + \text{R}_2\text{CHO}$$

or

$$\text{R}_1\overset{\overline{\text{O}}}{\underset{|}{\text{C}}}=\text{CHR}_2 \xrightarrow{\text{O}_2} \text{R}_1\overset{\text{O}}{\underset{\|}{\text{C}}}-\overset{\text{O}\overline{\text{O}}}{\underset{|}{\text{C}}}\text{HR}_2 \rightarrow \text{R}_1\overset{\overline{\text{O}}}{\underset{\backslash}{\text{C}}}_{\diagdown\text{O}} + \text{R}_2\text{CHO}$$

3.5 OXIDATION OF KETONES WITH OZONE

Oxidation of methyl ethyl ketone with a mixture of oxygen and ozone in CCl_4 at 20–50°C yields acetic acid, diacetyl (intermediate) and hydrogen peroxide [280]. The reaction is second order with a rate coefficient, $k = 3.6 \times 10^9 \exp(-17{,}000/RT)$ l mole^{-1} s^{-1}. The oxidation of the ketone under these conditions is a radical non-chain reaction. Peroxy radicals react faster by a termination reaction than by a propagation reaction at these temperatures. In aqueous solution, ozone oxidizes the ketone as well as the enol form of methyl ethyl ketone [281] and therefore acid accelerates the rate of oxidation.

Information has been obtained recently [282] on the oxidation of methyl ethyl ketone with ozone at low oxygen concentrations. The reaction rate was found to increase with decreasing concentration of oxygen in the ozone. The rate of ketone oxidation depended on $[O_2]$ and $[O_3]$ according to

$$W = \{k_o + (k'[O_3]/[O_2])\}[RH][O_3]$$

This result is explained by a chain reaction with an elementary propagation step $R\cdot + O_3 \rightarrow RO\cdot + O_2$, viz.

$$RH + O_3 \xrightarrow{k_o} R\cdot + O_2 + HO\cdot \overset{e}{\rightarrow} R\cdot + H_2O$$

$$R\cdot + O_3 \xrightarrow{k_1} RO\cdot + O_2$$

$$RO\cdot + RH \longrightarrow ROH + R\cdot \text{ (fast)}$$

$$R\cdot + O_2 \xrightarrow{k_2} RO_2\cdot$$

$$RO_2\cdot + RO_2\cdot \longrightarrow \text{molecular products}$$

The reaction rate according to the scheme is

$$W = k_o[RH][O_3] + 2k_ok_1k_2^{-1}e[RH][O_3]^2[O_2]^{-1}$$

and agrees with the experimental expression with $k' = 2k_ok_1 k_2^{-1}e$.

4. Oxidation of ethers

4.1 OXIDATION PRODUCTS

4.1.1 Aliphatic ethers

Hydroperoxides are the primary products of the oxidation of aliphatic ethers [186—203,283,284]. Under mild conditions (30—70°C), the yield of hydroperoxides is close to 100%. The α-C—H bond of ether is most reactive and the hydroperoxides formed are α-alcoxyhydroperoxides, as found by the synthesis of such hydroperoxides and analysis of their decomposition products [194,284—286].

Dihydroperoxides are formed in the oxidation of diethyl and diisopropyl ethers, together with the hydroperoxides [203,283,285]. Their production can be explained by isomerization of peroxy radicals.

The products of autoxidation and photo-oxidation of ethers are the same [187,287,288]. Aldehydes, alcohols, acids, and esters are the main products of hydroperoxide decomposition [186,187,202,283—285,289,290]. For example, ethanol, acetaldehyde, acetic acid, ethyl acetate, and ethyl formate were found in the products of diethyl ether oxidation [186,188, 202,203]. Their formation may be explained by the scheme

4.1.2 Cyclic ethers

α-Hydroperoxide is the primary intermediate in tetrahydrofuran oxidation [187,189,284,291]. It decomposes to α-hydroxytetrahydrofuran and γ-butyrolactone [187,189,202,291] according to

Oxidation of 2-methyltetrahydrofuran gives two hydroperoxides, 2-methyl-2-hydroperoxy and 2-methyl-5-hydroperoxyhydroperoxide, in a ratio of 3 : 2 [292]. The decomposition products of these hydroperoxides are γ-butyrolactone, γ-methyl-γ-butyrolactone, γ-acetopropanol and n-propyl acetate. Two hydroperoxides are formed in the oxidation of 2,5-dimethyltetrahydrofuran, a mono- and a dihydroperoxide [203].

The hydroperoxides are the primary products of oxidation of phthalane and isochromane [202,203,284,293]. Hydroperoxide is the main product of dioxan oxidation. The decomposition occurs according to

4.1.3 Ethers of benzyl alcohol

Dibenzyl ether is oxidized to hydroperoxide which decomposes under mild conditions to benzaldehyde and benzyl alcohol [186—188,294], viz.

$$C_6H_5CH_2OCH_2C_6H_5 \xrightarrow{O_2,RH} C_6H_5\overset{OOH}{\underset{|}{C}}HOCH_2C_6H_5$$

$$\rightarrow C_6H_5\overset{O\cdot}{\underset{|}{C}}HOCH_2C_6H_5 \rightarrow C_6H_5CHO + C_6H_5CH_2O\cdot \xrightarrow{RH} C_6H_5CH_2OH$$

$$C_6H_5CH_2O\overset{\cdot}{C}HC_6H_5 \rightarrow C_6H_5\overset{\cdot}{C}H_2 + C_6H_5CHO$$

Benzaldehyde and benzoic acid are formed in small amounts in parallel with the hydroperoxide, probably from the decomposition [197,203]

$$C_6H_5CH_2O\overset{\cdot}{C}HC_6H_5 \rightarrow C_6H_5\overset{\cdot}{C}H_2 + C_6H_5CHO$$

4.2 THE CHAIN MECHANISM OF ETHER OXIDATION

Ethers, as well as hydrocarbons and other organic compounds, oxidize by a chain mechanism. A specific feature of their oxidation is the propagation of chains by two reactions

$$RO_2 \cdot + RH \overset{k_p}{\to} ROOH + R \cdot$$

TABLE 12

Values of $k_p/\sqrt{2k_t}$ for ethers

Ether	Temp. (°C)	$k_p/\sqrt{2k_t}$ ($l^{1/2}$ mole$^{-1/2}$ s$^{-1/2}$)	Ref.
(i-Pr)$_2$O	30	3.7 $\times 10^{-3}$	203
	60	1.4 $\times 10^{-3}$	191
(n-Bu)$_2$O	30	0.1 $\times 10^{-3}$	203
	60	0.67 $\times 10^{-3}$	191
t-BuO-(i-Pr)	30	0.1 $\times 10^{-3}$	203
Me-O-cyclo-C$_6$H$_{11}$	50—81	1.6 $\times 10^3$ exp($-9800/RT$)	295
(C$_6$H$_5$CH$_2$)$_2$O	0—30	8.5 $\times 10^2$ exp($-6800/RT$)	197
(C$_6$H$_5$CH$_2$)$_2$O	30	(5.8—8.5) $\times 10^{-3}$	203
(C$_6$H$_5$CH$_2$)$_2$O	35—65	1.1 $\times 10^4$ exp($-9000/RT$)	197
(C$_6$H$_5$CH$_2$)$_2$O		2.3 $\times 10^3$ exp($-8000/RT$)	190
C$_6$H$_5$CH(CH$_3$)OCH$_2$C$_6$H$_5$	30	0.3 $\times 10^{-3}$	203
C$_6$H$_5$CH$_2$OC$_6$H$_5$	30	0.3 $\times 10^{-3}$	203
C$_6$H$_5$CH$_2$OC(CH$_3$)$_3$	30	(2.2—3.2) $\times 10^{-3}$	203
RC$_6$H$_4$CH$_2$OC$_6$H$_5$			
R = p-OCH$_3$	60	0.95 $\times 10^{-4}$	296
p-CH$_3$	60	1.75 $\times 10^{-4}$	296
m-CH$_3$	60	1.61 $\times 10^{-4}$	296
H	60	1.48 $\times 10^{-4}$	296
p-Cl	60	1.44 $\times 10^{-4}$	296
m-Cl	60	1.36 $\times 10^{-4}$	296
p-NO$_2$	60	1.55 $\times 10^{-4}$	296
m-NO$_2$	60	1.18 $\times 10^{-4}$	296
C$_6$H$_5$CH$_2$OR			
R = p-CH$_3$C$_6$H$_4$	60	1.67 $\times 10^{-4}$	296
p-NO$_2$C$_6$H$_4$	60	1.67 $\times 10^{-4}$	296
CH$_2$C$_6$H$_5$	60	6.95 $\times 10^{-4}$	296
C(CH$_3$)$_3$	60	6.05 $\times 10^{-4}$	296
CH$_2$CH$_3$	60	4.45 $\times 10^{-4}$	296
CH$_3$	60	4.15 $\times 10^{-4}$	296
1,3-Dioxan	50—95	3.1 $\times 10^3$ exp($-9700/RT$)	297
1,3-Dioxepan	50—95	4.8 $\times 10^3$ exp($-10,100/RT$)	297
4-Methyl-1,3-dioxan	50—95	1.3 $\times 10^4$ exp($-10,350/RT$)	297
2,2-Pentamethylen-4-methyl- -1,3-dioxan	50—95	5.1 $\times 10^3$ exp($-10,200/RT$)	297
2-Phenyl-1,3-dioxan	50—95	3.0 $\times 10^2$ exp($-7300/RT$)	297
2-Methyl-1,3-dioxan	50—95	2.5 $\times 10^2$ exp($-7500/RT$)	297
2,4-Dimethyl-1,3-dioxan	50—95	1.2 $\times 10^2$ exp($-6800/RT$)	297

TABLE 13

The rate coefficients k_p and k_t (l mol⁻¹ s⁻¹) for the oxidation of some ethers

Ether	Temp. (°C)	k_p (l mole⁻¹ s⁻¹)	$2k_t$ (l mole⁻¹ s⁻¹)	Ref.
(Et)₂O	30	0.47	1.4 × 10⁸	203
(i-Pr)₂O	30	1.20	1.4 × 10⁶	203
(n-Bu)₂O	30	0.02	1.4 × 10⁸	203
i-PrO(t-Bu)	30	1.5	4.3 × 10⁴	203
C₆H₅OCH₂C₆H₅	30	11.5	2.4 × 10⁷	203
t-BuOCH₂C₆H₅	30	84.0	(2.1—3.6) × 10⁷	203
(C₆H₅CH₂)₂O	30		(1.6—2.1) × 10⁸	203
Cyclo-C₆H₁₁OCH₃	50—81	2.9 × 10⁸ exp(−12,000/RT)	3.3 × 10⁷ exp(−4400/RT)	295
1,4-Dioxan	30	0.48	5.0 × 10⁷	203
1,3-Dioxan	30—70	5.0 × 10⁷ exp(−9900/RT)	1.7 × 10⁹ exp(−4800/RT)	297
4,4-Dimethyl-1,3-dioxan	30—70	5.0 × 10⁷ exp(−10,100/RT)	5.75 × 10⁸ exp(−2200/RT)	297
4-Methyl-1,3-dioxan	30—70	4.0 × 10⁷ exp(−9900/RT)	7.9 × 10⁸ exp(−2300/RT)	297
2,2-Pentamethylene-1,3-dioxan	70	2.3	1.1 × 10⁷	297
2,2-Pentamethylene-4-methyl-1,3-dioxan	30—70	6.9 × 10⁵ exp(−8800/RT)	1.0 × 10⁸ exp(−4100/RT)	297
2-n-Propyl-1,3-dioxan	30—70	1.7 × 10⁷ exp(−10,600/RT)	5.0 × 10⁹ exp(−4600/RT)	297
2-Methyl-1,3-dioxan	30	0.59	2.0 × 10⁶	297
2,4-Dimethyl-1,3-dioxan	30—70	1.6 × 10⁷ exp(−9600/RT)	1.7 × 10⁹ exp(−4100/RT)	297
2-Vinyldioxan	30	34.4	3.5 × 10⁷	297

and

$$-\overset{|}{\underset{|}{C}}-O-\overset{|}{\underset{|}{C}}-\ \overset{k'_p}{\to}\ -\overset{|}{\underset{|}{C}}-O-\overset{|}{\underset{|}{C}}-$$
$$\quad\ OO\cdot\quad H\qquad\ OOH$$

Therefore, the overall rate coefficient is given by

$$(k_p)_{exp} = k_p + \frac{k'_p}{[RH]}$$

Values of the ratio $k_p/\sqrt{2k_t}$ are given in Table 12, and the absolute rate coefficients k_p and k_t in Table 13.

5. Oxidation of acids

5.1 OXIDATIVE DECARBOXYLATION OF ACIDS

Oxidation of carboxylic acids alone and in hydrocarbon solution is accompanied by the production of CO_2 [204—215]. Carbon dioxide is formed from the carboxylic group, as established by the tracer technique [204—210]. Oxidation of $R^{14}COOH$ yields $^{14}CO_2$ [204—209] and that of $RC^{18}OOH$ produces $C^{18}O_2$ [210].

Carboxylic acids decarboxylate in oxidizing cumene at 80°C, whereas, in the absence of oxygen, they decarboxylate only above 300°C [208]. The reaction of peroxy radicals with carboxylic groups was suggested to account for CO_2 formation, viz.

$$RO_2\cdot + R_1CH_2COOH \longrightarrow ROOH + R_1CH_2COO\cdot$$

$$R_1CH_2COO\cdot \longrightarrow R_1CH_2\cdot + CO_2$$

But carboxylation can also proceed by reaction of the peroxy radical with the $-CH_2^-$ group, viz.

$$RO_2\cdot + R_1CH_2COOH \longrightarrow ROOH + R_1\overset{\cdot}{C}HCOOH$$

$$\overset{\cdot}{R}CHCOOH \xrightarrow{O_2} R\overset{\overset{\displaystyle OO\cdot}{|}}{C}HCOOH \longrightarrow CO_2 + products$$

n-Nonane was found in the products of capric acid oxidation in cumene at 135°C [298], giving evidence that decarboxylation occurs by abstraction from the OH group, viz.

$$R_1CH_2COOH \xrightarrow{RO_2\cdot} R_1CH_2COO\cdot \xrightarrow[-CO_2]{} R_1CH_2\cdot \xrightarrow{RH} R_1CH_3$$

Other evidence for peroxy radical reaction with the carboxylic group is the isotopic effect in the oxidative decarboxylation of R'COOH and R'COOD. In the case of trimethylacetic acid in oxidizing cumene, $k_H/k_D \approx 6$ at 135°C [299]; $k_H/k_D = 4.2$ (110°C) for acetic acid, and 2.1 (135°C) for n-butyric acid. However, i-butyric acid decarboxylates under the same conditions without an isotope effect ($k_H/k_D = 1.1$ [299]). Hence, two processes lead to carbon dioxide formation from acids in the oxidation, reaction of the peroxy radical with the carboxylic group and with the α-C—H group. The former is the main reaction with trimethylacetic acid and the second with i-butyric acid. Normal carboxylic acids were found by kinetic studies to decarboxylate by two parallel reactions [300], viz.

$$R_1CH_2COOH + RO_2\cdot \xrightarrow{k_1} R_1CH_2COO\cdot + ROOH \longrightarrow R_1CH_2\cdot + CO_2$$

$$R_1CH_2COOH + RO_2\cdot \xrightarrow{k_2} R_1\dot{C}HCOOH + ROOH$$

$$R_1\dot{C}HCOOH \xrightarrow{O_2,RH} R_1\overset{\overset{\displaystyle OOH}{|}}{C}HCOOH \xrightarrow{k_3} CO_2 + products$$

When the hydrocarbon oxidizes with a constant initiation rate, $[RO_2\cdot] =$ const. $= \sqrt{W_i/2k_t}$ and

$$[CO_2] = (k_1 + k_2)[R_1CH_2COOH][RO_2\cdot]t$$
$$+ k_2k_3^{-1}[R_1CH_2COOH][RO_2\cdot](e^{-k_3t} - 1)$$

$$W_0^{CO_2} = k_1[R_1CHCOOH][RO_2\cdot]$$

$$W_\infty^{CO_2} = (k_1 + k_2)[R_1CH_2COOH][RO_2\cdot]$$

The kinetic measurements are consistent with these relationships and the rate coefficients k_1, k_2, and k_3 are given in Table 14. In calculating rate coefficients, the dimerization of carboxylic acids through hydrogen bonding was taken into account. The ratio $k_1/(k_1 + k_2)$ increases with increasing temperature; for example, for stearic acid it is 0.20 at 125° and 0.24 at 145°C.

Dicarboxylic acids decarboxylate by attack of peroxy radicals on α-C—H bonds. The evidence for such a mechanism was obtained from data on the decarboxylation of adipic acid, with COOH and COOD groups, in oxidizing cumene, when the velocities of CO_2 production were found to be the same [299]. Carbon dioxide is produced from the acid in the initiated oxidation of cumene (Table 14) and cyclohexanol [215] after the induction period associated with the formation of an intermediate, probably α-hydroperoxide, after attack of peroxy radicals on α-C—H

TABLE 14

The rate coefficients for the reactions of cumylperoxy radicals with carboxylic groups (k_1) and α-C—H Bonds (k_2), and α-hydroperoxycarboxylic acid decomposition (k_3) with acids in oxidizing cumene (125—145°C) [215,299]

Acid	k_1 (l mol^{-1} s^{-1})	k_2 (l mol^{-1} s^{-1})	$k_3 \times 10^4$ (135°C) (s^{-1})
i-Butyric		1.05×10^9 exp(−15,500/RT)	12.5
n-Butyric	9.8×10^{11} exp(−22,900/RT)	9.5×10^8 exp(−16,000/RT)	6.3
n-Valeric	7.4×10^{11} exp(−22,600/RT)	2.0×10^9 exp(−16,700/RT)	6.4
Caprinic	8.3×10^{11} exp(−22,000/RT)	6.9×10^9 exp(−17,500/RT)	6.9
Stearic	1.4×10^{11} exp(−21,000/RT)	1.25×10^{10} exp(−17,900/RT)	7.4
Sebacic		0.58 (120°)	
Azelaic		0.44 (120°)	6.7
Pimelic		0.25 (120°)	

bonds [215]. For the decarboxylation of sebacic acid in oxidizing cyclo-hexanol [215] $k_2 = 3.8 \times 10^4$ exp(−12,000/RT) l mole^{-1} s^{-1} and $k_3 = 1.4 \times 10^{15}$ exp(−26,000/RT) s^{-1}. The peroxy radicals do not attack the carboxylic group because of the intermolecular hydrogen bond between these groups. Only oxalic acid, which does not form such hydrogen bonds, decarboxylates by the reaction of peroxy radicals with carboxylic groups [215] with a rate coefficient $k_1 = 1.9 \times 10^9$ exp(−16,200/RT) l mole^{-1} s^{-1}. Therefore, the mechanism of oxidative decarboxylation of a carboxylic acid depends on its structure and the initial step is the reaction of peroxy radicals either with the carboxylic group or with the α-C—H bond, or with both these reactions in parallel.

5.2 OXIDATION OF ACIDS

In addition to decarboxylation, the oxidation of acids yields hydro-peroxy, hydroxy, keto groups, lactones, and mono- and dicarboxylic acids of lower molecular weight. The mechanism of the oxidation of acids is similar to that for hydrocarbons. The reactivity of mono- [300] and dicarboxylic acids [216] with respect to cumylperoxy radicals was measured by oxidation in the presence of cumyl hydroperoxide as source of $RO_2\cdot$ (see Table 15). The reactivities of methylenic groups in mono- and dicarboxylic acids and in n-paraffin acids are close. For example, at 100°C, $k_{CH_2} \times 10^2$ (l mole^{-1} s^{-1}) = 4.8 (n-decane), 10.0 (glutaric, sebacic, β,γ groups), 6.4 (α-CH$_2$ of dibasic acids), 8.0 (for monocarboxylic acids), and 11.0 (>CH$_2$ for propionic acid).

In the presence of catalysts, the carboxylic group is attacked not only

TABLE 15

Rate coefficients for reactions of cumylperoxy radicals with mono- [300] and dicarboxylic [216] acids in chlorbenzene

Acid	Temp. (°C)	k (120°C) (l mole^{-1} s^{-1})	A (l mole^{-1} s^{-1})	E (kcal mole^{-1})
Acetic	100—110	0.19 (110°C)	2.24×10^8	16.2
Propionic	100—125	0.36	7.02×10^7	15.1
n-Butyric	100—125	0.49	6.76×10^7	14.8
n-Valeric	100—125	0.62	1.25×10^8	15.1
Enanthic	100—125	0.87	1.76×10^8	15.1
Caprinic	100—125	1.45	3.27×10^8	15.2
i-Butyric	100—125	0.81	1.93×10^7	13.4
Glutaric	125—145	0.98	2.73×10^6	11.4
Pimelic	130—145	1.79 (130°C)	2.21×10^6	11.0
Suberic	125—145	1.51	4.66×10^6	11.4
Azelaic	125—145	3.10 (135°C)	5.36×10^6	11.4
Sebacic	125—145	2.20	6.07×10^6	11.4

by peroxy radicals, but also by oxidizing ions [211,212], viz.

$$RCO_2H + Co(III) \longrightarrow RCOO—Co(III) + H^+$$

$$RCOO—Co(III) \longrightarrow RCOO\cdot + Co(II)$$

$$RCOO\cdot \longrightarrow R\cdot + CO_2$$

The kinetic parameters of the first step are shown in Table 16 for different acids.

When the acid undergoing oxidation has many CH_2 groups, peroxy radicals react both with the carboxylic group (CO_2 is formed) and with CH_2 groups (producing hydroxy, keto and hydroperoxy acids). The mechanism of oxidation of acids at CH_2 groups is similar to that of hydrocarbon oxidation.

TABLE 16

The values of E_a and $\Delta S^{\ddagger}$ for the oxidation of acids by Co(III) in water [211]

Acid	E_a (kcal mole^{-1})	$\Delta S^{\ddagger}$ (cal K^{-1} mole^{-1})
Propionic	26.7	+21.4
i-Butyric	22.6	+16.1
Phenylacetic	21.4	+13.1
Crotonic	23.6	+13.1
Cinnamic	26.5	+32.6

6. Oxidation of esters

6.1 THE PRODUCTS OF ESTER OXIDATION

The main oxidation products of the methyl esters of aliphatic acids containing n C atoms are methyl esters of dicarboxylic acids C_4–C_{n-3}, aliphatic acids C_1–C_{n-1}, and keto- and hydroxy compounds [301–307]. Oxidation of acetates (140–160°C) yield acids, carbon dioxide, hydroxy, and keto compounds (see Table 17). Hydroperoxide is the primary product of oxidation. Oxidation of dimethyl esters of dicarboxylic acids gives monoesters with a lower number of C atoms in the acidic group (see Table 17). Carbon dioxide is formed in parallel with acids and monoesters [308]. All monoesters C_{n-1}, C_{n-2}, etc. are also formed in parallel. This suggests several mechanisms of C—C bond scission in the oxidation, an α-mechanism with only one C—C bond broken to form C_m and C_{n-m} products, a β-mechanism with two C—C bonds broken in the β-position to form C_m, C_{n-m-1} and CO_2, etc. The α, β, and γ-mechanisms of C—C bond scission may be regarded as a result of peroxy radical isomerization to form labile dihydroperoxides, e.g.

$$CH_3OCCH_2CH_2CH_2CHCOCH_3 \rightarrow CH_3OCCHCH_2CH_2CHCOCH_3$$

$$CH_3OCCCH_2CH_2CHCCH_3$$

The probabilities of C—C bond rupture by different mechanisms (P_α, P_β ...), calculated by the analysis of the products formed in the oxidation of dimethyl esters (170°C) are [309]

$P_\alpha = 0.30$, $\quad P_\beta = 0.32$ for adipic acid

$P_\alpha = 0.17$, $\quad P_\beta = 0.33$, $\quad P_\gamma = 0.39$ for pimelic acid

$P_\alpha = 0.13$, $\quad P_\beta = 0.15$, $\quad P_\gamma = 0.18$ for sebacic acid

6.2 THE CHAIN MECHANISM OF ESTER OXIDATION

The oxidation of esters is a chain process with initiation by the reaction

$$RH + O_2 \xrightarrow{k_0} R\cdot + HO_2\cdot$$

where $k_0 = 5.0 \times 10^{11} \exp(-33{,}900/RT)$ for diethyl sebacate, $7.9 \times 10^{13} \exp(-39{,}300/RT)$ for dimethyl sebacate, and $1.0 \times 10^{12} \exp(-34{,}700/$

TABLE 17
The products of ester oxidation
Oxidation of acetates at 150°C for 4 h [308].

Acetate	Product (mole %)			
	Acetic acid	Propionic acid	n-Butyric acid	n-Valeric acid
n-Propyl	71.7	28.3		
n-Butyl	57.4	14.0	29.0	
n-Amyl	67.0	8.6	6.4	18.0
s-Butyl	86.5	13.5		

Oxidation of dimethyl esters of dicarboxylic acids at 170°C [310].

Acid	Monomethyl esters (mole%) of dicarbonic acids with number of C atoms							
	C_4	C_5	C_6	C_7	C_8	C_9	C_{10}	C_{11}
Apidic	0.56	11.0	3.9	2.7				
Pimelic	0.59	11.0	8.3	2.7	12.0			
Azelaic	0.36	16.0	15.5	10.0	4.1	1.7	8.5	
Sebacic	0.50	15.2	18.0	9.2	5.8	4.0	1.5	10.3

RT) l mole^{-1} s^{-1} for diisopropyl sebacate [311]. However, in the oxidation of methyl esters of oleic, linolenic, and linoleic acids [312], the chains are initiated by a third-order reaction

$$2 \, RH + O_2 \longrightarrow R\cdot + H_2O_2 + R\cdot$$

The rate of initiated oxidation of esters obeys the equation

$$W = \frac{k_p}{\sqrt{2k_t}} \, [RH]\sqrt{W_i}$$

Values of the ratio $k_p/\sqrt{2k_t}$ are given in Table 18. The absolute rate coefficients of cumylperoxy radical reactions with various esters were measured by Agabekov et al. (see Table 19). It is of interest to note that the reactivity of the esters of monocarboxylic acids is much greater than that of dicarboxylic esters (compare, for example, propionates and glutarates).

Carbon dioxide formation accompanies the oxidation of esters. In the initiated oxidation of cyclohexanol [319,320] and cumene [320], CO_2 is formed from dimethyl esters of dicarboxylic acids through an intermediate, probably hydroperoxide, by

$$\text{Ester} \xrightarrow[RO_2\cdot]{k} \text{intermediate} \xrightarrow{k_m} CO_2 + \text{products}$$

The values of k and k_m are given in Table 20.

TABLE 18

The values of $k_p/\sqrt{2k_t}$ for ester oxidation

Ester	Temp. (°C)	$k_p/\sqrt{2k_t}$ ($l^{1/2}$ $mol^{-1/2}$ $s^{-1/2}$)	Ref.
n-Butyl acetate	80— 95	$88 \exp(-8600/RT)$	269
n-Amyl acetate	80— 95	$2.26 \times 10^2 \exp(-9200/RT)$	269
n-Decyl acetate	80— 95	$7.80 \times 10^2 \exp(-9600/RT)$	269
Methyl adipate	145—170	$1.66 \times 10^4 \exp(-13,000/RT)$	313
Dimethyl adipate	145—170	$5.74 \times 10^9 \exp(-24,900/RT)$	313
Ethyl adipate	145—170	$1.25 \times 10^4 \exp(-12,300/RT)$	313
Diethyl adipate	145—170	$2.39 \times 10^9 \exp(-22,100/RT)$	313
Di-n-propyl adipate	145—170	$8.23 \times 10^9 \exp(-23,200/RT)$	313
i-Propyl adipate	145—170	$2.28 \times 10^5 \exp(-10,500/RT)$	313
Di-i-propyl adipate	145—170	$2.85 \times 10^9 \exp(-15,800/RT)$	313
Di-i-butyl adipate	145—170	$8.75 \times 10^9 \exp(-16,900/RT)$	313
Methyl oleate		$4.4 \times 10^4 \exp(-10,800/RT)$	312
Methyl linoleate		$1.6 \times 10^2 \exp(-5700/RT)$	312
Methyl linolenate		$1.3 \times 10^4 \exp(-8000/RT)$	312

TABLE 19

The rate coefficients of reaction of cumylperoxy radicals with some esters

Ester	Temp. (°C)	k (l $mole^{-1}$ s^{-1})	Ref.
Methyl propionate	50— 70	$4.3 \times 10^5 \exp(-7000/RT)$	314
Ethyl propionate	70— 90	$8.5 \times 10^5 \exp(-8900/RT)$	314
i-Propyl propionate	80—100	$7.4 \times 10^6 \exp(-10,700/RT)$	314
t-Butyl propionate	90—110	$4.5 \times 10^7 \exp(-12,800/RT)$	314
t-Butyl i-butyrate	85—115	$1.15 \times 10^9 \exp(-16,200/RT)$	314
p-Methyl toluate	130—160	$9.0 \times 10^6 \exp(-12,700/RT)$	315
Dimethyl oxalate	75—100	$8.6 \times 10^3 \exp(-5500/RT)$	316
Dimethyl malonate	110—145	$3.4 \times 10^9 \exp(-16,000/RT)$	316
Dimethyl succinate	130—145	$4.15 \times 10^{10} \exp(-18,700/RT)$	316
Dimethyl glutarate	130—145	$3.9 \times 10^9 \exp(-16,500/RT)$	316
Dimethyl glutarate	125—145	$3.8 \times 10^9 \exp(-16,500/RT)$	317
Diethyl glutarate	125—145	$3.7 \times 10^9 \exp(-16,500/RT)$	317
Di-n-propyl glutarate	125—145	$1.4 \times 10^8 \exp(-13,700/RT)$	317
Di-n-butyl glutarate	125—145	$9.7 \times 10^7 \exp(-13,300/RT)$	317
Di-t-butyl glutarate	135—145	$4.5 \times 10^9 \exp(-17,400/RT)$	318
Dimethyl adipate	140	7.57	316
Methyl-t-butyl adipate	140—160	$4.0 \times 10^8 \exp(-17,900/RT)$	317
Ethyl-t-butyl adipate	140—160	$1.9 \times 10^8 \exp(-17,300/RT)$	317
i-Propyl-t-butyl adipate	140—160	$9.3 \times 10^7 \exp(-16,700/RT)$	317
Di-t-butyl adipate	140—160	$1.4 \times 10^8 \exp(-16,700/RT)$	314
Dimethyl pimelate	130—145	$2.55 \times 10^9 \exp(-16,000/RT)$	316
Dimethyl azelate	130—145	$3.2 \times 10^8 \exp(-14,200/RT)$	316
Methyl sebacate	135—145	$1.4 \times 10^9 \exp(-16,000/RT)$	318
Dimethyl sebacate	130—145	$1.9 \times 10^8 \exp(-13,700/RT)$	318

TABLE 20

The rate coefficients for CO_2 formation in the cooxidation of esters with cyclohexanol (I) and cumene (II) [319,320] by the reactions

$$\text{Ester} \xrightarrow{k} P \xrightarrow{k_m} CO_2 + \text{products}$$

Ester, oxidizing substrate	Temp. (°C)	k ($l\,mole^{-1}\,s^{-1}$)	k_m (s^{-1})
Dimethyl oxalate (I)	85— 95	$7.0 \times 10^6 \exp(-12,200/RT)$	$1.8 \times 10^{13} \exp(-27,800/RT)$
Dimethyl adipate (I)	135—145	$1.4 \times 10^7 \exp(-12,600/RT)$	$4.5 \times 10^{11} \exp(-28,700/RT)$
Dimethyl succinate (II)	135	0.30	2.2×10^{-4}
Dimethyl adipate (II)	135	0.18	2.1×10^{-4}
Dimethyl sebacate (II)	135	0.02	2.2×10^{-4}

7. Oxidation of phenols

7.1 OXIDATION OF PHENOLS IN HYDROCARBON SOLUTIONS

Phenols react with O_2 in hydrocarbon solution by non-chain reactions [217], e.g.

$$PhOH + O_2 \rightarrow PhO\cdot + HO_2\cdot$$

$$PhOH + HO_2\cdot \rightarrow PhO\cdot + H_2O_2$$

$$PhO\cdot + PhO\cdot \rightarrow products$$

The first bimolecular step is rate-limiting. Its activation energy is close to the endothermicity of the reaction, and the pre-exponential factor is close to the collision frequency (Table 21). A linear dependence is observed between $\log k$ and σ with $\rho = -5.0$ at $180°C$.

The compensation effect is observed: $\log k_o = 8.6 + 0.35E_a$ at $180°C$. A termolecular initiation process was suggested for the oxidation of α-naphthol in n-decane [218], but under conditions when α-naphthol was consumed by reaction with peroxy radicals as well as by reaction with O_2. A bimolecular mechanism for this reaction in benzene and cyclohexane was established recently with the same rate coefficients for both solvents [217].

Phenols are widely used as oxidation inhibitors, but the reaction

$$ArOH + O_2 \rightarrow ArO\cdot + HO_2\cdot$$

is a source of free radicals. It is interesting to estimate its role in oxidation. Let us take readily oxidized α-naphthol as inhibitor and compare the rates of formation of free radicals by the reaction $ArOH + O_2$ and by decomposition of hydroperoxide (W_i). Assuming W_i to be 10^{-6} mole l^{-1} s^{-1} at $180°C$ and $[O_2] = [\alpha\text{-naphthol}] = 3 \times 10^{-3}$ mole l^{-1}, then $W(ArOH + O_2) = 3.4 \times 10^{-3} \times 3 \times 10^{-3} \times 3 \times 10^{-3} = 3 \times 10^{-8}$ mole l^{-1} s^{-1} which represents only 3% of W_i. At a lower temperature, this fraction would be still less due to the relatively higher activation energy for the reaction $ArOH +$

TABLE 21

The kinetic parameters of the reaction $PhOH + O_2$ in benzene [217]

Phenol	A (l mole^{-1} s^{-1})	E_a (kcal mole^{-1})	k (180°C) (l mole^{-1} s^{-1})
α-Naphthol	2.2×10^9	25.0	3.44×10^{-3}
p-Methoxyphenol	7.5×10^9	27.0	7.46×10^{-4}
β-Naphthol	4.5×10^{10}	29.0	4.20×10^{-4}
p-Cresol	3.0×10^{11}	32.0	3.25×10^{-4}
Phenol	7.0×10^{12}	35.0	3.16×10^{-5}

O_2. The value of W_i assumed seems to be an underestimate since such values are usual for the oxidation of hydrocarbons at 110—130°C. Therefore the reaction $ArOH + O_2$, being very slow due to its relatively high activation energy, plays a part neither in initiation of chains nor in overall consumption of the inhibitor when the latter is added to the hydrocarbon.

7.2 OXIDATION OF PHENOLS IN POLAR SOLUTIONS

Oxidation of phenols with O_2 in aqueous solutions is faster than that in hydrocarbon solutions. Oxygen attacks both phenol molecules and phenolate ions. Therefore the rate of oxidation depends on the concentration of hydrogen ions. The rate of phenol oxidation in aqueous solutions (at 14—21°C) is [219]

$$W = k[\text{PhOH}][O_2], \quad k = k_1 + k_2 K[\text{H}^+]^{-1}$$

$$k_1 = 4.9 \times 10^8 \exp(-25{,}600/RT) \, \text{l mole}^{-1} \, \text{s}^{-1}$$

$$k_2 = 3.5 \times 10^{12} \exp(-17{,}000/RT) \, \text{l mole}^{-1} \, \text{s}^{-1}$$

assuming [220] that $K = 8 \times 10^{-10}$ mole l^{-1}. The suggested mechanism is

$$\text{PhOH} + O_2 \rightarrow \text{PhO·} + \text{HO}_2·$$

$$\text{PhOH} \rightleftharpoons \text{PhO}^- + \text{H}^+$$

$$\text{PhO}^- + O_2 \rightarrow \text{PhO·} + O_2·$$

$$O_2^-· + \text{H}^+ \rightleftharpoons \text{HO}_2·$$

$$\text{HO}_2· + \text{PhOH} \rightarrow \text{H}_2O_2 + \text{PhO·}$$

Photochemical oxidation of phenol in aqueous solution produces dimers, $\text{HOC}_6\text{H}_4\text{C}_6\text{H}_4\text{OH}$, and dihydroxyphenols [221]. It is suggested that these products are formed by reactions of phenoxyl radicals.

Oxidation of phenols is faster in alkaline than in neutral solutions because the reaction $\text{ArO}^- + O_2$ is much faster than $\text{ArOH} + O_2$. The dependence of oxidation rates in alkaline aqueous alcohol solutions on phenol structure was studied [222,223]. A linear correlation between $\log W$ and σ was observed with $\rho = 3.55$ (40°C, m- and p-substituted phenols).

Oxidation of phenols of the type

in aqueous alcohol solution produces semiquinone ion-radicals

(as established by the ESR technique) [224—227], i.e. oxidation of such phenols is accompanied by elimination of a group in the *para*-position.

When phenols react with oxygen in polar alkaline solutions, semi-quinone radicals are formed; the oxidation proceeds autocatalytically producing hydroperoxides and quinols.

7.3 CATALYTIC OXIDATION OF PHENOLS

Oxidation of hydroquinone and pyrocatechol in aqueous solution is accelerated by quinones [228,229], which seems to be due to the formation of semiquinone radicals and their fast reaction with oxygen. Duro-hydroquinone oxidation is autocatalytic [229], due to the accumulation of quinone and formation of semiquinone radicals [230—233].

Complexes of transition metals accelerate the oxidation of phenols in polar solvents [234]. Formation of phenoxy radicals was established by the ESR technique [235—241]. Molecular products are produced as a result of the reactions of ArO·. The initial steps suggested are

$$ArOH + Me(Ox) \rightarrow ArO\cdot + H^+ + Me(Red)$$

$$Me(Red) + O_2 \rightarrow Me(Ox)$$

Phenoxyl radicals are converted to quinones and products of phenol dimerization. Phenol oxidizes in water giving pyrocatechol and hydroquinone in the presence of metal ions [242]. The best yields were obtained with Cu^{2+} and Fe^{3+} ions (100—150°C).

Catalytic oxidation of phenols in non-aqueous solutions under the action of certain catalysts is accompanied, together with the formation of quinones and dimers, by oxidative polymerization [243,244], viz.

Selective polymerization can be produced by proper choice of the solvent and catalyst.

The reaction of phenols with oxygen in $CHCl_3$ and CH_3OH is catalyzed by bis(salicylidene)ethylenediiminocobalt(II). Intramolecular oxidation with recombination of radicals is suggested [245,246]. The composition of the products of oxidation depends on experimental conditions and the structure of phenol.

8. The role of hydrogen bonds in oxidation

8.1 HYDROGEN BONDING AND THE FORMATION OF FREE RADICALS

Molecules of hydroperoxide in solution are linked by hydrogen bonds, forming dimers

$$ROOH + ROOH \overset{K}{\rightleftharpoons} (ROOH)_2$$

The heat of formation of one hydrogen bond was estimated as 2 kcal mole^{-1} for cumene hydroperoxide in CCl_4 [247] and as 2.8 kcal mole^{-1} for t-butyl hydroperoxide [248]. Hydroperoxide dimers decompose to free radicals by the reaction [150]

$$\overset{\overset{\text{H}}{|}}{ROOH...OOR} \rightarrow RO_2 \cdot + H_2O + RO \cdot$$

This mode of decay is more rapid than unimolecular decomposition with scission of the O—O bond as it is energetically more favorable. When the concentration of hydroperoxide is low, $[ROOH] \gg [(ROOH)_2]$, and the rate of decomposition is

$$W_i = k_1[ROOH] + k_2[(ROOH)_2] \cong k_1[ROOH] + k_2K[ROOH]^2$$

If $k_2K[ROOH] \gg k_1$, $W_i \cong k_2K[ROOH]^2$ and so long as the rate of oxygen uptake (W_{O_2}) is proportional to $\sqrt{W_i}$, $W_{O_2} \propto [ROOH]$. Such a dependence has been established for the oxidation of a number of hydrocarbons [249—251] and interpreted as the result of bimolecular decomposition of hydroperoxide [150]. If this bimolecular decomposition occurs via preliminary formation of a dimer, then at a high concentration of hydroperoxide (when almost all molecules are associated) the rate of initiation will increase linearly with the concentration of hydroperoxide (taken as ROOH), i.e. $W_i = k_2[(ROOH)_2] = 0.5 k_2[ROOH]_\Sigma$. This dependence, has been established for t-butyl hydroperoxide in n-heptane [252] (Fig. 5). The rate coefficient for decomposition of the dimer to radicals appears to be $k_2 = 1.14 \times 10^8 \exp(-23,000/RT)$ s^{-1}. The equilibrium constant of t-butyl hydroperoxide association estimated from kinetic data (0.8 l mole^{-1} at 90°C) is very close to K found from spectroscopic measurements (0.74 l mole^{-1} at 90°C) [253].

Alcohols are produced in the oxidation of hydrocarbons and form

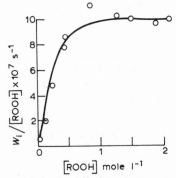

Fig. 5. Dependence of $W_i/[ROOH]$ on $[ROOH]$ for t-BuOOH in n-heptane (90°C).

hydrogen bonds with the hydroperoxides formed

$$\text{ROOH} + \text{R'OH} \rightleftharpoons \text{ROO}\overset{\displaystyle H}{\underset{\displaystyle /}{...}}\text{HOR'}$$

Such adducts decompose to free radicals [252], as do dimers of hydroperoxides, probably by the reaction

$$\text{ROO}\overset{\displaystyle H}{\underset{\displaystyle /}{...}}\text{HOR'} \overset{k}{\rightarrow} \text{RO·} + \text{H}_2\text{O} + \text{R'O·}$$

This mode of decay is again energetically more favorable than unimolecular fission of the O—O bond (by 16 kcal mole^{-1}). An increase in the rate of free radical formation with increase of alcohol concentration is observed

$$W_i = kK[ROOH][R'OH]$$

However, at high concentrations of alcohol, W_i attains a maximum and then decreases with increasing alcohol concentration. Such a dependence is ascribed to the effect of the solvent dielectric constant ϵ on the rate coefficient k. At a sufficient alcohol concentration, all of the hydroperoxide molecules are bound in adducts with alcohol. Further increase in $[R'OH]$ changes ϵ and decreases the rate coefficient of the $\text{ROO}\overset{\displaystyle H}{\underset{\displaystyle |}{}} ... \text{HOR'}$ adduct decomposition to free radicals [252]. The relationship between $\log k$ and $(\epsilon - 1)/(2\epsilon + 1)$ is linear. The rate coefficient of t-butyl hydroperoxide decomposition in n-butanol is [252] $k = 5.0 \times 10^6 \exp(-20,000/RT)$ s^{-1}. The product kK for the system cyclohexyl hydroperoxide—cyclohexanol—cyclohexane [169] is $kK = 4.5 \times 10^8 \exp(-22,000/RT)$ l mole^{-1} s^{-1} and for the system t-amyl hydroperoxide—t-amyl alcohol—i-octane [254] $kK = 3.2 \times 10^8 \exp(-22,300/RT)$ l mole^{-1} s^{-1}

Carboxylic acids accelerate the decay of hydroperoxides to free radicals by forming associates of two molecules of acid with one molecule of hydroperoxide [255]

$$\text{ROOH} + 2\ \text{R}'\text{COOH} \overset{K'}{\rightleftharpoons} \text{X} \overset{k'}{\rightarrow} \text{free radicals}$$

The rate and equilibrium constants were estimated for n-decyl hydroperoxide and caproic acid in n-decane to be [255] $k' = 2.0 \times 10^5$ $\exp(-16{,}100/RT)$ s^{-1} and $K' = 5.8 \times 10^{-4} \exp(10{,}200/RT)$ l^2 mole^{-2}.

8.2 FORMATION OF HYDROGEN BONDS WITH AND ABSTRACTION OF HYDROGEN ATOMS FROM O—H AND N—H BONDS

Inhibitors, such as phenols and aromatic amines, which inhibit oxidations by reaction with peroxy radicals

$$\text{InH} + \text{RO}_2\cdot \rightarrow \text{In}\cdot + \text{ROOH}$$

form hydrogen bonds of the types —O—H ... O and >N—H ... O with polar molecules such as alcohols, ketones, acids, hydroperoxides, and H$_2$O. The inhibiting group is then blocked by the hydrogen bond and the hydrogen-bonded inhibitor molecules do not react with peroxy radicals. Therefore, the activity of inhibitors is markedly diminished in the presence of polar molecules [73,256—258]. For example, α-naphthol reacts with RO$_2\cdot$ in cyclohexane with a rate coefficient of 2×10^5 l mole^{-1} s^{-1}, whereas that for the reaction in cyclohexanol is 1.3×10^3 l mole^{-1} s^{-1} (75°C). The equilibrium constant may be estimated from the dependence of the effective rate coefficient $k_{\text{eff}}(\text{RO}_2\cdot + \text{InH})$ on concentration of the added compound forming hydrogen bonds, e.g.

$$\text{InH} + \text{O} \overset{R}{\underset{H\cdot}{\diagdown\diagup}} \overset{K}{\rightleftharpoons} \text{InH...O} \overset{R}{\underset{H}{\diagdown\diagup}}$$

$$k_{\text{ef}} = \frac{K[\text{InH}]}{[\text{InH}]_\Sigma}$$

where $[\text{InH}]_\Sigma = [\text{InH}] + [\text{InH...ROH}]$. Thus

$$[\text{InH...ROH}] = K[\text{InH}]$$

and

$$k_{\text{eff}} = k/(1 + K[\text{ROH}])$$

or

$$\frac{1}{k_{\text{eff}}} = \frac{1}{k} + \frac{K[\text{ROH}]}{k}$$

Values of K are collected in Table 22.

TABLE 22

Equilibrium constants (K) for production of hydrogen bonded adducts determined from kinetic data

Inhibitor	Polar compound	Solvent	Temp. (°C)	K (l mole^{-1})
Propagating radical in system, RCH⟨OH⟩⟨OO·⟩				
α-Naphthol	Cyclohexanol	Cyclohexanol	75	13
α-Naphthol	Cyclohexanol	Chlorobenzene	75	22
α-Naphthylamine	n-BuOH	n-Heptane	74	2.5
Propagating radical in system, $CH_3COCH(OO·)CH_3$ [256]				
Hydroquinone	t-BuOH	Methyl ethyl ketone	60	12
Hydroquinone	H_2O	Methyl ethyl ketone	60	30
Trimethylphenol	t-BuOH	Methyl ethyl ketone	60	15.8
Trimethylphenol	H_2O	Methyl ethyl ketone	60	82
α-Naphthol	t-BuOH	Methyl ethyl ketone	60	43
α-Naphthol	H_2O	Methyl ethyl ketone	60	88
2,6-Di-t-butyl-4-methyl-phenol	t-BuOH	Methyl ethyl ketone	60	8.1
2,6-Di-t-butyl-4-methyl-phenol	H_2O	Methyl ethyl ketone	60	11.1
2,4,6-Tri-t-butyl phenol	t-BuOH	Methyl ethyl ketone	60	11.8
2,4,6-Tri-t-butylphenol	H_2O	Methyl ethyl ketone	60	20
2,6-Di-t-butylphenol	t-BuOH	Methyl ethyl ketone	60	6.3
2,6-Di-t-butylphenol	H_2O	Methyl ethyl ketone	60	12.7

Addition of a hydroperoxide ROOH to a hydrocarbon R′H induces an exchange reaction [43,259]

$$R'OO· + ROOH \rightarrow R'OOH + ROO·$$

In the presence of polar compounds (for example alcohol), this exchange becomes slower due to the formation of hydrogen bonds (Table 23) [260].

Carboxylic acid products of the hydrocarbon oxidation are attacked by peroxy radicals at the carboxylic group, and become decarboxylated

$$RO_2· + R'COOH \rightarrow ROOH + R'COO·$$

$$R'COO \rightarrow R'· + CO_2$$

The rate of decarboxylation of labelled caproic acid in octadecane oxidation decreases as other acids accumulate in the hydrocarbon [207]. This is due to dimerization of the acids via hydrogen bonds, preventing

TABLE 23

Effect of t-butanol on the rate coefficient of the exchange reaction between tetrallyl hydroperoxide and cumylperoxy radicals at 30°C [260]

[Cumene] (mole l^{-1})	[t-BuOH] (mole l^{-1})	k (l mole^{-1} s^{-1})
7.2		600
6.6	0.8	180
3.9	4.9	30
2.2	7.3	10

decarboxylation, viz.

$$R'COOH + RCOOH \rightleftharpoons R'-C \underset{O-H...O}{\overset{O...H-O}{\diagup\diagdown}} C-R''$$

Dicarboxylic acids (except oxalic) are not attacked to any appreciable extent at the carboxylic groups as both groups are connected by intramolecular hydrogen bonds [215]. Thus hydrogen bonds between molecules always protect O—H bonds from attack by peroxy radicals.

8.3 HYDROGEN BONDING WITH PEROXY RADICALS

The occurrence of hydrogen bonding between peroxy radicals and hydroxyl-containing molecules was first suggested as an explanation of the decrease in cyclohexanone oxidation rate in the presence of water [261]. Oxidation of 2,4-disubstituted pentanes, $CH_3CHX_1CH_2CHX_2CH_3$ produces dihydroperoxide in a high yield (70—90%) only when X_1 and X_2 are not hydroxyl groups [262] but are, for example, CH_3 or OCH_3. Monohydroperoxide is formed when one or two hydroxyl groups are present. This is accounted for by formation of a hydrogen bond of the type

$$\underset{X \quad CH_2}{\overset{CH_3 \quad O-O...H-O CH_3}{\diagup\diagdown \qquad \diagup\diagdown}}{\underset{H}{C \qquad C}}$$

hindering intramolecular abstraction of the H atom from the β-C—H bond with formation of dihydroperoxide. Quantitative investigation of hydrogen bonding of the type $RO_2\cdot$ ... HOR' was carried out by Zaikov et al. [263—267].

For methyl ethyl ketone—water and hydrocarbon—alcohol mixtures, the rate coefficients $k_{p(eff)}$ and $k_{t(eff)}$ determined were found to decrease with increasing concentration of H_2O (R'OH) in the system. The results obtained are analyzed on the basis of the scheme

$$RO_2 \cdot + RH \xrightarrow{k_p} ROOH + R \cdot$$

$$RO_2 \cdot + R'OH \overset{K}{\rightleftharpoons} RO_2 \cdot ...HOR'$$

$$RO_2 \cdot ...HOR' + RH \xrightarrow{k'_p} ROOH + R \cdot + HOR'$$

$$RO_2 \cdot + RO_2 \cdot \xrightarrow{k_t}$$

$$RO_2 \cdot ...HOR' + RO_2 \cdot \xrightarrow{k'_t} \left.\begin{array}{c} \\ \\ \\ \end{array}\right\} \text{molecular products}$$

$$RO_2 \cdot ...HOR' + RO_2 \cdot ...HOR' \xrightarrow{k'_t}$$

The measured $k_{p(eff)}$ and $k_{t(eff)}$ represent combinations of relevant elementary rate coefficients, viz.

$$k_{p_{(eff)}} = \frac{k_p + k'_p K[R'OH]}{1 + K[R'OH]} \approx \frac{k_p}{1 + K[R'OH]}$$

when $k'_p K[R'OH] \ll k_p$

$$k_{t_{(eff)}} = \frac{k_t + Kt'_t[R'OH] + K^2 k''_t[R'OH]^2}{1 + K[R'OH] + K^2[R'OH]^2}$$

The individual rate coefficients and equilibrium constants found are collected in Table 24.

The formation of hydrogen bonds in 2-methylpentene-2 oxidation makes the peroxy radicals more reactive (Table 24). These radicals are usually of a low reactivity due to their intramolecular π-bonds. It is suggested that the latter are broken by the formation of hydrogen bonds, the radical activity thus increasing, viz.

$$\begin{array}{cc} \overset{\cdot}{O}-O & \overset{\cdot}{O}O...HOR' \\ \overset{..}{(CH_3)_2C=CH-CH-X} + R'OH \rightleftharpoons (CH_3)_2C=CH-CH-X \end{array}$$

A hydrogen bond usually lowers the peroxy radical activity since the approach of the peroxide group and the molecule is hindered. Hydroxyperoxy radicals formed during the oxidation of alcohols apparently possess intramolecular hydrogen bonds; these lower the reactivity of such radicals.

The formation of hydrogen bonds between polar molecules and peroxy radicals permits a new interpretation of reactions involving $RO_2 \cdot$ with molecules having O—H and N—H bonds. The peroxy radical first forms a

TABLE 24
Rate coefficients of reactions of peroxy radicals, free and hydrogen-bonded, and equilibrium constants for formation of the adducts

Reaction	K (l mole^{-1} s^{-1}) at 60°C	log A	E_a or Q (kcal mole^{-1})
RH = methyl ethyl ketone [263—265]			
$RO_2^. + RH$	3.9×10^{-1}	5.10	8.4 ± 0.5
$RO_2^. ... H_2O + RH$	2.0×10^{-2}	9.11	16.5 ± 1
$RO_2^. + RO_2^.$	1.8×10^6	7.30	1.6 ± 0.8
$RO_2^. ... H_2O + RO_2^. ... H_2O$	7.0×10^4	7.86	4.6 ± 1.0
$RO_2^. + H_2O \rightleftharpoons RO_2^. ... H_2O$	1.1 [a]	$(\overline{4.88})$	4.8 ± 0.4
$RO_2^. ... HOCH_3 + RH$	6.7×10^{-2}		
$RO_2^. ... HOCH_3 + RO_2^.$	8×10^5		
$RO_2^. ... HOCH_3 + RO_2^. ... HOCH_3$	7×10^5		
$RO_2^. + HOCH_3 \rightleftharpoons RO_2^. ... HOCH_3$	0.7 [a]		
$RO_2^. ... HOC(CH_3)_3 + RH$	9.0×10^{-2}		
$RO_2^. ... HOC(CH_3)_3 + RO_2^.$	8×10^5		
$RO_2^. ... HOC(CH_3)_3 + RO_2^. ... HOC(CH_3)_3$	7×10^5		
$RO_2^. + HOC(CH_3)_3 \rightleftharpoons RO_2^. ... HOC(CH_3)_3$	0.64		
$RO_2^. ... H_2O + HO$—⬡—OH	4.0×10^4		
$RO_2^. ... H_2O + \alpha$-Naphthol	3.4×10^4		
$RO_2^. ... H_2O + HO$—⬡(H_3C, H_3C)—CH_3	2.2×10^4		
$RO_2^. ... H_2O + HO$—⬡(X, X)—CH_3	0.30×10^4		
$RO_2^. ... H_2O + HO$—⬡(X, X)	0.15×10^4		
$RO_2^. ... H_2O + HO$—⬡(X, X)—X	0.61×10^4		
$RO_2^. ... H_2O + HO$—⬡(X, X)—azolactone	0.12×10^4		
RH = cyclohexane [266]			
$RO_2^. ... HOC(CH_3)_3 + RH$	0.22		
$RO_2^. + RH$	0.53		
$RO_2^. + HOC(CH_3)_3 \rightleftharpoons RO_2^. ... HOC(CH_3)_3$	0.9 [a]	$(\overline{3.69})$	3.5
$RO_2^. + RO_2^.$	5.0×10^6	7.724	$1.6 \dagger 0.6$
$RO_2^. ... HOC(CH_3)_3 + RO_2^.$	4.0×10^6	8.518	3.0 ± 0.8
$RO_2^. ... HOC(CH_3)_3 + RO_2^. ... HOC(CH_3)_3$	1.6×10^6	9.740	5.4 ± 0.8

TABLE 24 (continued)

Reaction	K (l mole^{-1} s^{-1}) at 60°C	log A	E_a or Q (kcal mole^{-1})
RH = 2-methylpentene-2 [267]			
$RO_2^{\cdot}$ + RH	1.77(40°C)	1	
$RO_2^{\cdot}$ ... HOC(CH$_3$)$_3$ + RH	13.3(40°)		
$RO_2^{\cdot}$ + HOC(CH$_3$)$_3$ ⇌ $RO_2^{\cdot}$ ... HOC(CH$_3$)$_3$	0.76 [a]	($\overline{3}$.22)	4.2
$RO_2^{\cdot}$ + $RO_2^{\cdot}$	4.3 × 10^6		
$RO_2^{\cdot}$ ... HOC(CH$_3$)$_3$ + $RO_2^{\cdot}$	9.0 × 10^6		
$RO_2^{\cdot}$ ... HOC(CH$_3$)$_3$ + $RO_2^{\cdot}$ ... HOC(CH$_3$)$_3$	13.2 × 10^6		

[a] l mole^{-1}.

hydrogen bond with the molecule and then abstracts H, viz.

$$RO_2 \cdot + HOR' \overset{K}{\rightleftharpoons} RO_2 \cdot ...HOR' \overset{k}{\to} ROOH + R'O\cdot$$

$$RO_2 \cdot + AmH \rightleftharpoons RO_2 \cdot ...HAm \to ROOH + Am\cdot$$

$$RO_2 \cdot + HOOR' \rightleftharpoons RO_2 \cdot ...HOOR' \to ROOH + R'OO\cdot$$

The measured rate coefficient of the reaction between a peroxy radical and such a molecule in a non-polar medium is the product kK and the experimental activation energy is $E_{obs} = E_a - Q$, where Q is the heat of hydrogen bond formation.

9. Reactions of peroxy radicals with polyfunctional molecules

The reaction of peroxy radicals with C—H bonds seems to be the most important in the mechanism of oxidation of O-containing compounds as well as of hydrocarbons. The rate coefficient of this reaction for hydrocarbons depends on the dissociation energy of C—H bond and on steric hindrance at constant temperature and solvent. The neighbouring groups influence both factors. Consequently, all molecules of hydrocarbon may be divided into a number of groups. Every group is supposed to react with its own partial rate coefficient, so that the molecular rate coefficient is equal to the sum of all partial rate coefficients (the rule of additivity of partial rate coefficients). This rule is valid for hydrocarbon fragments of monofunctional compounds too. For example, the rate coefficient of cumylperoxy radicals with dimethyl esters, $CH_3OCOCH_2(CH_2)_n$-CH_2COOCH_3, may be represented in the form

$$k = 2k_o + n\Delta$$

where Δ is the partial rate coefficient of the CH$_2$ group (from Table 19).

TABLE 25

Comparison of rate coefficients, experimental (k) and calculated according to the additivity rule (Σk_i), of cumylperoxy radical reactions with esters at 140°C [322,323]

Compound	k (l mole^{-1} s^{-1})	Number of groups	Σk_i (l mole^{-1} s^{-1})	E (kcal mole^{-1})	log A	$\Delta G_n^{\ddagger}$ (kcal mole^{-1})	$\Delta S_n^{\ddagger}$ (cal K^{-1} mole^{-1})	$L_G \times 10^2$ (cal K^{-1} mole^{-1})
Ethyl propionate	17	1	17	8.8	5.97	0	0	
Diethyl glutarate	6.8	2	34.7	16.5	9.70	1.2	17	1.8
Diethylene glycol dicaprylate	10.8	2 + 2 a	68	23.4	12.63	1.6	29	2.2
Pentaerythritol tetravalerate	4.0	4	74	23.9	13.05	2.4	32	1.6

The question is whether this additivity rule is valid with polyfunctional molecules. Recent experimental data on the reactivity of polyfunctional esters and alcohols with peroxy radicals provides evidence that the additivity rule cannot be applied to polyfunctional compounds [321]. The experimental rate coefficients of cumylperoxy radical reactions with ethyl propionate and 3 polyfunctional esters are compared with the sums of the partial rate coefficients in Table 25. The latter were calculated from k for ethyl propionate and $k_{CH_2} = 0.75$ l mole^{-1} s^{-1} at $140°C$ (Table 19). It is seen that the calculated rate coefficients, Σk_i, are several times greater than the experimental values of k for all three polyfunctional esters. It is interesting to note that the greater the number of ester groups in the molecule, the greater is the difference between experimental and calculated rate coefficients. Another peculiarity of peroxy radical reactions with polyfunctional compounds is a strong compensation effect, i.e. an increase of the pre-exponential factor A with increasing E (Table 25).

Both peculiarities may be explained by the concept of the transition state as a multidipole system [323]. A polyfunctional molecule may be treated as number of n interacting dipoles $n_1, ... n_n$. These dipoles are orientated at restricted angles due to the motion of molecular segments. If two functional groups are separated by two or more C—C bonds, the distance between them r_{ij} is more than the charge separation in the dipoles. In this case, the Gibbs free energy of interaction of n dipoles G_n is

$$G_n = \frac{1}{\epsilon} N_A \sum_i^n \frac{n_i n_j}{r_{ij}^3} f(\theta_i, \theta_j, \Delta\phi_{ij})$$

where N_A is Avogadro's number, ϵ is dielectric constant of the medium, θ_i and θ_j are the angles of inclination of the dipole axes to the line connecting their centres, $\Delta\phi_{ij}$ is the angle between the planes in which the two dipoles lie and every plane passes through the line connecting the centres of the dipole. Accordingly, the transition state of the reaction $R'O_2\cdot + RH$ has $n + 1$ interacting dipoles and the contribution of multidipole interaction to the Gibbs free energy of activation is

$$\Delta G_n^{\ddagger} = G_n^{\ddagger} - G_n = \frac{1}{\epsilon} N_A \left(\sum^{n+1} \frac{n_i n_j}{r_{ij}^3} f_{ij} - \sum^n \frac{n_i n_j}{r_{ij}^3} \right)$$

Thus, for the rate coefficient, one will have the equation

$$\ln k = \ln k_o - \frac{\Delta G_n^{\ddagger}}{RT}$$

The reaction between $RO_2\cdot$ and a monofunctional compound may be taken as the reaction for comparison. In this case, $\Delta G_n^{\ddagger}$ will be the difference between $\Delta G_n^{\ddagger}$ for two and polydipole systems, i.e. the free energy of dipole—dipole interaction in multidipole systems. The value of $\Delta G_n^{\ddagger}$ may

TABLE 26
Rate coefficients for the reaction of *sec*-hexadecylperoxy radicals with alcohols at 130°C
Values of Σk_i are calculated from data on monohydroxy alcohols [321].

Alcohol	k (l mole^{-1} s^{-1})	Σk_i (l mole^{-1} s^{-1})	$\Delta G_n^{\ddagger}$ (kcal mole^{-1})
1-Decanol	15.1		0
2-Octanol	20.7		0
1,2 Ethanediol	6.0	30.2	1.3
1,4-Butanediol	6.5	30.2	1.2
1,3-Butanediol	7.7	35.8	1.2
1,2-Dioxypropane ether	12.0	71.6	−0.4
1,1,1-Trimethylpropanol	63.3	45.2	−0.3
2-Methyl-3,4-dioxytetrahydrofuran	145.0	56.5	−0.8
2,2-Dimethyl-1,3-propanediol	206.0	30.2	−1.6

be found from the experimental data using the equation $\Delta G_n^{\ddagger} = RT \ln(k_0/k)$. One can see from Table 25 that $\Delta G_n^{\ddagger}$ increases from 0 to 2.5 kcal mole^{-1} for the reaction of the cumyl peroxy radical with polyfunctional esters. The value of $\Delta G_n^{\ddagger}$ may be positive ($G_n^{\ddagger} > G_n$) or negative ($G_n^{\ddagger} < G_n$). We can see both cases for the reaction of *sec*-hexadecyl peroxy radicals with polyatomic alcohols (Table 26). The rate coefficients (k) for the reactions were estimated from the results of experiments on the co-oxidation of *n*-hexadecane with different alcohols in the presence of benzoquinone as a selective inhibitor (see Sect. 2.5.3) [321].

The compensation effect (Table 25) has been mentioned earlier. This effect can be easily explained by the concept of multidipole interaction [323]. The distance r_{ij} and angles θ_i, θ_j, $\Delta\phi_{ij}$ in a polyfunctional molecule and in the transition state depend on the temperature due to deformational vibrations. The amplitude of vibration increases with temperature and one can expect that $\Delta G_n^{\ddagger}$ decreases. The temperature dependence of $\Delta G_n^{\ddagger}$ can be supposed analogous to that of the dielectric constant ($\epsilon = \epsilon_0 e^{-LT}$), viz.

$$\frac{d \ln \Delta G_n^{\ddagger}}{dT} = -L_G$$

As $\ln k = \ln A - E/RT$, $\ln k = \ln k_0 - \Delta G_n^{\ddagger}/RT$, and $\ln k_0 = \ln A_0 - E_0/RT$ then

$$E = RT^2 \, d \ln k/dT = E_0 + \Delta G_n^{\ddagger} - T\frac{d\Delta G_n^{\ddagger}}{dT}$$

$$= E_0 + \Delta G_n^{\ddagger} + (L_G - L)T\Delta G_n^{\ddagger}$$

$$\Delta S_n^{\ddagger} = (L_G - L) \, \Delta G_n^{\ddagger}$$

$$\ln A = \ln A_0 + \frac{\Delta G_n^{\ddagger}(L_G - L)}{R}$$

and

$$E = E_0 + \Delta G_n^{\ddagger} + RT \ln \frac{A}{A_0}$$

the latter relationship explains the compensation effect. Values of $\Delta G_n^{\ddagger}$, $\Delta S_n^{\ddagger}$, and L_G, calculated from the experimental data, are given in Table 25.

When a peroxy radical attacks a monofunctional molecule, the rate coefficient depends on the dissociation energy of the C—H bond as well as on dipole—dipole interaction. The influence of this interaction may be included in the partial rate coefficient. When a polyfunctional molecule is attacked, the situation with respect to dipole—dipole interaction becomes more complicated. The attacking peroxy radical interacts, not only with the dipole moment of the nearest functional group, but with all functional groups of the molecule. Therefore, every polyfunctional molecule reacts individually depending on the number of functional groups and the structure of the molecules. Thus, the partial rate coefficient additivity rule is not valid in these cases. So the electrostatic interaction between polar groups in the transition state changes the reactivity of every C—H bond in a polyfunctional molecule. This phenomenon must also be present in the reactions of other polar radicals.

References

1 N.M. Emanuel, E.T. Denisov and Z.K. Maizus, Liquid-phase Oxidation of Hydrocarbons, Plenum Press, New York, 1967.
2 L.V. Schibaeva, D.I. Metlitsa and E.T. Denisov, Zh. Fiz. Khim., 44 (1970) 2793.
3 E.A. Blyumberg, G.E. Zaikov, Z.K. Maizus and N.M. Emanuel, Dokl. Akad. Nauk SSSR, 133 (1960) 144.
4 G.E. Zaikov and Z.K. Maizus, Izv. Akad. Nauk SSSR, Otd. Khim. Nauk, (1962) 1175.
5 E.A. Blyumberg, G.E. Zaikov, Z.K. Maizus and N.M. Emanuel, Kinet. Katal., 1 (1960) 510.
6 G.E. Zaikov, Z.K. Maizus and N.M. Emanuel, Neftekhimiya, 4 (1964) 91.
7 G.E. Zaikov and Z.K. Maizus, Dokl. Akad. Nauk SSSR, 150 (1963) 116.
8 G.E. Zaikov and Z.K. Maizus, M.I. Vinnik and N.M. Emanuel, Neftekhimiya, 7 (1967) 260.
9 G.E. Zaikov, Z.K. Maizus, M.I. Vinnik and N.M. Emanuel, Neftekhimiya, 7 (1967) 260.
10 G.E. Zaikov, Neftekhimiya, 3 (1963) 381.
11 D.C. Hull, U.S. Pat., 2,522,175 (1951).
12 Kodak Ltd., Br. Pat. 534,633 (1941).
13 D.C. Hull, U.S. Pat., 2,287,803 (1942)
14 D.C. Hull, U.S. Pat., 2,425,878 (1947).
15 R.M. Deansly, U.S. Pat., 2,456,683 (1948).

196

16 A.I. Il'in and V.G. Nedavnjaja, Neftekhimiya, 6 (1966) 440.
17 W.G. Lloyd, J. Am. Chem. Soc., 78 (1956) 72.
18 R. Cantieni, Ber. Dtsch. Chem. Ges. B 69 (1936) 1101, 1386.
19 N.A.Milas, U.S. Pat., 2,115,206 (1938).
20 G.O. Schenck and H.D. Becker, Ger. Pat., 1,076,688 (1961).
21 N.V. de Bataafsche Petroleum Maatschapij, Brit. Pat., 708,339 (1954).
22 N. Brown, M.J. Hartig, M.J. Roedel, A.W. Anderson and C.E. Schweitzer, J. Am. Chem. Soc., 77 (1955) 1756.
23 E.T. Denisov and V.V. Kharitonov, Zh. Fiz. Khim., 35 (1961) 444.
24 T. Kunugi, N. Mazura and S. Oguni, J. Jpn. Pet. Inst., 6 (1963) 26.
25 E.T. Denisov and V.M. Soljanikov, Neftekhimiya, 4 (1964) 458.
26 E.T. Denisov and V.V. Kharitonov, Neftekhimiya, 2 (1962) 760.
27 N.A. Milas, S.A. Harris and P.C. Panagiotakos, J. Am. Chem. Soc., 61 (1939) 2430.
28 R. Criegee, Fortschr. Chem. Forsh., 1 (1950) 508.
29 W. Cooper and W.H.T. Davison, J. Chem. Soc., (1952) 1180.
30 M.S. Kharash and G. Sosnovsky, J. Org. Chem., 23 (1959) 1322.
31 V.L. Antonovsky and V.A. Terent'ev, Zh. Fiz. Khim., 39 (1965) 621, 2901; 40 (1966) 19.
32 E.T. Denisov, V.V. Kharitonov and E.N. Raspopova, Kinet. Katal., 5 (1964) 981.
33 N.A. Milas and A. Golubović, J. Am. Chem. Soc., 81 (1959) 5824.
34 N.A. Milas and A. Golubović, J. Am. Chem. Soc., 81 (1959) 3361.
35 Energy of Rupture of Chemical Bonds. Ionization Potentials and Electron Affinities, Izd. Akad. Nauk SSSR, Moscow, (1962).
36 M.S. Kharash, J.L. Rowe and W.H. Urry, J. Org. Chem., 16 (1951) 905.
37 A.I. Brodskij, V.I. Franchuk, M.M. Aleksankin and V.A. Lunenok-Burmakina, Dokl. Akad. Nauk SSSR, 123 (1958) 117.
38 Ju.B. Krjukov, R.M. Smirnova, V.A. Seleznev, V.V. Kamzolkin and A.N. Bashkirov, Neftekhimiya, 3 (1963) 238.
39 F. Mashio and S. Kato, Mem. Fac. Ind. Arts Kyoto Tech. Univ. Sci. Technol., (1957) 67; Chem. Abstr., 52 (1958) 16178c.
40 E.T. Denisov and V.M. Soljanikov, Neftekhimiya, 3 (1963) 360.
41 C. Parlant, I. Seree de Roch and J.-C. Balaceanu, Bull. Soc. Chim. Fr., (1963) 2452.
42 C. Parlant, Rev. Inst. Fr. Pet., 19 (1964) 1.
43 J.R. Thomas and C.A. Tolman, J.Am. Chem. Soc., 84 (1962) 2079.
44 R.L. Vardanjan, V.V. Kharitonov and E.T. Denisov, Izv. Akad. Nauk SSSR, Ser. Khim., (1970) 1536.
45 A.L. Alexandrov and E.T. Denisov, Izv. Akad. Nauk SSSR, Ser. Khim., (1969) 2322.
46 B.Ja. Ladygin and V.V. Saraeva, Kinet. Katal., 7 (1966) 967.
47 G. Hughes and H.A. Makada, Advan. Chem. Ser., 75 (1968) 102.
48 A.L. Alexandrov and E.T. Denisov, Izv. Akad. Nauk SSSR, Ser. Khim., (1966) 1737.
49 A.L. Alexandrov, T.I. Sapacheva and V.F. Shuvalov, Izv. Akak. Nauk SSSR, Ser. Khim., (1969) 955.
50 K.A. Zhavnerko, Cand. Sci. Thesis, Institute of Phys. Org. Chem., Akad. Nauk Beloruss SSR, Minsk, 1969.
51 R.L. McCarthy and A. MacLachlan, Trans. Faraday Soc., 57 (1961) 1107.
52 E.T. Denisov, Dokl. Akad. Nauk SSSR, 130 (1960) 1055.
53 P. Gray, Trans. Faraday Soc., 55 (1959) 408.
54 E.T. Denisov, Dokl. Akad. Nauk SSSR, 141 (1961) 131.
55 D.J. Carlsson and J.C. Robb, Trans. Faraday Soc., 62 (1966) 3403.
56 E.T. Denisov and V.V. Kharitonov, Kinet. Katal., 5 (1964) 781.

57 V.V. Kharitonov and E.T. Denisov, Neftekhimiya, 6 (1966) 235.
58 E.T. Denisov and V.V. Kharitonov, Neftekhimiya, 3 (1963) 558.
59 V.V. Kharitonov, Zh. Fiz. Khim., 40 (1966— 2699.
60 H.J. Bäckström, The Svedberg (1944) 45; Chem. Abstr., 39 (1945) 1105.
61 J.L. Bolland and H.R. Cooper, Nature (London), 173 (1953— 413.
62 M. Kamija, Ju. Fudzita and T. Kan, Shokubai, 6 (1964) 15.
63 G.G. Jayson, G. Scholes and J. Weiss, J. Chem. Soc., (1957) 1358.
64 A. Hummel and A.O. Allen, Radiat. Res., 17 (1962) 302.
65 S. Up Choi and N. Lichtin, J. Am. Chem. Soc., 86 (1964) 3948.
66 P.N. Komarov, E.V. Barelko and M.A. Proskurnin, Neftekhimiya, 3 (1963) 609; 5 (1965) 715.
67 P.N. Komarov, Neftekhimiya, 5 (1965) 721, 863.
68 E. Hayon and J.J. Weiss, J. Chem. Soc., (1961) 3970.
69 G.G. Jayson, G. Scholes and J. Weiss, J. Chem. Soc. A, (1968) 662.
70 E.T. Denisov and V.V. Kharitonov, Dokl. Akad. Nauk SSSR, 132 (1960) 595.
71 E.T. Denisov and V.V. Kharitonov, Zh. Fiz. Khim., 38 (1964) 639.
72 E.T. Denisov and A.L. Alexandrov, Zh. Fiz. Khim., 38 (1964) 491.
73 E.T. Denisov, A.L. Alexandrov and V.P. Shcheredin, Izv. Akad. Nauk SSSR, Otd. Khim. Nauk, (1964) 1583.
74 C.E. Boozer and G.S. Hammond, J. Am. Chem. Soc., 76 (1954) 3861; G.S. Hamond, C.E. Boozer, C.E. Hamilton and J.N. Sen, J. Am. Chem. Soc., 77 (1955) 3238.
75 J.S. Thomas, J. Am. Chem. Soc., 85 (1963) 2166.
76 E.T. Denisov and V.V. Kharitonov, Iz. Akad. Nauk SSSR, Ser. Khim., (1963) 2222.
77 E.T. Denisov and V.P. Shcheredin, Izv. Akad. Nauk SSSR, Ser. Khim., (1964) 919.
78 R.L. Vardanjan, V.V. Kharitonov and E.T. Denisov, Neftekhimiya, 11 (1971) 247.
79 V.V. Kharitonov and E.T. Denisov, Izv. Akad. Nauk SSSR, Ser. Khim., (1967) 2764.
80 E.T. Denisov, Izv. Akad. Nauk SSSR, Ser. Khim., (1969) 328.
81 V.Ja. Shljapintokh, O.N. Karpukhin, L.M. Postnikov, I.V. Zakharov, A.A. Vichutinskii and V.F. Tsepalov, Chemiluminestsentnye Metody Issledovanija Khimicheskikh Protsessov, Izd. Nauka, Moskva, 1966.
82 R.L. Vardanjan, V.V. Kharitonov and E.T. Denisov, Kinet. Katal., 12 (1971) 903.
83 A.L. Alexandrov and E.T. Denisov, Izv. Akad. Nauk SSSR, Ser. Khim., (1969) 1652.
84 A.L. Alexandrov, T.I. Sapacheva and E.T. Denisov, Kinet. Katal., 10 (1970) 711.
85 G. Czapski and B.H.J. Bielski, J. Phys. Chem., 67 (1963) 2180.
86 A.L. Alexandrov and E.T. Denisov, Dokl. Akad. Nauk SSSR, 178 (1968) 379.
87 A. Kato, N. Takeyama, J. Mizoguchi and T. Seiyame, J. Chem. Soc. Jpn., Ind. Chem. Sect., 67 (1964) 1214.
88 T. Seiyama, A. Kato, N. Watamori and N. Takeyama, J. Chem. Soc. Jpn., Ind. Chem. Sect., 68 (1965) 1576.
89 V.M. Soljanikov and E.T. Denisov, Izv. Akad. Nauk SSSR, Ser. Khim., (1968) 1504.
90 E.T. Denisov, Izv. Akad. Nauk SSSR, Ser. Khim., (1967) 1608.
91 V.M. Solyanikov and E.T. Denisov, Dokl. Akad. Nauk SSSR, 173 (1967) 1106.
92 V.M. Solyanikov and E.T. Denisov, Neftekhimiya, 9 (1969) 116.
93 E.T. Denisov, V.M. Solyanikov and A.L. Alexandrov, Adv. Chem. Ser., 75 (1968) 112.
94 K.A. Zhavnerko and B.V. Erofeev, Vestn. Akad. Nauk Beloruss. SSR, Ser. Khim. Nauk, 3 (1966) 123.

198

95 V.M. Solyanikov and E.T. Denisov, Iz. Akad. Nauk SSSR, Ser. Khim., (1968) 1391.
96 A.V. Nikiforova, I.I. Moiseev and Ja.K. Syrkin, Zh. Obshch. Khim., 33 (1963) 3239.
97 W.G. Lloyd, J. Org. Chem., 32 (1967) 2816.
98 M. Crimeaux, C.R. Acad. Sci., 104 (1887) 1276.
99 A. Sftreker, Ann. Chim. (Paris), 93 (1855) 370.
100 R.A. Sheeden and R.B. Turner, J. Am. Chem. Soc., 77 (1955) 130.
101 I.I. Ioffe, Ju.T. Nkolaev and M.S. Brodskii, Kinet. Katal., 1 (1960) 125.
102 I.I. Ioffe, N.V. Klimova and A.G. Makeev, Kinet. Katal., 3 (1962) 107.
103 I.I. Ioffe and Ju.T. Nikolaev, Kinet. Katal., 2 (1961) 245.
104 Ja.B. Gorokhvatskii, E.N. Popova, A.I. Pjatnitskaja and E.I. Jaremenko, Symposium Kataliticheskie Reaktsii v Zhidkoi Phaze, Izd. Akad. Nauk Kaz. SSR, Alma-Ata, 1967, p. 531.
105 V.V. Shalja, B.I. Kolotusha, Ph.A. Jampolskaja and Ja.B. Gorokhvatskii, Kinet. Katal., 10 (1969) 1090.
106 A. Le Berre, C.R. Acad. Sci., 252 (1961) 1341; Bull. Soc. Chim. Fr., (1961) 1543.
107 A. Etienne and A. Le Berre, C.R. Acad. Sci., 252 (1961) 1166.
108 A. Le Berre and P. Goasguen, Bull. Soc. Chim. Fr., (1962) 1682.
109 M.N. Shchukina, V.G. Ermolaeva and A.E. Kolmanson, Dokl. Akad. Nauk SSSR, 158 (1964) 436.
110 G.A. Russell, E.G. Janzen and E.T. Strom, J. Am. Chem. Soc., 84 (1962) 4155.
111 V.V. Pereshein, N.A. Sokolov, V.A. Shushunov and G.A. Abakumov, Tr. Khim. Khim. Tekhnol., 1 (1966) 167; Zh. Obshch. Khim., 37 (1967) 386.
112 A. Le Berre, Bull. Soc. Chim. Fr., (1961) 1198.
113 G.A. Russell, A.G. Bemis, E.J. Geels, E.G. Janzen and A.J. Moye, Adv. Chem. Ser., 75 (1968) 174.
114 W. Brackman and C.J. Gaasbeek, Rec. Trav. Chim., 85 (1966) 242.
115 W. Brackman and C.J. Gaasbeek, Rec. Trav. Chim., 85 (1966) 257.
116 L.V. Shibaeva, D.I. Metelitsa and E.T. Denisov, Neftekhimiya, 10 (1970) 682.
117 L.V. Shibaeva, D.I. Metelitsa and E.T. Denisov, Neftekhimiya, 11 (1971) 92.
118 C.R. Jefcoate, J.R. Smith and R.O. Norman, J. Chem. Soc. B, (1969) 1013.
119 R. Cantieni, Chem. Ber., 69 (1936) 2282.
120 P.E. Frankenburg and W.A. Noyes, J. Am. Chem. Soc., 75 (1953) 2847.
121 G. Ciamician and P. Silber, Chem. Ber., 46 (1913) 3077.
122 E.T. Denisov, Kinetika i Kataliz, Izd. Akad. Nauk SSSR, Moskow, 1960, p. 95.
123 G.E. Zaikov and Z.K. Maizus, Kinet. Katal., 3 (1962) 846.
124 G.E. Zaikov, Z.K. Maizus and N.M. Emanuel, Neftekhimiya, 7 (1967) 82.
125 D.B. Sharp, S.E. Whitcomb, L.W. Patton and A. Moorhead, J. Am. Chem. Soc., 74 (1952) 1802.
126 C. Paquot, Bull. Soc. Chim. Fr., (1945) 450.
127 D.B. Sharp, L.W. Patton and S.E. Whitcomb, J. Am. Chem. Soc., 73 (1951) 5600.
128 K. Ito, S. Sakai and I. Isij, J. Chem. Soc. Jpn., Ind. Chem. Sect., 68 (1965) 2403.
129 R.V. Kozlenkova, V.V. Kamzolkin and A.N. Bashkirov, Neftekhimiya, 9 (1969) 586.
130 W. Pritzkow, Chem. Ber., 87 (1954) 1668.
131 A. Robertson and W.A. Waters, J. Chem. Soc., (1948) 1574.
132 R. Schölner and W. Treibs, Chem. Ber., 94 (1961) 2978.
133 A.I. Kamneva and L.A. Muzychenko, Tr. Mosk. Khim. Tekhnol. Inst., 23 (1956) 61.
134 S.S. Jenkins, J. Am. Chem. Soc., 57 (1935) 2733.
135 G.E. Zaikov and Z.K. Maizus, Zh. Fiz. Khim., 40 (1966) 211.

136 I.G. Tishchenko and L.S. Stanishevskii, Zh. Obshch. Khim., 33 (1963) 3751; I.G. Tishchenko, L.S. Stanislevskii and L.S. Novikov, Zh. Org. Khim., 5 (19690 301.
137 R.C. Fuson and H.L. Jackson, J. Am. Chem. Soc., 72 (1950) 1637.
138 A. Rieche, Angew. Chem., 50 (1937) 520; 51 (1938) 707.
139 W. Pritzkow, Chem. Ber., 88 (1955) 572.
140 B.I. Makalets, Cand. Sci. Thesis, Moscow State University, 1960.
141 L.K. Obukhova and N.M. Emanuel, Izv. Akad. Nauk SSSR, Otd. Khim. Nauk, (1960) 1545.
142 A.E. Robson and D. Joung, Br. Pat., 771,992 (1957).
143 A. Elhil, A.E. Robson and D. Joung, Br. Pat., 771,991 (1957).
144 I.V. Berezin and B.I. Makalets, Zh. Fiz. Khim., 33 (1959) 2351.
145 E.T. Denisov, Iz. Akad. Nauk SSSR, Otd. Khim. Nauk, (1960) 812.
146 G.E. Zaikov, A.A. Vichutinskii and Z.K. Maizus, Kinet. Katal., 8 (1967) 675.; G.E. Zaikov, Kinet. Katal., 9 (1968) 1166.
147 A.L. Alexandrov and E.T. Denisov, Kinet. Katal., 10 (1969) 904.
148 A.E. Semenchenko, V.M. Soljanikov and E.T. Denisov, Neftekhimiya, 11 (1971) 555.
149 A.L. Buchachenko, K.Ja. Kaganskaya, M.B. Neuman and A.A. Petrov, Kinet. Katal., 2 (1961) 44.
150 L. Bateman, Q. Rev., 8 (1954) 147.
151 E.T. Denisov, Izv. Akad. Nauk SSSR, Otd. Khim. Nauk, (1967) 2396.
152 G.E. Zaikov, A.A. Vichutinskii, Z.K. Maizus and N.M. Emanuel, Dokl. Akad. Nauk SSSR, 168 (1966) 1096.
153 G.E. Zaikov, Izv. Aakd. Nauk SSSR, Ser. Khim., (1967) 1692.
154 G.E. Zaikov, Z.K. Maizus and N.M. Emanuel, Teor. Eksp. Khim., 3 (1967) 612.
155 G.E. Zaikov and Z.K. Maizus, Zh. Fiz. Khim., 43 (1969) 115.
156 G.E. Zaikov, Kinet. Katal., 9 (1968) 511.
157 L.M. Andronov, G.E. Zaikov, Z.K. Maizus and N.M. Emanuel, Izv. Akad. Nauk SSSR, Ser. Khim., (1968) 1748.
158 E.T. Denisov and L.N. Denisova, Izv. Akad. Nauk SSSR, Ser. Khim., (1964) 1108.
159 L.K. Obukhova, A.A. Boldin and N.M. Emanuel, Neftekhimiya, 1 (1961) 70.
160 G.A. Russell, J. Am. Chem. Soc., 79 (1957) 3871.
161 J.A. Howard and K.U. Ingold, Can. J. Chem., 43 (1965) 2737; J. Am. Chem. Soc., 90 (1968) 1056.
162 G.E. Zaikov, Z.K. Maizus and N.M. Emanuel, Neftekhimiya, 4 (1964) 91.
163 T.E. Denisov, Kinet. Katal., 4 (1963) 53.
164 E.T. Denisov and L.N. Denisova, Izv. Akad. Nauk SSSR, Ser. Khim., (1963) 1731.
165 G.E. Zaikov, Z.K. Maizus and N.M. Emanuel, Izv. Akad. Nauk SSSR, Ser. Khim., (1968) 53.
166 G.E. Zaikov, Z.K. Maizus and N.M. Emanuel, Dokl. Akad. Nauk SSSR, 140 (1961) 405.
167 V.L. Antonovskii and V.A. Terent'ev, Zh. Fiz. Khim., 43 (1969) 2549; 2727.
168 E.T. Denisov, Dokl. Akad. Nauk SSSR, 146 (1962) 394; Zh. Fiz. Khim., 37 (1963) 1896.
169 A.E. Semenchenko, V.M. Soljanikov and E.T. Denisov, Neftekhimiya, 10 (1970) 864.
170 S. Chubachi, H. Matsui, K. Yamamoto and S. Ishimoto, Bull. Chem. Soc. Jpn., 42 (1969) 789.
171 J.J. den Hertog and E.C. Kooyman, J. Catal., 6 (1966) 357.
172 R. van Helden and E.C. Kooyman, Rec. Trav. Chim., 80 (1961) 57.
173 V.D. Kossarov and E.T. Denisov, Neftekhimiya, 7 (1967) 420; 8 (1968) 595; Kinet. Katal., 10 (1969) 513; Zh. Fiz. Khim., 43 (1969) 769; 44 (1970) 390.

200

174 E.T. Denisov, V.D. Komissarov and D.I. Metelitsa, Discuss. Faraday Soc., 46 (1968) 127.
175 W. Miller and X. Rohede, Chem. Ber., 25 (1892) 2095.
176 V. Bogdanowska, Chem. Ber., 25 (1892) 1271.
177 C. Graebe and E. Gfeller, Ann. Chim. (Paris), 276 (1893) 12; 290 (1896) 199.
178 C. Harries, Chem. Ber., 34 (1901) 2105.
179 C. Harries and Stähler, Ann. Chim. (Paris), 330 (1904) 264.
180 A.H. Salway and E.S. Kipping, J. Chem. Soc., 95 (1909) 166.
181 W.Treibs, Chem. Ber., 63 (1930) 2423; 64 (1931) 2178, 2545; 65 (1932) 163, 1314; 66 (1933) 610, 1483; 68 (1935) 1049.
182 J. Rigaudy, C.R. Acad. Sci., 228 (1949) 253.
183 W.v.E. Doering and R.M. Haines, J. Am. Chem. Soc., 76 (1954) 482.
184 R. Hanna and X. Ourisson, Bull. Soc. Chim. Fr., (1967) 3742.
185 T.J. Wallace, H. Pobiner and A. Schriesheim, J. Org. Chem., 30 (1965) 3768.
186 A.M. Clover, J. Am. Chem. Soc., 44 (1922) 1107.
187 N.A. Milas, J. Am. Chem. Soc., 53 (1931) 221.
188 A. Rieche and R. Meister, Angew. Chem., 49 (1936) 101.
189 A. Robertson, Nature (London), 162 (1948) 153.
190 F. Reimers, Q.J. Pharm. Pharmacol., 19 (1946) 27.
191 L. Sajus, Adv. Chem. Ser., 75 (1968) 59.
192 N.A. Bakh, V.I. Medvedovsky and V.V. Saraeva, 2nd U.N. Int. Conf. Peaceful Uses At. Energy, Geneva, Vol. 29, 1958, p. 128.
193 V.V. Saraeva, N.A. Bakh, V.I. Dakin and P. Dillinger, Kinet. Katal., 3 (1962) 865.
194 G.O. Shenck, H. Becker, K. Schulte-Elte and C.H. Krauch, Chem. Ber., 96 (1963) 509.
195 E.K. Varfolomeeva and Z.G. Zolotova, Ukr. Khim. Zh., 25 (1959) 708.
196 D.B. Sharp and T.M. Patrick, J. Org. Chem., 26 (1961) 1389.
197 L. Debiais, M. Niclause and M. Letort, J. Chim. Phys., 56 (1959) 41, 54, 63, 69.
198 P. Grosborne and I. Seree de Roch, Bull. Soc. Chim. Fr., (1967) 2260.
199 V.I. Stenberg, R.D. Olson, Chiou Tong Wang and N. Kulevsky, J. Org. Chem., 32 (1967) 3227.
200 R. Lombard and Blum, Bull. Soc. Chim. Fr., (1957) 1310.
201 M. Galin-Vacherot, Bull. Soc. Chim. Fr., (1969) 765.
202 N. Kulevsky, Chiou Tong Wang and V.I. Stenberg, J. Org. Chem., 34 (1969) 1345.
203 J.A. Howard and K.U. Ingold, Can. J. Chem., 47 (1969) 3809; 48 (1970) 873.
204 I.V. Berezin, B.I. Makalets and L.G. Chuchukina, Zh. Obshch. Khim., 28 (1958) 2718.
205 I.V. Berezin, L.G. Berezkina and T.A. Nosova, Okislenie Uglevodorodov v Zhidkoi Faze, Izd. Akad. Nauk SSSR, Moscow, 1959, p. 101.
206 I.V. Berezin, A.M. Ragimova and N.M. Emanuel, Izv. Akad. Nauk SSSR, Otd. Khim. Nauk, (1959) 1733.
207 I.V. Berezin and A.M. Ragimova, Zh. Fiz. Khim., 36 (1962) 581.
208 N.I. Mitskevich, D. Sc. Thesis, Institute of Physical Organic Chemistry, Akademii Nauk Berlorusskoi SSR, Minsk, 1963.
209 N.I. Mitskevich, B.V. Erofeev and V.A. Lashitskii, Neftekhimiya, 5 (1965) 381.
210 H. Hübner, Kernenergie, (1960) 839.
211 W.A. Waters, Discuss. Faraday Soc., 46 (1968) 158.
212 V.M. Goldberg, Cand. Sci. Thesis, Institute of Chemical Physics, Akademii Nauk SSSR, Moscow, 1965.
213 N.I. Mitskevich and V.E. Agabekov, Neftekhimiya, 6 (1966) 867.
214 V.E. Agabekov, Cand. Sci. Thesis, Institute of Physical Organic Chemistry, Akademii Nauk Belorusskoi SSR, Minsk, 1969.

215 V.E. Agabekov, E.T. Denisov and N.I. Mitskevich, Izv. Akad. Nauk SSSR, Ser. Khim., (1968) 2254.
216 V.E. Agabekov, E.T. Denisov and N.I. Mitskevich, Kinet. Katal., 10 (1969) 731.
217 L.N. Denisova, E.T. Denisov and D.I. Metelitsa, Izv. Akad. Nauk SSSR, Ser. Khim., (1969) 1657.
218 Z.K. Maizus, N.M. Emanuel and V.N. Jakovleva, Dokl. Akad. Nauk SSSR, 143 (1962) 366.
219 L.V. Shibaeva, D.I. Metalitsa and E.T. Denisov, Kinet. Katal., 10 (1969) 1020, 1239.
220 D.T.Y. Chen and K.J. Leidler, Trans. Faraday Soc., 58 (1962) 480.
221 H.-I. Joschek and S.I. Miller, J. Am. Chem. Soc., 88 (1966) 3273.
222 U.E. Kirso, M.Ja. Gubergrits and K.A. Kuiv, Organic Reactivity (USSR), 3 (1966) 33.
223 U.E. Kirso and M.Ja. Gubergrits, Izv. Akad. Nauk Est. SSR, 16 (1967) 26.
224 L.M. Strigun, A.I. Prokof'ev, F.N. Pirnazarova and N.M. Emanuel, Izv. Akad. Nauk SSSR, Ser. Khim., (1968) 59, 2242; (1969) 1462.
225 N.M. Emanuel and L.M. Strigun, Usp. Khim., 37 (1968) 969.
226 M. Adams, M.S. Bois and R.H. Sands, J. Chem. Phys., 28 (1958) 774.
227 J.K. Bicconcal, S. Clough and G. Scott, Proc. Chem. Soc. (London), (1959) 308.
228 M. Augustin, J. Schneider and W. Langenbeck, J. Prakt. Chem., 13 (1961) 245.
229 T.H. James and A. Weissberger, J. Am. Chem. Soc., 60 (1938) 98.
230 A. Hantzsch, Chem. Ber., 49 (1916) 511; 54 (1921) 1276.
231 E. Weitz, Angew. Chem., 66 (1954) 664; Z. Elektrochem., 34 (1928) 538.
232 B. Eleme, Rec. Trav. Chim., 50 (1931) 807; 52 (1933) 569.
233 L. Michaelis, M.P. Schubert and S. Granick, J. Am. Chem. Soc., 61 (1939) 1981.
234 R.D. Korpusova and L.A. Nikovaev, Zh. Fiz. Khim., 30 (1956) 2831.
235 L.V. Gorbunova, M.L. Khidekel and G.A. Razuvaev, Dokl. Akad. Nauk SSSR, 147 (1962) 368.
236 E. Ochiai, Tetrahedron, 20 (1964) 1831.
237 Y. Ogata and T. Morimoto, Tetrahedron, 21 (1965) 2791.
238 V.V. Karpov and M.L. Khidekel, Zh. Org. Khim., 3 (1967) 1669; 4 (1968) 861.
239 T. Yonezawa, T. Tsuruya, S. Tsuchiya and T. Kawamura, J. Chem. Soc. Jpn., Ind. Chem. Sect., 71 (1968) 1007.
240 L.P. Il'icheva, K.B. Jatsimirskii, Izv. Vyssh. Uchebn. Zaved., Khim. Khim. Tekhnol., (1968) 520.
241 C.M. Orlando, J. Org. Chem., 33 (1968) 2516.
242 B.I. Makalets and L.G. Ivanova, Neftekhimiya, 9 (1969) 280.
243 G.F. Endres, A.S. Hay and J.W. Eustance, J. Org. Chem., 28 (1963) 1300.
244 H. Finkbeiner, A.S Hay, H.S. Blanchard and C.F. Endres, J. Org. Chem., 31 (1966) 549.
245 H.M. van Dort and H.J. Geursen, Rec. Trav. Chim., 86 (1967) 520.
246 L.H. Vogt, Jr., J.G. Wirth and H.L. Finkbeiner, J. Org. Chem., 34 (1969) 273.
247 I.I. Shabalin, Cand. Sci. Thesis, Kazan' State University, 1968.
248 V.V. Zharmov and N.K. Rudnevskii, Opt. Spektrosk., 12 (1962) 479.
249 J.L. Bolland and G. Gee, Trans. Faraday Soc., 42 (1946) 236.
250 W. Kern and H. Willersinos, Makromol. Chem., 1 (1955) 1; Angew. Chem., 67 (1955) 573.
251 L. Bateman, H. Hughes and A.L. Morris, Discuss. Faraday Soc., 14 (1953) 190.
252 E.T. Denisov, Zh. Fiz. Khim., 38 (1964) 2085.
253 E.T. Denisov, D. Sc. Thesis, Moscow State University, 1964.
254 T.G. Degtyareva, V.M. Solyanikov and E.T. Denisov, Neftekhimiya, 12 (1972) 857.
255 L.G. Privalova and Z.K. Maizus, Izv. Akad. Nauk SSSR, Otd. Khim. Nauk, (1964) 281.

202

256 L.M. Andronov and G.E. Zaikov, Izv. Akad. Nauk SSSR, Ser. Chim., (1968) 2261.
257 L.M. Andronov, G.E. Zaikov and Z.K. Maizus, Teor. Eksp. Khim., 3 (1967) 620.
258 L.M. Andronov, G.E. Zaikov, Z.K. Maizus and N.M. Emanuel, Zh. Fiz. Khim., 41 (1967) 2002.
259 J.R. Thomas and C.A. Tolman, J. Am. Chem. Soc., 84 (1962) 2079.
260 J.A. Howard, W.J. Schwalm and K.U. Ingold, Adv. Chem. Ser., 75 (1968) 6.
261 E.T. Denisov, Izv. Akad. Nauk SSSR, Otd. Khim. Nauk, (1960) 53.
262 F.F. Rust and E.A. Youngman, J. Org. Chem., 27 (1962) 3778.
263 G.E. Zaikov, Z.K. Maizus and N.M. Emanuel, Neftekhimiya, 8 (1968) 217; Dokl. Akad. Nauk SSSR, 173 (1967) 859.
264 L.M. Andronov, G.E. Zaikov and Z.K. Maizus, Zh. Fiz. Khim., 41 (1967) 1122.
265 G.E. Zaikov, L.M. Andronov, Z.K. Maizus and N.M. Emanuel, Dokl. Akad. Nauk SSSR, 174 (1967) 127.
266 G.E. Zaikov and Z.K. Maizus, Izv. Akad. Nauk SSSR, Ser. Khim., (1969) 598.
267 G.E. Zaikov, Z.K. Maizus and N.M. Emanuel, Izv. Akad. Nauk SSSR, Ser. Khim., (1969) 2265.
268 B.Ya. Ladygin, M.S. Fupman and V.I. Mogilev, Khim. vys. Energ., 6 (1972) 447.
269 V.M. Potekhin, D. Sc. Thesis, University of Leningrad, 1972.
270 R.L. Vardanyan, E.T. Denisov and V.I. Zozulya, Izv. Akad. Nauk SSSR, Ser. Khim., (1972) 611.
271 A.Ya. Gerchikov, E.P. Kuznetsova and E.T. Denisov, Kinet. Katal., 15 (1974) 509.
272 L.G. Galimova, Cand. Sci. Thesis, University of Upha, 1975.
273 A.A. Alexandrov, G.I. Solov'ev and E.T. Denisov, Izv. Akad. Nauk SSSR, Ser. Khim., (1972), 1527.
274 N.G. Zubareva, E.T. Denisov and A.V. Ablov, Kinet. Katal., 14 (1973) 346.
275 V.E. Agabekov, E.T. Denisov, N.I. Mitskevich and I.I. Korsak, Neftekhimiya, 13 (1973) 845.
276 I.I. Korsak, V.E. Agabekov and N.I. Mitskevich, Vesti Akad. Navuk BSSR, Ser. Khim. Navuk, 1 (1972) 30.
277 L.K. Obukhova, Neftekhimiya, 5 (1965) 97.
278 V.A. Itskovich, Cand. Sci. Thesis, University of Leningrad, 1970.
279 I.I. Korsak, V.E. Agabekov and N.I. Mitskevich, Neftekhimiya, 15 (1975) 130.
280 A.Ya. Gerchtkov, V.D. Kossarov, E.T. Denisov and G.B. Kochemasova, Kinet. Katal., 13 (1972) 1126.
281 A.Ya. Gerchikov, E.M. Kuramshin, V.D. Komissarov and E.T. Denisov, Kinet. Katal., 15 (1974) 230.
282 V.D. Komissarov, A.Ya. Gerchikov, L.G. Galimova and E.T. Denisov, Dokl. Akad. Nauk SSSR, 213 (1973) 881.
283 K.I. Ivanov, V.K. Savinova and E.G. Mikhailova, Zh. Obshch. Khim., 16 (1946) 65, 1003, 1015.
284 A. Rieche, Angew. Chem., 70 (1958) 251.
285 A. Rieche and K. Koch, Chem. Ber., 75 (1942) 1016.
286 N.A. Milas, R.L. Peeler and O.L. Mageli, J. Am. Chem. Soc., 76 (1954) 2322.
287 H. Wieland and A. Wingler, Ann. Chem., 431 (1923) 317.
288 H. King, Nature (London), 20 (1927) 843.
289 L.M. Kaliberdo, R.I. Suprun and L.B. Shuvalova, Neftepererab. Neftekhim. (Moscow), 6 (1972) 25.
290 R.D. Olson and V.J. Stenberg, Proc. N.D. Acad. Sci., 17 (1963) 50.
291 H. Rein and R. Criegee, Angew. Chem., 62 (1950) 120.
292 G.I. Nikishin, V.G. Glukhovtsev, M.A. Peikova and A.V. Ignatenko, Izv. Akad. Nauk SSSR, Ser. Khim., (1971) 2323.

293 S. Siegel and S. Coburn, J. Am. Chem. Soc., 73 (1951) 5494.

294 F.C. Eichel and D.F. Othmer, Ind. Eng. Chem., 41 (1949) 2623.

295 G.A. Kovtun and A.L. Alexandrov, Izv. Akad. Nauk SSSR, Ser. Khim., (1973) 1946.

296 G.A. Russel and R.C. Williamson, J. Am. Chem. Soc., 86 (1964) 2357, 2364.

297 S.A. Agisheva, A.L. Alexandrov, V.S. Martemyanov, S.S. Zlotskii and D.L. Rakhmankulov, Neftekhimiya, 15 (1975) 742.

298 V.E. Agabekov, N.I. Mitskevich, V.A. Azazko and N.L. Budeiko, Dokl. Akad. Nauk BSSR, 17 (1973) 826.

299 Azarko, V.A., V.E. Agabekov and N.I. Mizkevich, Dokl. Akad. Nauk BSSR, 17 (1973) 340.

300 V.E. Agabekov, N.I. Mitskevich, E.T. Denisov, V.A. Azarko and N.L. Budeiko, Dokl. Akad. Nauk BSSR, 18 (1974) 38.

301 H. Thaler and W. Saumweber, Fette Seifen Anstrichm., 63 (1961) 945, 1045.

302 L.I. Ermolaeva, I.B. Blanshtein, Yu.L. Moskovich and B.G. Freidin. Zh. Prikl. Khim., 43 (1970) 1589.

303 H. Thaler and W. Saumweber, Nahrung, 7 (1963) 106.

304 C. Paquot and F. Goursac, Bull. Soc. Chim. Fr., 2 (1950) 172.

305 V. Ramanathan, T. Sakuragi and F.A. Kummerow, J. Am. Oil Chem. Soc., 36 (1959) 244.

306 V. Pritzkow and K. Dietzsch, Chem. Ber., 93 (1960) 1733.

307 N.I. Mitskevich, B.V. Erofeev and V.A. Laschitskii, Neftekhimiya, 5 (1966) 381.

308 M.K. Novozhilova, V.M. Potekhin, V.A. Proskuryakov, A.E. Drabkin and E.P. Tarasenkova, Zh. Prikl. Khim., 43 (1970) 2313.

309 V.E. Agabekov, T.G. Kosmacheva and N.I. Mitskevich, Dokl. Akad. Nauk BSSR, 16 (1972) 38.

310 T.G. Kosmachova, V.E. Agabekov and N.I. Mitskevich, Vestsi Akad. Navuk BSSR, 5 (1971) 37.

311 G.V. Butovskaya, V.E. Agabekov and N.I. Mitskevich, React. Kinet. Catal. Lett., 4 (1976) 105.

312 N. Yanishlieva, I.P. Skibida, Z.K. Maizus and A. Popov, Izv. Otd. Khim. Nauki, Bulg. Akad. Nauk., 4 (1) (1971) 1.

313 T.G. Kosmacheva, V.E. Agabekov, N.I. Mitskevich and M.N. Fedorishcheva, Vestsi Akad. Navuk BSSR, Ser. Khim. Navuk, 2 (1976) 102.

314 V.E. Agabekov, N.I. Mitskevich, G.V. Butovskaya and T.G. Kosmacheva, React. Kinet. Catal. Lett., 2 (1975) 123.

315 N.I. Mitskevich, N.G. Ariko, V.E. Agabekov and H.M. Karmilova, React. Kinet. Catal. Lett., 1 (1974) 467.

316 V.E. Agabekov, E.T. Denisov, N.I. Mitskevich, T.G. Kosmacheva and G.V. Butovskaya, Kinet. Katal., 15 (1974) 883.

317 V.E. Agabekov, N.I. Mitskevich and M.N. Fedorishcheva, Kinet. Katal., 15 (1974) 1149.

318 V.E. Agabekov, M.N. Fedorishcheva and N.I. Mitskavich, Vestsi Akad. Navuk BSSR, Ser. Khim. Navuk, 4 (1973) 24.

319 V.E. Agabekov, T.G. Kosmacheva and N.I. Mitskevich, Vestsi Akad. Navuk BSSR, Ser. Khim. Navuk, 3 (1972) 31.

320 V.E. Agabekov, T.G. Kosmacheva, N.I. Mitskevich and I.P. Stremok, Vestsi Akad. Navuk BSSR, Ser. Khim. Navuk, 4 (1972) 38.

321 E.T. Denisov, Neftekhimiya, 18 (1978) 525.

322 G.G. Agliullina, V.S. Martemyanov, E.T. Denisov and T.I. Eliseeva, Izv. Akad. Nauk SSSR, Ser. Khim., (1977) 50.

323 E.T. Denisov, Izv. Akad. Nauk SSSR, Ser. Khim., (1978) 1746.

324 E.T. Denisov, Zh. Fiz. Khim., 52 (1978) 1585.

Chapter 4

The Liquid Phase Oxidation of Sulphur, Nitrogen, and Chlorine Compounds

D.L. TRIMM

1. Introduction

Interest in the oxidation of organic compounds containing nitrogen, sulphur, and chlorine has arisen both in its own right and as a result of the fact that these compounds are often used to inhibit the autoxidation of hydrocarbons in solution [1,2]. The present chapter considers both aspects, even though this leads to some small duplication of Chap. 1. Here, however, attention is primarily focused on the reactions and fate of the inhibitor [3—5].

This chapter is focused on the oxidation of organic compounds containing nitrogen, sulphur, and chlorine by molecular oxygen. Oxidation with other reagents is not considered except insofar as it throws light on the reaction with molecular oxygen. In general, oxidation of the three types of compound are considered separately, dealing first with their co-oxidation with other hydrocarbons and then with the oxidation of the pure compounds. Similarities emerge, as, for example, with the role of electron-directing substituents in determining the nature and the rate of oxidation, but the nature of the three types of compound is such that they are best considered with respect to individual systems.

2. Autoxidation of organic compounds containing nitrogen

2.1 CO-OXIDATION WITH HYDROCARBONS

Although it is necessary to consider the importance of individual reactions in particular systems, there is general agreement that the autoxidation of hydrocarbons (RH) in the presence of inhibitors (AH) may be represented by the overall scheme

$$\text{Initiation} \rightarrow \text{R} \cdot \tag{1}$$

$$\text{ROOH} \rightarrow \text{R} \cdot \text{ or } \text{RO}_2 \cdot \tag{1a}$$

$$\text{AH} + \text{O}_2 \rightarrow \text{A} \cdot + \text{HO}_2 \cdot \tag{1b}$$

$$\text{R} \cdot + \text{O}_2 \rightarrow \text{RO}_2 \cdot \tag{2}$$

$$\text{RO}_2 \cdot + \text{RH} \rightarrow \text{ROOH} + \text{R} \cdot \tag{3}$$

$$2 \, RO_2 \cdot \rightarrow products \tag{4}$$

$$RO_2 \cdot + AH \rightarrow ROOH + A \cdot \tag{5}$$

$$A \cdot + RO_2 \cdot \rightarrow RO_2A \tag{6}$$

$$A \cdot + RH \rightarrow AH + R \cdot \tag{7}$$

$$2 \, A \cdot \rightarrow products \tag{8}$$

$$RO_2 \cdot + AH \rightarrow (AH \rightarrow RO_2) \tag{9}$$

$$(AH \rightarrow RO_2) + RO_2 \cdot \rightarrow products \tag{10}$$

$$ROOH + AH \rightarrow R'CO + R''OH + AH \tag{11}$$

The oxidation of hydrocarbons, reactions (1), (1a), (2)—(4), is inhibited to an extent that depends on the efficiency of chain termination, reactions (5), (6), (8)—(10), on the possibility of chain transfer and regeneration, reactions (1b), (5) and (7), and on the possibility of degradation of hydroperoxides to inert products, reaction (11). Amines and phenols are known to be efficient chain breaking inhibitors, while sulphides promote reaction (11).

The relative importance of individual reactions depends on the nature of the system and the inhibitor. This importance is usually assessed by developing rate expressions on the basis of different assumptions, and comparing predictions with experimental results. The most satisfactory method of doing this was applied by Mahoney [3], who used a computer to handle a complex interactive reaction scheme. However, in order to identify and to discuss alternative reactions, the present article adopts the classical approach of considering inhibition reactions individually. Where the situation demands a more complex approach, individual reactions are considered in the light of possible interactions.

This is particularly pertinent in the context of nitrogen-containing inhibitors, which can interfere with the autoxidation of hydrocarbons at various points. Indeed, provided that their redox potential is sufficiently low, they may even initiate oxidation via reaction (1b) [6]. Simple and complex inhibition is discussed below: the influence of different factors on these reactions is discussed at the end of this section.

2.1.1 Inhibition by complex formation

Although inhibition of autoxidation by donation of hydrogen to peroxy radicals, reaction (5), is an important reaction, Boozer and Hammond [7] have suggested that inhibition by complex formation may also be important. Assuming that the major termination step involves reactions (9) and (10), and that the reaction is initiated by azobisisobutyronitrile (AIBN), then the rate of initiation is

$$r_i = 2k_i[AIBN] \tag{I}$$

where k_i includes a term for the efficiency of initiation. Then, at steady state conditions, the rate of initiation equals the rate of termination $(2k_{10}[RO_2]^2[AH])$ and the overall rate becomes

$$r = -\frac{d[O_2]}{dt} - 2k_i[AIBN] = k_3[RH]\left\{\frac{k_i[AIBN]}{k_{10}[AH]}\right\}^{1/2} \tag{II}$$

Under these circumstances, a plot of $r/[AIBN]^{1/2}$ versus $1/[AH]^{1/2}$ should be linear for initial oxidation rates, as has been observed experimentally [8—10].

Difficulties arise, however, in that a similar rate expression can also be obtained if reactions (9) and (10) are not used. If an overall scheme is used in which these reactions are replaced by chain transfer, reaction (7), by biradical termination, reaction (6), and by termination by hydrogen abstraction, reaction (5), then a similar relationship can be developed. Thus, the assumption that chain transfer is more efficient than reaction (6), leads [9] to the equation

$$r = (k_3[RH] + k_5[AH])\left(\frac{2k_7k_i[AIBN]}{2k_5k_6}\right)^{1/2}\left(\frac{[RH]}{[AH]}\right)^{1/2} \tag{III}$$

Experimental evidence has been obtained to support this mechanism. Thus, for example, Mahoney and Ferris [10] and Lloyd and Lange [11] report a three halves dependency of rate upon substrate concentration in the presence of inhibitor. In addition, the above equation can be rearranged to show that there should be a minimum oxidation rate at $[AH] = (k_3/k_5)[RH]$, and an inflection point at $[AH] = 3(k_3/k_5)[RH]$ at constant $[RH]$: both have been observed experimentally [9].

However, Boozer and Hammond [7] have obtained additional support for the postulated chain termination via reaction (9) as opposed to reaction (5) on the basis of isotopic labelling experiments and from the inhibiting effect of amines not possessing a labile N—H function in their structure [8]. The use of deuterium-labelled methyl aniline and diphenylamine as inhibitors in the oxidation of tetralin and cumene did not show the isotope effect which would be expected if reaction (5) was important. Similarly, both N-dimethylaniline and N,N'-tetramethyl-p-phenylenediamine have measurable inhibitory activity despite the fact that neither has a labile hydrogen [8]. However, it has been argued [12] that neither a kinetic isotope effect nor a labile hydrogen is necessary if inhibition results from an electron transfer reaction of the type

$$RO_2\cdot + AH \rightarrow RO_2^- + AH^+ \tag{9a}$$

Hausser [13] reports that such a reaction occurs between diphenylpicrylhydrazyl free radicals and N,N'-tetramethyl-p-phenylenediamine with a heat of reaction in toluene at $-40°C$ equal to -20 ± 5 kcal mole^{-1}, but this is a system which would be expected to favour electron transfer. The position with kinetic isotope effects is equally complex, both positive and negative effects being observed [14,15], sometimes in the same system [16]!

Attempts to clarify the picture have been made by isolating intermediates of the type (AH → RO₂) but here, too, the possibility of biradical termination, reaction (6), leading to similar intermediates throws doubt on the evidence. Polymeric peroxides have been identified [17] but Mahoney et al. [18,19], in some excellent studies of the thermochemistry of peroxy and phenoxy radicals in solution, have no hesitation in ascribing products of this type to the reaction of two radicals.

Although the bulk of the evidence in favour of termination by reaction (9) and (10) can be seen to be questionable, it does seem that these reactions can be of importance in certain special cases. For phenolic inhibitors, there seems little doubt that reaction (5) alone is important: for amines, and particularly for those amines that do not possess a labile hydrogen atom, termination via complex formation may not be negligible.

2.1.2 Inhibition by hydrogen abstraction

The evidence in support of the importance of reaction (5) is fairly conclusive. In addition to the kinetic isotope effects discussed above, experimental observations are in good agreement with predictions obtained on the basis of steady state treatments of the overall reaction scheme above. The evidence for the reaction has been summarised in detail elsewhere [1,2,5] and since it is not directly pertinent to the subject of this chapter, the interested reader is referred to these reviews.

Of more importance here is the fate of the inhibitor and the possibility that reaction (5) can be accompanied by chain transfer, reactions (5), (7) etc. Autoxidation of the additive itself will be dealt with later: evidence for the occurrence of reactions (6)—(8) and the nature of the products formed thereby are of more interest in the present context.

The behaviour of amine-inhibited autoxidation systems has been clarified as a result of the observation of Thomas et al. [14,20,21] that amines can react with peroxy radicals to produce nitroxides, which themselves can influence the course of subsequent reactions. The importance of this observation can be illustrated by reference to studies of the inhibited autoxidation of hydrocarbons carried out by Ingold and his coworkers [22—27].

In comparing the effectiveness of N-alkylanilines and N-arylanilines as inhibitors for the autoxidation of styrene, Brownlie and Ingold consider a number of cases of increasing complexity [22,23]. Considering first the reaction sequence

$$I_2 \rightarrow 2e \ I\cdot \tag{1}$$

$$R\cdot + O_2 \rightarrow RO_2\cdot \tag{2}$$

$$RO_2\cdot + RH \rightarrow ROORH \tag{3'}$$

$$RO_2 \cdot + AH \to ROOH + A \cdot \tag{5}$$

$$RO_2 \cdot + A \cdot \to \text{products} \tag{6}$$

then the rate of oxidation at steady state can be shown to be

$$r = \frac{xek_1k_3'[I_2][RH]}{k_5[AH]} \tag{IV}$$

where x is 1 or 2 depending on whether the inhibitor can stop two or one chains, respectively. In the initial stages of reaction, conventional amine inhibitors stop only one chain and $x = 2$.

If, following Thomas et al., amino radicals react with peroxides to produce nitroxides

$$RO_2 \cdot + A \cdot \to AO \cdot + RO \cdot \tag{6'}$$

then the rate equation depends on the subsequent reactions

$$A \cdot + ROOH \to AH + RO_2 \cdot \tag{12}$$

$$AO \cdot + ROOH \to AOH + RO_2 \cdot \tag{12'}$$

and can expand to

$$r = \frac{ek_1k_3'[I_2][RH]}{2k_5[AH]} \left\{ 1 + \left(\frac{1 + 4k_5k_{12}[AH][ROOH]}{ek_1k_6[I_2]} \right)^{1/2} \right\} \tag{V}$$

The inclusion of chain initiation by amino radicals and of biradical termination

$$A \cdot + RH \to AH + R \cdot \tag{7}$$

$$2 A \cdot \to \text{inactive products} \tag{8}$$

leads to an even more complex rate equation

$$r = \frac{2ek_1k_3'[I_2][RH]}{k_5[AH]} \left(\frac{1 + k_7[RH]}{2(ek_1k_8[I_2])^{1/2}} \right) + k_7[RH] \left(\frac{ek_1[I_2]}{k_8} \right)^{1/2} \tag{VI}$$

Brownlie and Ingold were able to relate the behaviour of many of the amine inhibitors used to one or other of these rate equations, although some of their experimental data were not sufficiently precise to make an absolute attribution. In general, with most N-arylanilines, eqn. (V) was obeyed, while with N,N'-diphenyl-p-diphenylenediamine, eqn. (IV) appeared to be obeyed, although the accuracy was not very good. Reaction (5) was shown to be of importance in nearly every case by the kinetic isotope effects observed on replacing H by D in the inhibitor.

In addition to the study of the effect of nitroxides on the reaction rate (see below), Ingold and his coworkers [24] have also turned their attention to the possibility of nitroxide formation from different amines in the context of the possibility of complex formation [7,8], reactions (9) and (10), in the systems. ESR studies have been used to show that, for example, diphenylnitroxides are produced from diphenylamine radicals [24] via formation of the amino radical followed by oxygen transfer, reaction (6′). Maximal amounts of nitroxide were produced in oxidations involving tertiary peroxy radicals, largely because of the possibility of disproportionation within the solvent cage between a primary or secondary alkoxy radical and the nitroxide to produce a carbonyl compound and diphenyl hydroxylamine: this reaction cannot occur with a tertiary peroxide.

The investigation was also extended to consider the AIBN-initiated decomposition of t-butylhydroperoxide in the presence of various amine inhibitors [25—29]. Primary and secondary aliphatic amines, and tertiary amines which are not usually used as antioxidants, were found not to affect the ESR signal corresponding to the peroxy radical, while diaryl and alkaryl amines gave signals associated with the nitroxide radicals. Primary aromatic amines did not give simple aryl nitroxides. The situation was complicated by the observation that the overall effect of the addition of some inhibitors was to increase chain inhibition. Thus, for example, on adding $N,N,N'N'$-tetramethyl-p-phenylenediamine, a butoxy radical was produced in the solution which initiated a fresh chain. However, by analogy with the autoxidation of anilines [30], they were able to resolve all their results in terms of inhibitor action involving the intermediate production of nitroso compounds, viz.

$$RO_2 \cdot + ArNH_2 \rightarrow RO_2H + ArNH \cdot \tag{5}$$

$$Ar\overset{\bullet}{N}H + RO_2 \cdot \rightarrow \{ArNHOOR\} \rightarrow ArNO + ROH \tag{13}$$

The similarity to the ideas of Boozer and Hammond [7,8] is amusing in many respects.

Brownlie and Ingold [23] have also examined the kinetic implications of the production of nitroxides or of hydroxylamines in the system, using an aliphatic nitroxide, 2,2,6,6-tetramethyl-4-piperidone nitroxide (A), two stable aromatic nitroxides, 4,4′-dimethoxydiphenyl (B), and 4,4′-dinitrodiphenyl (C), and the hydroxylamines produced therefrom. Using the sequence

$$I_2 \rightarrow 2e \; I \cdot \tag{1}$$

$$R \cdot + O_2 \rightarrow RO_2 \cdot \tag{2}$$

$$RO_2 \cdot + RH \rightarrow RO\overset{\bullet}{O}RH \tag{3'}$$

$$RO_2 \cdot + AH \rightarrow \text{inactive products} \tag{5}$$

$$R \cdot + AH \rightarrow \text{inactive products} \tag{14}$$

the rate of reaction was shown to be

$$r = r_i \frac{k_2[O_2](k_3'[RH] + k_5[AH])}{[AH](k_3'k_{14}[RH] + k_5k_{14}[AH] + k_2k_5[O_2])} \tag{VII}$$

where r_i is the rate of initiation. This reduces to

$$r = r_i \frac{k_3'[\text{RH}]}{k_5[\text{AH}]} + r_i \qquad\qquad\text{(VIII)}$$

if $k_{14} \to 0$, and to

$$r = r_i \frac{k_2[\text{O}_2]}{k_{14}[\text{AH}]} \qquad\qquad\text{(IX)}$$

if $k_5 \to 0$.

Aromatic nitroxides were found to attack both alkyl and alkylperoxy radicals, but aliphatic nitroxides attacked only alkyl radicals, the rate obeying equation (IX) above. The kinetics of reaction involving nitroxide B were close to predictions made on the basis of equation (VII) and it is interesting to note that the rate of reaction was affected by solvent, possibly via the formation of a loose solvent—nitroxide complex. Hydroxylamines A and B reacted only with peroxy radicals, but hydroxylamine C was involved in a complex reaction sequence producing first strong inhibition and then equally strong catalysis of autoxidation!

As a result of these studies, Ingold has summarised the overall reaction scheme needed to explain the experimental results as

$$\text{I}_2 \to 2e\ \text{I} \qquad\qquad\text{(1)}$$

$$\text{R}\cdot + \text{O}_2 \to \text{RO}_2\cdot \qquad\qquad\text{(2)}$$

$$\text{RO}_2\cdot + \text{RH} \to \text{ROOH} + \text{R}\cdot \qquad\qquad\text{(3)}$$

$$\text{RO}\cdot + \text{RH} \to \text{ROH} + \text{R}\cdot \qquad\qquad\text{(3')}$$

$$\text{RO}_2\cdot + \text{Ph}_2\text{NH} \rightleftharpoons \text{ROOH} + \text{Ph}_2\text{N}\cdot \qquad\qquad\text{(5)}$$

$$\text{RO}_2\cdot + \text{Ph}_2\text{N}\cdot \to \text{RO}\cdot + \text{Ph}_2\text{NO}\cdot \qquad\qquad\text{(6')}$$

$$\text{RO}_2\cdot + \text{Ph}_2\text{N}\cdot \to \text{products} \qquad\qquad\text{(6)}$$

$$2\ \text{RO}_2\cdot \to \text{products} \qquad\qquad\text{(4)}$$

$$\text{RO}_2\cdot + \text{Ph}_2\text{NO}\cdot \to \text{products} \qquad\qquad\text{(6'')}$$

$$\text{R}\cdot + \text{Ph}_2\text{NO}\cdot \to \text{products} \qquad\qquad\text{(13')}$$

$$\text{PhN}\cdot \to ? \qquad\qquad\text{(8')}$$

$$\text{ROOH} + \text{Ph}_2\text{NO}\cdot \rightleftharpoons \text{RO}_2\cdot + \text{Ph}_2\text{NOH} \qquad\qquad\text{(12')}$$

Reaction (5) becomes of major importance in the presence of appreciable quantities of amine, the radical being removed from the system via reaction (8'). With small amounts of amine, only small amounts of nitroxide were identified, reactions such as (6'') and (12') becoming much more important. Reaction (—5) was thought to be of importance in given systems, but the difficulty of differentiating (5) from (12′) precluded an absolute measure of its importance.

The whole problem of inhibitor regeneration and of negative catalysis has also been recently examined by Denisov [31], using quinones and α-naphthylamine. The quinone can inhibit the oxidation of cyclohexanol by abstraction of H, viz.

$$>C(OH)OO\cdot + Q \rightarrow \cdot QH + >C=O + O_2 \tag{15}$$

or, with α-naphthylamine

$$
\begin{array}{c}
\text{OO} \\
\backslash\,/ \\
\text{C}_{\delta^- \;\; \delta^+} + \bar{\text{N}}< \rightarrow \text{InH} + >C=O + O_2 \\
/\,\backslash \\
\text{O..H}
\end{array}
\tag{16}
$$

the reaction being favoured by the electron distribution in the molecules (see below). Denisov explains the synergistic effect of using both phenol and amine in terms of replacement of the reactive amino radical (which was found to attack the fuel under some circumstances) by a phenoxy radical

$$A\cdot + HOPh \rightarrow AH + PhO\cdot \tag{17}$$

The situation with alcohols is, perhaps, more complex than would be suggested by this scheme. Emanuel [32] quotes some Russian work in which the effect of alcohols on α-naphthylamine inhibitors is suggested to involve the formation of a hydrogen bond between the two, viz.

$$
\begin{array}{c}
\qquad\qquad\text{R} \\
\qquad\qquad/ \\
A\!-\!H \cdots O \\
\qquad\qquad\backslash \\
\qquad\qquad\text{H}
\end{array}
$$

which hinders the detachment of H from the inhibitor. A kinetic scheme is suggested on the basis of the formation of such compounds, which is found to give quite a good agreement with experiment if the alcohol concentrations are not too high. Quantitative comparisons of the efficiency of α-naphthol, α-naphthylamine, and phenyl-α-naphthylamine with various alcohols is given.

Compounds of this type have been postulated in many cases and they may well be important in alcohol systems. Unfortunately, however, the authors have not considered the possibility of the generation of alkoxy radicals in solution and the values that they quote do indicate that RO· radicals may be interfering with the reaction in some cases.

Factors controlling the reactivity of amine inhibitors are discussed below, but in general, the efficiency of an inhibitor is increased by an increase in the electron density at the reactive centre. Many direct qualitative comparisons of the efficiency of various additives have been made by comparing the induction periods for oxygen adsorption or hydroperoxide build up [33—36] and by investigation of the rate of reaction of amines with peroxides [37—39]. Although the reactivity of individual antioxidants depends on many different factors, the energy of activation of the reaction of anilines with benzoyl peroxide [39] typically lies in the range $E_a = 11—16$ kcal mole^{-1}.

The efficiency of a particular amine must depend not only on the rate of the initial hydrogen abstraction, but also on the nature and subsequent reactions of the radical produced. The free radical produced by H transfer may well be stabilised by resonance and may be insufficiently reactive to start a new oxidation chain [40], particularly when the amino group is surrounded by bulky substituents [9]. If the radical does react, then the subsequent rate and nature of the reaction will depend upon the intermediates and on the relative importance of chain termination and chain transfer reactions. Some formal grouping of the factors affecting the efficiency of a given inhibitor and the kinetics of the inhibited reaction is possible.

2.2 FACTORS AFFECTING INHIBITOR EFFICIENCY

2.2.1 Electron directing and steric effects: solvents

In many respects, the behaviour of amino and phenolic inhibitors can be predicted on the same grounds, because of the fact that the peroxy radical is an electron acceptor and prefers to react with a centre of high electron density [41]. As a result, electron-releasing substituents, such as alkoxy or alkyl groups, where they do not sterically hinder the reaction, will improve antioxidant performance. Bulky *ortho* substituents, which involve steric influences, may retard the rate of reaction [42], although this will obviously depend on individual systems. The effect of phenolic inhibitors has been dealt with elsewhere [1], but it is of interest to consider some factors that throw light on the mechanism and kinetics of amino inhibition.

Attempts to relate the efficiency of inhibition with the Hammett— Taft equation were soon abandoned in favour of more accurate correlations with Brown's electrophilic substituent constants, σ^+, as given by [43]

$$\log_{10} \frac{k_5}{(k_5)_o} = \sigma^+ \rho + a \tag{X}$$

where $(k_5)_o$ was obtained by extrapolation to zero inhibitor concentration and ρ makes allowance for steric effects. Thus, for example, Howard and Ingold [27] obtained an excellent correlation between the relative rates of *meta*- and *para*-substituted phenols (where the steric hindrance was low) and their σ^+ values. In some cases, values of the Hammett σ factor and

Brown's σ^+ factor are very similar [2], and it was possible for Boozer and his coworkers [8] to relate the action of phenols and amines in these simpler terms.

To reinforce the point that a similar electron directing effect operates, results published by Emanuel [32] for the autoxidation of a mixture of hydrocarbons in the presence of substituted N,N'-dialkyl-p-phenylene-diamines show that substituents raising the electron density at the inhibitor reaction centre increase the efficiency of the antioxidant, while substituents that decrease electron density decrease inhibitory powers. No attempt was made to quantify these observations in terms of σ or σ^+ factors.

The importance of electron-directing effects in amines has been studied recently by Brownlie and Ingold [22]. The experimental observations discussed above were largely resolved in terms of a transition step consisting of the resonance forms

$$XC_6H_4\underset{R}{N} : HOOR' \rightleftharpoons XC_6H_4\overset{+}{\underset{R}{N}}\overline{H}OOR' \rightleftharpoons XC_6H_4\underset{R}{N}H:OOR'$$

(I)   (II)   (III)

which were very similar to those suggested for substituted phenols [27]. The reaction rate coefficients measured for the autoxidation of styrene in the presence of diphenylamines and N-methylanilines was found to be in reasonable agreement with that predicted on the basis of both positive and negative σ^+ constants (in the absence of *ortho* groups) implying that polar effects were playing a significant role in reactions involving the transition state. Hammett ρ factors, obtained from the slopes of the $\log k_5$ versus σ^+ plots were found to be -0.89 (diphenylamines) and -1.6 (N-methylanilines), compared with -1.1 (di-t-butylphenols), -1.36 (2,6-dimethylphenols), and -1.49 (phenols with no *ortho* substituent effects).

There seems little doubt that, in the absence of steric effects, relationships of this type offer a good estimate of the kinetic parameter k_5 for a given member of a series. The particular effect of a given substituent has been examined in detail elsewhere [44,45] and need not be discussed here save to stress the point that inhibition is favoured by electron-supplying substituents on the amine.

Brownlie and Ingold [22,46] reported that, for both amines and phenols, the value of ρ decreased as the steric protection given by substituents to the reaction centre increased. Although the effect of electron-directing substituents could be important in determining the relative importance of complexes (I) and (III), above, the major effect determining the importance of (III), at least, was the steric hindrance for the approach of the peroxy radical to the H atom, particularly as a result of *ortho* substitution. In fact, structure (III) was deemed to be of little importance in the absence of steric effects [46].

Ortho substitution has not been investigated in detail for amine inhibitors, but Ingold and his coworkers [27,47] report that steric effects need not be overwhelmingly important unless two bulky *ortho* groups are substituted into a phenolic molecule. For amines, it may well be possible that initial donation of a hydrogen atom is possible, but that bulky substituent groups may protect the free radical produced from further reaction [9]. The fact that electron-directing influences can be less important under suitable circumstances is confirmed by studies of the oxidation of acrylonitrile, methyl vinyl ketone, and methacrylonitrile, initiated by persulphate [48], in which the relative rates of oxidation were found to be in closer agreement with predictions based on Q factors (which are responsive to resonance stabilisation) rather than those made on the basis of polarity effects in the molecule.

It is also possible to relate the effect of "inert" solvents with steric effects. Neglecting the possibility of solvent interaction of the type [49]

$$ROOH + SH \rightarrow RO_2 \cdot + S \cdot + H_2O \tag{11'}$$

there still appears to be definite effects of solvent on the rate of reaction. This has been elegantly explained in several papers by Ingold and coworkers in terms of complex formation in solution. Considering the autoxidation of styrene in the presence of various solvents [25], the effect was correlated with the dielectric constants of the solvents. Similar effects were observed during studies of the reaction of *t*-butoxy radicals with phenols in carbon tetrachloride and chlorobenzene [46]. Differences in reactivity between the two solvents was ascribed to the possibility of complex formation between alkoxy radicals and the aromatic solvent. A decrease in the Hammett ρ factor on going from carbon tetrachloride to chlorobenzene was ascribed to the increasing importance of steric effects and of transition state complexes of type (III), above. This explanation

$$RCH_2NHR' + R''OOH \rightarrow RCH_2NH(OH)R' + R''O^-$$

$$\downarrow \qquad\qquad\qquad\qquad \downarrow {-H^+}$$

$$\overset{+}{R}CH_2N(OH)(OH)R' + R''O^- \leftarrow RCH_2N(OH)R') \quad (+R''O_2 \cdot)$$

$$\downarrow \qquad\qquad\qquad\qquad \uparrow {R''OOH}$$

$$\cdot OH + RCH_2\overset{+}{N}(OH)R' \xrightarrow{-H^+} RCH_2N(O\cdot)R'$$

$$\qquad {R''OOH} \quad \downarrow {RCH_2NHR'} \tag{18}$$

$$R\overset{\cdot}{C}HNHR' + RCH_2N(OH)R'$$

$$\diagup {R''OOH} \qquad \diagdown$$

$$\cdot H + RCH = NR' \qquad\qquad R\overset{\cdot}{C}HNHR' + R''O$$

$$\qquad\qquad\qquad\qquad\qquad |$$

$$\qquad\qquad\qquad\qquad\quad OH$$

was also advanced for the styrene case [25] and extended to cover the reactions of nitroxides produced in the oxidation of amine inhibitors [23]. The observation that nitroxides can form (II) complexes with aromatic compounds [50] was quoted in support of this argument.

To summarise, then, it is possible to predict kinetic characteristics from Hammett-type relationships, provided that electron directing and steric characteristics are similar in a series of substituted amines. The mechanism proposed by De La Mere [29] for the oxidation of, for example, a secondary amine, see reaction (18), underlines the complexity of the kinetics that can be expected, unless one particular transition state and one reaction path is the major reaction route.

2.2.2 Oxidation—reduction potentials

Since the efficiency of any inhibitor is known to be enhanced by an increase in the electron density at the reactive centre, it should be possible to quantify this effect via the redox potentials of inhibitors. Thus the efficiency of the antioxidant should be increased by a decrease in the redox potential or by a decrease in the A—H bond strength. Correlations between inhibitors have been made successfully in terms of bond strength [45] and redox potentials although, if the redox potential is decreased to too low a value, then the inhibitor becomes liable to self-oxidation via reaction (1b) [35,40]. Much less work has been completed on the amine inhibitors than on phenols [2,5], but inhibition correlates well with redox potentials where sufficient data are available.

2.3 AUTOXIDATION OF NITROGEN-CONTAINING COMPOUNDS

The oxidation of nitrogen-containing molecules with molecular oxygen has been the subject of an authoritative review by Höft and Schulze [51]. The authors concentrate primarily on the products of reaction and the reaction mechanisms, probably as a result of the paucity of kinetic data for the systems. Reaction is suggested to occur generally via the formation of peroxy compounds, although such intermediates have been isolated in only a few cases. Primary attack occurs at the free electron pair of the nitrogen, leading to the formation of a peroxy compound at the α-carbon atom. These may be, or may form, radical intermediates but such radicals will be polarised by the neighbouring nitrogen atom and can often react differently to the corresponding purely carbon-containing radicals. The review discusses reaction mechanisms in depth, particularly in the light of the importance of such oxidations in biological systems.

Investigation of the kinetics of the reactors is sparse, with the possible exception of the production of nitroxides from amines discussed above. What results are available show that the autoxidation of the nitrogen-containing molecule is similar to that of hydrocarbons, bearing in mind the

TABLE 1

Rate coefficients of chain propagation and termination reactions measured by the rotating sector technique

Oxidised substrate	Temp. (°C)	$\log(k_p/\text{l mole}^{-1}\text{ s}^{-1})$ [a]	$\log(2k_t/\text{l mole}^{-1}\text{ s}^{-1})$	Ref.
Cyclo-$C_6H_{11}NH_2$	50— 90	$9.04—13,300/\theta$	$9.25—3000/\theta$	53,54
n-Bu_2NH	50— 75	$9.80—13,500/\theta$	$9.24—2600/\theta$	53,54
$C_6H_5CH_2NH_2$	50— 80	$9.83—12,200/\theta$	$9.90—2200/\theta$	55
$(CH_3)_2NCOCH_3$	50— 90	$8.11—11,000/\theta$	$8.146(35—55°)$	56
n-BuNHCOCH$_3$	50—125	$7.98—12,000/\theta$	$7.362(70—80°)$	56
iso-PrNHCOCH$_3$	60—120	$8.04—13,000/\theta$	$6.94(80°)$	56

[a] $\theta = 4.57T$ kcal mole^{-1}.

electron-directing influence of the nitrogen atom. However, there is some evidence that more attention is being focused on these systems, particularly from some elegant work of Denisov. Thus, for example, considering the reaction sequence for amines

$$\text{Initiation} \rightarrow \text{R}\cdot \tag{1}$$

$$\text{R}\cdot + \text{O}_2 \rightarrow \text{RO}_2\cdot \tag{2}$$

$$\text{RO}_2\cdot + \text{RH} \overset{k_p}{\rightarrow} \text{ROOH} + \text{R}\cdot \tag{3}$$

$$\text{RO}_2\cdot \overset{k_t}{\rightarrow} \text{X}\cdot \xrightarrow[\text{RO}_2\cdot]{\text{fast}} \text{products} \tag{4'}$$

where reaction (4') is known to be important for tertiary amines [52], rate coefficients have been measured by the rotating sector method for reactions (3) and (4') (Table 1).

The peroxy radicals involved in this reaction sequence are interesting in that they possess both oxidative and reductive capabilities. As a result, molecules such as aromatic amines can undergo a cyclic reaction such as [52,55,57,58]

$$\text{RCH(OO}\cdot\text{)NHR}' + \text{InH} \rightarrow \text{RCH(OOH)NHR}' + \text{In}\cdot \tag{5'}$$

$$\text{RCH(OO}\cdot\text{)NHR}' + \text{In}\cdot \rightarrow \text{InH} + \text{RCH}=\text{NR}' + \text{O}_2 \tag{7'}$$

In contrast, the peroxy radicals of amides show no reducing activity and the stoichiometric coefficient of inhibition for phenols and aromatic amines is close to 2 [56,59,60].

Again, similar to hydrocarbons, the autoxidation of amines is affected by the presence of metal ions. This appears to have been first reported by Bacon [61], who used $Ag^+/S_2O_8^{2-}$ to oxidise primary and secondary amines. Subject to the availability of an α-hydrogen atom, good yields of ketones could be obtained

$$RR'CHNH_2 \rightarrow RR'C{:}NH \quad + \quad RR'CHNH_2$$

$$RR'C{:}NCHRR' + NH_3 \tag{19}$$

$$RR'CO + RR'CHNH_2 \cdot$$

A similar reaction was reported by Meth-Cohn and Suschitzky [62], who obtained aldehydes from the aerobic oxidation of amines in neutral solution in the presence of a manganese dioxide catalyst.

Study of a similar reaction, the oxidation of α-naphthylamine in the presence of copper stearate [63], reveals that the reaction proceeds via oxygen insertion into a copper—amine complex. It was suggested that the electron involved in the bonding of the nitrogen to the copper was transferred to the molecular oxygen, subsequently followed by migration of a proton from the amine to form an amine—copper hydroperoxide intermediate, which breaks down to form products. The activation energy and entropy of the overall oxidation was found to be 14.7 kcal mole^{-1} and -35.3 eu, respectively. The formation of this type of complex was confirmed by ESR examination of the intermediates formed during the ferricyanide ion-catalysed oxidation of alkaline hydroxylamine [64] and during the metal-catalysed autoxidation of substituted hydrazines [65]. The kinetic parameters observed are at least consistent with the proposed oxidation route, and the observation that a copper stearate—N-substituted aniline system is an extremely efficient inhibitor for the autoxidation of decane [66] would indicate that any competition favours oxidation of the nitrogen-containing compound.

As with hydrocarbons, the presence of a metal salt capable of undergoing oxidation—reduction can also inhibit oxidation by the sequence [55,58]

$$RO_2 \cdot + M^{n+} \rightarrow RO_2^- + M^{(n+1)+}$$

$$M^{(n+1)+} + RCH(OO \cdot)NHR' \rightarrow M^{n+} + O_2 + H^+ + RCH{=}NR'$$

Rate coefficients for termination, measured by the Russian school, are summarised in Table 2.

The kind of complexity that can arise in these systems is well illustrated by the study of the oxidation of N-alkylamides [68,69]. Double reactions occurring in the presence of metal ion catalysts are laid out in Scheme 1, the relative importance of each reaction depending on reaction conditions and on electron-directing and steric effects in the molecule. Overall kinetics are given for different reactions [68,69], but the complexity is such that detailed analysis is difficult or impossible.

In general, the factors that influence the efficiency of the compounds as inhibitors also affect their oxidisability. Thus, for example, studies of

TABLE 2

Rate coefficients (k_t) of reactions of radicals with inhibitors and transition metals salts, determined from the rates of inhibited oxidation

Oxidised substrate	Inhibitor	Temp. (°C)	k_t (l mole^{-1} s^{-1})	Ref.
Cyclo-$C_6H_{11}NH_2$	1-Naphthylamine	75	7.2×10^2	53,54
	4,4'-Dimethoxydiphenylamine	75	2.0×10^4	53,54
	N-Phenyl-2-naphthylamine	75	1.3×10^4	53,54
	2,4,6-Tri-t-butylphenol	75	8.2×10^3	53,54
	1-Naphthol	75	1.0×10^3	53,54
	2,2,6,6-Tetramethyl-4-oxy-piperidine-1-oxyl	75	2.7×10^4	54,57
	Dianisilnitroxyl	75	8.1×10^4	54,57
	$Co(C_{17}H_{35}COO)_2$	75	3.0×10^4	54,58
	$CoCl_2 \cdot 6 H_2O$	75	6.1×10^4	54,58
	$Co(cyclo-C_6H_{11}COO)_2$	75	8.7×10^4	54,58
	$Co(CH_3COO)_2$	75	1.2×10^5	54,58
	$Ce(C_{17}H_{35}COO)_3$	75	8.6×10^5	54,58
	$Fe(C_{17}H_{35}COO)_3$	75	1.2×10^6	54,58
	$Cu(C_{17}H_{35}COO)_3$	75	9.4×10^7	54,58
	$Mn(C_{17}H_{35}COO)_2$	75	1.6×10^8	54,58
	$Cu(acetylacetone)_2$	75	7.2×10^5	67
	Cu-porfirine	75	7.3×10^5	67
n-Bu_2NH	1-Naphthol	75	1.5×10^3	53,54
	4,4'-Dimethoxydiphenylamine	75	4.2×10^4	53,54
	2,4,6-Tri-t-butylphenol	75	3.0×10^4	53,54
	1-Naphthol	75	1.8×10^3	53,54
	$Cu(C_{17}H_{35}COO)_2$	75	4.1×10^6	54,58
	$Mn(C_{17}H_{35}COO)_2$	75	3.5×10^7	54,58
	$Cu(acetylacetonate)_2$	75	4.8×10^5	67
	Cu-porfirine	75	9.3×10^4	67
	2,2,6,6-Tetramethyl-4-oxy-piperidine-1-oxyl	75	1.5×10^5	57,54
$C_6H_5CH_2NH_2$	1-Naphthylamine	65	1.6×10^3	55
	4,4'-Dimethoxydiphenylamine	65	8.2×10^3	55
	N-Phenyl-2-naphthylamine	65	2.6×10^4	55
	2,4,6-Tri-t-butylphenol	65	1.3×10^4	55
	1-Naphthol	65	3.3×10^4	55
	2,2,6,6-Tetramethyl-4-oxy-piperidine-1-oxyl	65	1.1×10^5	55
	Dianisilnitroxyl	65	3.4×10^5	55
	$Cu(CH_3COO)_2$	65	1.5×10^8	55
	$Mn(CHCOO)_3$	65	2.3×10^8	55
	$Cu(acetylacetone)_2$	65	1.3×10^7	67
	Cu-porfirine	65	1.5×10^7	67
$(CH_3)_2NCOCH_3$	1-Naphthol	70	7.8×10^2	56,60
	2,4,6-Tri-t-butylphenol	70	1.6×10^3	56,60
	Hydroquinone	70	3.5×10^3	56,60

TABLE 2 (continued)

Oxidised substrate	Inhibitor	Temp. (°C)	k_t (l mole^{-1} s^{-1})	Ref.
	1-Naphthylamine	70	9.8×10^3	56,60
	N-Phenyl-2-naphthylamine	70	7.6×10^3	56,60
	4,4'-Dimethoxydiphenylamine	70	1.0×10^5	56,60
n-BuNHCOCH$_3$	1-Naphthol	90	1.2×10^4	56,60
	Hydroquinone	90	5.0×10^4	56,60
iso-PrNHCOCH$_3$	1-Naphthol	90	2.0×10^4	56,60
	2,4,6-Tri-t-butylphenol	90	2.4×10^4	56,60

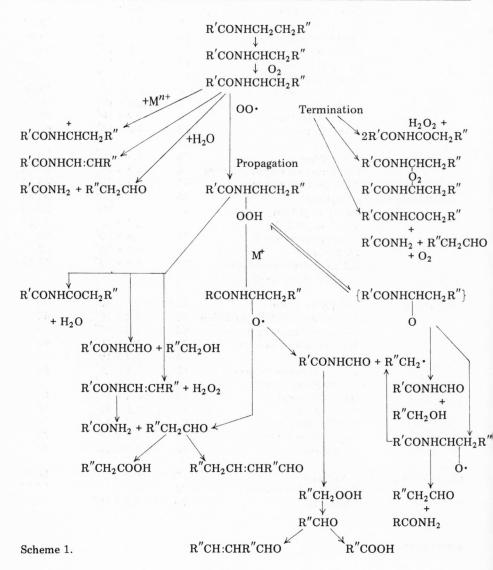

Scheme 1.

the oxidation of p-nitrotoluene show that substituents that increase the rate of ionisation also increase the rate of oxidation [70]. Substituents such as α-cyano or 2,6-dinitro stabilise the p-nitrobenzyl anions to such an extent that oxidation is slow. Alternatively, stabilisation can lead to differing products depending on the solvent and on the possibility of solvent—nitrotoluene complex formation [71].

Similar observations have been reported for the oxidation of aliphatic hydroxylamines, where di-ethyl, di-n-propyl, and di-n-butyl hydroxylamine have been found to react with relative rates equal to 1 : 5.3 : 3.4 [72]. The reaction may be represented by

$$(RCH_2)_2NO\cdot + (RCH_2)_2NOH \rightarrow (RCH_2)_2NOH + RCH_2N(OH)\overset{\cdot}{C}HR$$
$$\downarrow O_2$$

$$RCH_2N(OH)CH(OOH)R \xleftarrow{(RCH_2)_2NOH} RCH_2N(OH)\overset{\overset{\displaystyle OO\cdot}{|}}{C}HR$$
$$\downarrow$$
$$H_2O_2 + (RCH_2N(OH)\underset{\underset{\displaystyle R}{|}}{C}HO{-})_2$$

With radicals formed from hydroxylamine showing high stability as a result of resonance between the forms [73]

$$RCH_2\underset{\underset{\displaystyle R}{|}}{\overset{\overset{\displaystyle O^-}{|}}{N}}_{\underset{\displaystyle CH\cdot}{}} \rightleftharpoons RCH_2\underset{\underset{\displaystyle R}{|}}{\overset{\overset{\displaystyle O^-}{|}}{N^+}}_{\underset{\displaystyle HC^-}{}} \rightleftharpoons RCH_2\underset{\underset{\displaystyle R}{|}}{\overset{\overset{\displaystyle O\cdot}{|}}{N}}_{\underset{\displaystyle HC^-}{}}$$

The relative rates were suggested to reflect the differing importance of differing isomers depending on the molecular structure of the substrate. However, there is some question as to whether the authors avoided complications due to metal ion impurities and Hughes et al. [74] were careful to investigate the autoxidation of hydroxylamine in the presence of EDTA. The kinetics of reaction were found to be in good agreement with the sequence

$$NH_2OH + OH^- \rightleftharpoons NH_2O^- + H_2O \tag{22}$$

$$NH_2O^- + O_2^- \rightarrow products \tag{23}$$

but, unfortunately, relative rates of oxidation of organic hydroxylamines were not established under similar conditions. However, Cowley and Waters [75] established that both the resonance-stabilised ion and short-lived free radicals could be important in the oxidation of N,N-dibutylhydroxylamine and it seems well-established that rates of oxidation do depend on the degree of ionisation and the possibility of resonance stabilisation in oxidation intermediates.

The evidence that electron-directing groups also affect the rate and nature of autoxidation of nitrogen-containing compounds has been well reviewed by Höft and Schultze [51], particularly for the oxidation of amyl phenylhydrazines where it has been possible to relate rates of oxidation with predictions based on the Taft—Hammett relationship [76]. Similar relationships have been established with phenylhydrazones, which have been found to oxidise to produce a hydroperoxide [77]

$$PhCH=N-NH-Ph \rightarrow HOOPhCHN=NPh \qquad (24)$$

$$\text{A} \qquad\qquad \text{B} \qquad \text{A} \qquad \text{B}$$

The substitution of any group in ring A resulted in the compound oxidising more slowly than the parent compound, with *meta* groups having more effect than *para* [78]. Electron-donating groups substituted into ring B accelerated the reaction, while electron-accepting groups retarded oxidation. The results were stated to be related to predictions based on the Hammett relationship, but no quantitative assessment was attempted. Similar semi-quantitative assessments of electron-directing effects have also been completed for dialkylanilines [79] and β-naphthylamine [80,81].

Although kinetic measurements have not been reported to any extent and the situation can be complicated by oxidative attack on the organic side chain, the oxidation of nitrogen-containing molecules does seem to reflect the results of the comparable oxidation of hydrocarbons, bearing in mind the electron-directing capabilities of the nitrogen atom.

3. Autoxidation of organic compounds containing sulphur

3.1 CO-OXIDATION WITH HYDROCARBONS

Although the oxidation of sulphur-containing substrates is of interest in its own right, the role of such compounds as possible inhibitors for the autoxidation of hydrocarbons has also generated considerable scientific attention. This has been focused mainly on the reactions of sulphides and of products derived therefrom, which interfere with the hydrocarbon oxidation chain mainly at the hydroperoxide, viz.

$$R_2S + R'OOH \rightarrow R_2SO + R'OH \qquad (25)$$

The product sulphoxide can also react with hydroperoxide to produce disulphide which, in addition, is itself an inhibitor. Clearly, the potential of the system as an antioxidant is high and considerable effort has been expended on establishing the reaction mechanisms involved.

Reaction of sulphide with hydroperoxide has been suggested to occur in two ways, represented by [82,83]

$$
\begin{array}{ccc}
\text{R--O--O} \quad \text{SR'R''} & \text{R--O} \quad \text{OSR'R''} & \\
\quad | \quad & \quad | & \\
\quad \text{H} \quad \rightarrow & \quad \text{H} & (26) \\
\quad \text{H} & \quad \text{H} & \\
\quad \backslash & \quad / & \\
\quad \text{O} & \quad \text{O} & \\
\quad | & \quad | & \\
\quad \text{O--R} & \quad \text{O--R} &
\end{array}
$$

and

$$
\begin{array}{ccc}
\text{R--O--O} \quad \text{SR'R''} & \text{R--O} \quad \text{OSR'R''} & \\
\quad \text{H H} \quad \rightarrow & \quad \text{H} \quad \text{H} & (27) \\
\quad \backslash & \quad / & \\
\quad \text{O} & \quad \text{O} & \\
\quad | & \quad | & \\
\quad \text{R'''} & \quad \text{R'''} &
\end{array}
$$

The reaction is remarkably clean with secondary hydroperoxides, being second order with respect to hydroperoxide and first order in sulphide. However, several side reactions can occur with, for example, t-butylhydroperoxide [84], largely as a result of degradation reactions producing peroxy and alkoxy radicals.

$$
\text{RO}_2\cdot + \text{R'R''S} \rightarrow \text{RO}\cdot + \text{R'R''SO} \tag{28}
$$

$$
\text{RO}\cdot + \text{ROOH} \rightarrow \text{ROH} + \text{RO}_2\cdot \tag{29}
$$

Such radicals may be trapped by the use of unsaturated sulphides or unsaturated solvents to give a return to the more simple kinetic behaviour.

Both mono- and disulphides owe a large part of their inhibitory action to products formed by their oxidation, i.e. sulphoxides and thiosulphinates [84,85]. Thus, for example, Hargrave [84] has reported that the oxidation of organic sulphides involves two distinct steps, the formation of the sulphoxide and the subsequent reactions of the sulphoxide with hydroperoxide to produce disulphide, water, and an unidentified peroxide. In addition, the disulphide itself can act as an inhibitor for the hydrocarbon oxidation [84,85]. In fact, as shown by Barnard et al. [85] in a more detailed investigation of inhibition by mono- and disulphides, the antioxidant efficiency of the sulphides depends directly on their ability to form sulphoxides or thiosulphinates. Sulphides oxidising to sulphones, thiosulphonates, and disulphones were found to be inefficient inhibitors, but sulphides oxidising to sulphoxides gave an overall antioxidant activity of the order of that observed for conventional antioxidants such as phenyl-β-naphthylamine. The activity of thiosulphinates was effectively independent of organic substituents, but alkyl-substituted sulphoxides were more effective than those substituted with aryl groups.

Although sulphoxides are more efficient inhibitors than sulphides, it is preferable to use the latter compounds as inhibitors [85]. Sulphoxides are not thermally stable and, indeed, there is evidence that their antioxidant activity is related to their ease of degradation [85]. In this case, it is preferable to use sulphides, which react with hydroperoxide to produce sulphoxides, as a "reservoir" for the more active inhibitor.

Sulphoxides are interesting inhibitors in that their mode of action seems to be related only to the reaction with hydroperoxides [86]. Evidence obtained from the study of the inhibition of peroxide decomposition in mineral oil [86] shows that the reaction is probably free radical at low temperatures, but may change to an ionic process at high temperatures, viz.

$$RR'R''COOH + A \rightarrow RR'C{=}O + R''OH + A \qquad (30)$$

where A is a Lewis acid present in solution and possibly originating from the acidic end products of the oxidation of sulphoxide.

However, there is some evidence that the acidity or alkalinity of the solution has a profound effect even on the low temperature reaction. Thus Ogata and Suyama [91] report that sulphoxides react with organic hydroperoxides in non-aqueous alkaline solution to produce sulphones, which have no antioxidant activity. In acids, as stated above, disulphides are the major sulphur-containing products.

The concept of a more active peroxide decomposer generated from a sulphide has received kinetic support. On this mechanistic concept, it is possible to postulate the scheme

$$RH + O_2 \rightarrow R{\cdot} + HO_2{\cdot} \qquad (1)$$

$$R{\cdot} + O_2 \rightarrow RO_2{\cdot} \qquad (2)$$

$$RO_2{\cdot} + RH \rightarrow ROOH + R{\cdot} \qquad (3)$$

$$2\,RO_2{\cdot} \rightarrow products \qquad (4)$$

$$ROOH + R'SR'' \rightarrow R'SOR'' + ROH \qquad (14)$$

$$ROOH + R'SOR'' \rightarrow products + R'SOR'' \qquad (31)$$

$$ROOH + R'SOR'' \rightarrow inert\ products \qquad (32)$$

$$ROOH \rightarrow RO{\cdot} + OH{\cdot} \qquad (33)$$

Assuming a stationary state and high chain lengths, then

$$k_{14}[ROOH][R'SR''] = k_{31}[ROOH][R'SOR'']$$

and, in the presence of added hydroperoxides

$$k_{32}[ROOH] \gg k_1[RH][O_2]$$

Assuming a steady state concentration of hydroperoxide, we may derive the equation

$$-\frac{d[RSR]}{dt} = k_{14}[ROOH][R'SR'']$$

$$= \frac{k_{32}^2 k_3^2 k_{33}[RH]^2}{k_{31}^2 k_4 k_{14}[R'SR'']} \tag{XI}$$

Integrating between the limits $[R'SR''] = [R'SR'']_0$ at $t = 0$ and $[R'SR''] = 0$ at $t = t_i$ (induction period), gives

$$t_i = [R'SR'']_0^2 \frac{k_{31}^2 k_4 k_{14}}{2k_3^2 k_{33} k_{32}^2 [RH]^2} \tag{XII}$$

Verification of this equation from plots of $\log t_i$ versus $\log[RSR]_0$ has been obtained for several inhibitors of this type [87].

Recognition of the role of sulphoxides in the inhibition chain has resulted in several studies of the autoxidation of mono- and disulphides (see later). In addition to the work by Hargrave [84], Bateman et al. [88–90] have studied the oxidation of monosulphides, unsaturated mono-sulphides and cyclo-hex-2-enyl methyl sulphide. Allylic and vinylic sulphides were found to be much more reactive than saturated sulphides, although the methylene group, rather than the allylic double bond, was suggested to be the primary reaction centre. Reactivities were compared for the percentage yield of sulphoxide (A) and for the yield of hydro-peroxide (B), viz.

Sulphide	A	B
1,3-Dimethylallyl n-butyl	5	3
1,3-Dimethylallyl methyl	5.5	8
Cyclohexenyl methyl	13	44
3-Methylallyl methyl	18	55
n-Butyl cinnamyl	24	93

At least some part of the inhibition by disulphides was suggested to be due to the formation of a complex with peroxy radicals aided by the readily available electrons at the sulphur atoms.

A certain amount of doubt as to the overall effectiveness of sulphoxides vis-à-vis other inhibitors has arisen as a result of one of the most interesting studies of recent years [92]. This has involved the study of the decomposition of cumene hydroperoxide by a series of mono-, di-, tri-, and tetra-sulphides, using high performance liquid chromatography (HPLC) to identify the products.

Cumene hydroperoxide was found to decompose to produce mainly phenol, acetophenone, (α,α-dimethyl)benzyl alcohol and α-methylstyrene and the amounts of each product indicate the importance of different breakdown paths, viz.

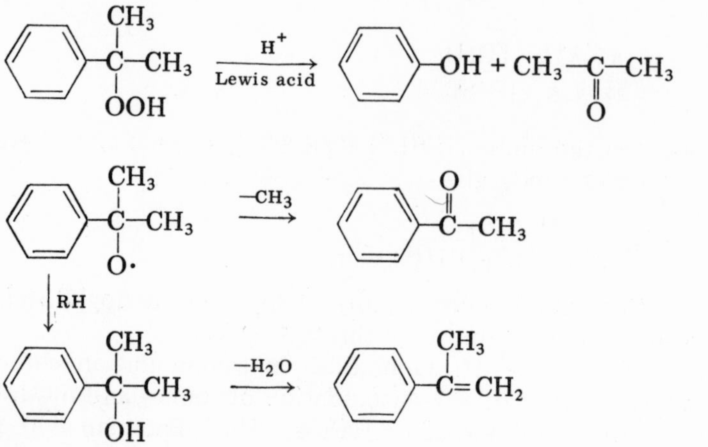

$$\underset{\underset{\text{OOH}}{\overset{\text{CH}_3}{\vert}}}{\underset{\vert}{\text{C}}}-\text{CH}_3 \xrightarrow[\text{Lewis acid}]{\text{H}^+} \text{Ph}-\text{OH} + \text{CH}_3-\overset{\overset{\text{O}}{\parallel}}{\text{C}}-\text{CH}_3 \qquad (35)$$

$$\underset{\underset{\text{O·}}{\vert}}{\overset{\overset{\text{CH}_3}{\vert}}{\text{C}}}-\text{CH}_3 \xrightarrow{-\text{CH}_3} \text{Ph}-\overset{\overset{\text{O}}{\parallel}}{\text{C}}-\text{CH}_3 \qquad (36)$$

$\downarrow$ RH

$$\underset{\underset{\text{OH}}{\vert}}{\overset{\overset{\text{CH}_3}{\vert}}{\text{C}}}-\text{CH}_3 \xrightarrow{-\text{H}_2\text{O}} \text{Ph}-\underset{\underset{\text{C}}{\vert}}{\overset{\overset{\text{CH}_3}{\vert}}{\text{C}}}=\text{CH}_2 \qquad (37)$$

Comparing the efficiency of hydroperoxide decomposing catalysts of the form

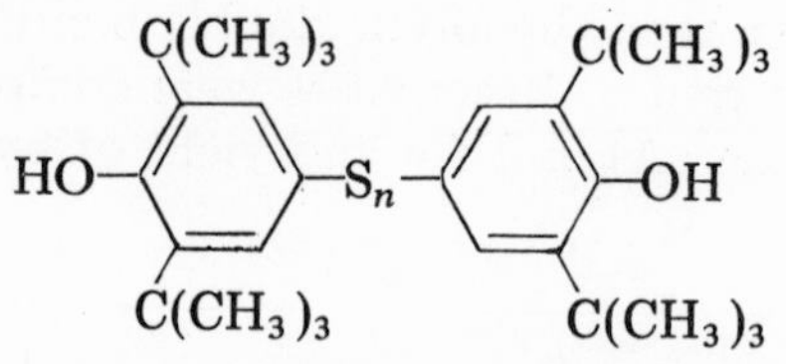

$$\text{HO}-\underset{\underset{\text{C(CH}_3)_3}{}}{\overset{\overset{\text{C(CH}_3)_3}{}}{\bigcirc}}-\text{S}_n-\underset{\underset{\text{C(CH}_3)_3}{}}{\overset{\overset{\text{C(CH}_3)_3}{}}{\bigcirc}}-\text{OH}$$

where $n = 1-4$, two very interesting observations were made. First, when the compounds were used at the same sulphur level, the products of reaction were found to have a similar distribution, as given in Table 3. Secondly, measurement of the kinetics of decomposition of the hydroperoxide showed that, when $n = 2-4$, the rate of decomposition was con-

TABLE 3

Reaction products from the decomposition of hydroperoxide with sulphide at the same sulphur level

Sulphide n	Product		
	Phenol (%)	Alcohol + styrene (%)	Acetophenone (%)
1	48.5	23.5	18.0
2	53.5	26.5	16.8
3	51.0	25.6	19.1
4	52.0	23.7	15.3

trolled only by the concentration of sulphur and was independent of the sulphide used. As a result, the conclusion was drawn that all sulphur atoms are equivalent in terms of their peroxide decomposition ability and that the results indicate the production of a common intermediate, which is the active species for decomposition. By elimination, this was suggested to be sulphur dioxide, which can catalyse the hydroperoxide decomposition, viz.

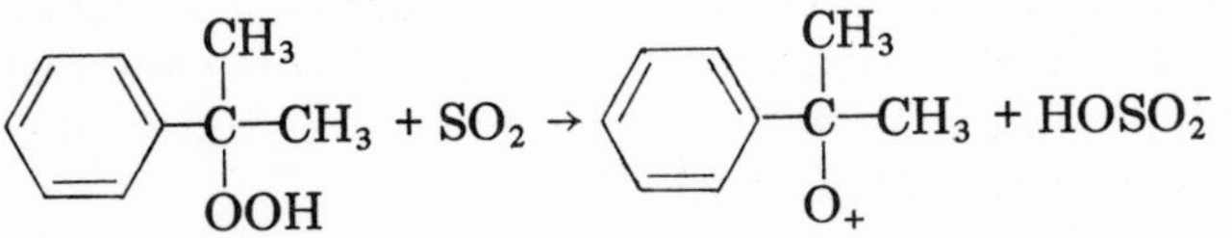

$$\underset{\underset{\text{OOH}}{\overset{\overset{\text{CH}_3}{|}}{\underset{|}{\text{C}}}}{\bigcirc}\text{—CH}_3 + \text{SO}_2 \rightarrow \underset{\underset{\text{O}_+}{\overset{\overset{\text{CH}_3}{|}}{\underset{|}{\text{C}}}}{\bigcirc}\text{—CH}_3 + \text{HOSO}_2^- \tag{38}$$

Experimental evidence confirmed this, showing that a molecule of sulphur dioxide can decompose approximately 2×10^5 molecules of hydroperoxide with a product distribution similar to that obtained from the sulphides.

Attention was then focused on the production of sulphur dioxide from the sulphides and inspection of the literature producing the following reaction scheme for mono- and disulphides, viz.

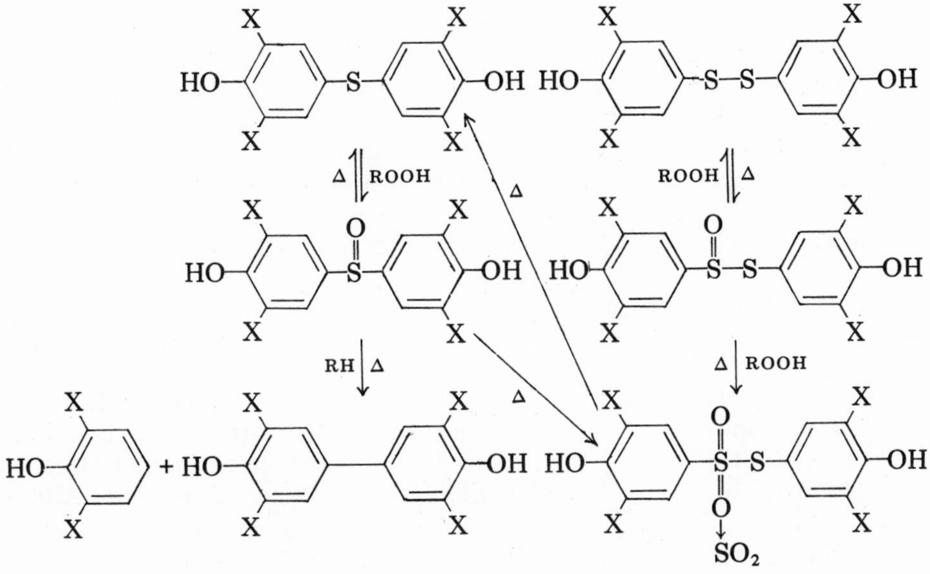

$$\tag{39}$$

Investigation of the proposed intermediates and their reactions, as well as of the kinetics of individual reactions, showed that this scheme was entirely in agreement with the overall picture obtained from a study of cumene hydroperoxide decomposition catalysed by the sulphides.

Extension of these ideas to alkyl sulphides indicated that they also decompose hydroperoxides by liberating sulphur dioxide. However, sulphides that can be oxidised and pyrolysed to yield sulphuric acid can also act as free radical scavengers.

As a result of these very recent findings, some reassessment of the role of organic sulphur antioxidants may be necessary. It seems probable that, with simple molecules, sulphoxides play a major role as inhibitors, but that sulphur dioxide is an active inhibitor in many cases.

Industrial interest has also focused on the petroleum sweetening process where one other co-oxidation reaction has attracted some interest, the co-oxidation of thiols and olefins with oxygen. In general, the reaction has been suggested to involve free radicals [93], viz.

$$RSH \rightarrow RS \cdot \xrightarrow{CH_2=CHR'} RS-CH_2-CHR' \xrightarrow{O_2} RSCH_2CHR'O_2 \cdot$$

$$\downarrow RSH$$

$$\underset{O}{\overset{\parallel}{R \cdot S \cdot CH_2 \cdot \underset{OH}{CHR'}}} \leftarrow RSCH_2-\underset{OOH}{CHR'} + RS \cdot$$

(40)

Hydroperoxides have been isolated from the systems [94,95], even though they react readily with excess thiol. Mono-olefins were found to lead to 2-sulphinyl-ethanol secondary products [95], while the secondary products of di-olefins depended on the relative reactivity of the two double bonds [96]. An interesting review of the detailed chemistry involved in recognising the reaction mechanism is given by Oswald and Wallace [97]: some of the more pertinent details are discussed below.

3.2 AUTOXIDATION OF SULPHUR-CONTAINING SUBSTRATES IN THE ABSENCE OF OTHER HYDROCARBON FREE RADICAL CHAIN REACTIONS

Studies of the autoxidation of sulphur compounds alone has been largely concerned with thiols because of their importance in biological systems [98] and of the commercial interest in petroleum sweetening [99—101] by the oxidation of thiols to less noxious disulphides. Although this process is gradually being outdated with the increasing importance of hydrodesulphurisation, considerable interest still remains.

Investigation of the kinetics and mechanism of the oxidation was eased by the early observation that oxidation of the thiol anion was considerably easier than of the parent thiol. Indeed, Kharasch [102] has found that the relative rates of thiol reactions can alter drastically between the unionised and ionised molecule. Recognition of the importance of the ion led immediately to the suggestion that oxidation is primarily by electron transfer, and to the recognition of a range of electron transfer catalysts.

Although most attention has been paid to metal-catalysed systems, amine-based catalysts are potentially important in petroleum systems. N-Alkylaromatic amines in alkaline solution, in particular, have a marked catalytic effect on the oxidation of thiols [103—105], apparently via the formation of amino anions [97], viz.

In the presence of reactive olefins, co-oxidation of the thiol with the olefin [94,95,106] increases the rate of disappearance of the thiol by the sequence

$$RSH \xrightarrow{O_2} RS\cdot \xrightarrow{H_2C=CR_2} RS\!-\!H_2C\!-\!CR_2\cdot \xrightarrow{O_2} RS\!-\!CH_2\!-\!CR_2\!-\!O_2\cdot$$

$$\text{RSH} \downarrow \text{R'H} \qquad (41)$$

$$RSCH_2\!-\!CR_2OOH$$

The hydroperoxide then reacts immediately with excess thiol to produce disulphide, hydroxyethyl sulphide and water [107].

It is difficult to discuss the uncatalysed autoxiation of thiols, in that traces of metal catalyst can have a very profound effect on the rate. Much of the earlier work is open to question on these grounds, even though useful mechanistic pointers have emerged. Thus, for example, the autoxidation of three carboxythiols (glutathione, cysteine, and thioglycolic acid) has been studied by Dixon and Tunnicliffe [98] over a wide range of pH. Autoxidation of glutathione was highest between pH 7 and 7.5, dropping to ca. 10% of the maximum value at pH 13.8, while the autoxidation of thioglycolic acid increased steadily with pH. Benesch and Benesch [112] suggested that an increase in pH resulted in an increase in the thiol anion concentration, but that this effect was offset above pH 7 with glutathione and cysteine by the loss of a proton from the substituted amino group.

The autoxidation of simple thiols is also very dependent on the basicity of the solvent and high polar solvents such as dimethylformamide [108—110] and tetramethylguanidine [108] accelerate the reaction. The first detailed investigation was carried out with the thiols dissolved in aqueous sodium hydroxide solution [111]. Although the apparatus was crude, stoichiometric conversion to the disulphide was observed with the ease of oxidation of different thiols decreasing in the order n-propyl > n-butyl > n-amyl > benzyl > phenyl.

An extensive investigation of the base-catalysed oxidation of simple thiols has been carried out by Wallace and Schriesheim [109], who suggested that the oxidation proceeds according to the scheme

$$RSH + B \rightleftharpoons RS^- + BH \tag{42}$$

$$RS^- + O_2 \rightarrow RS\cdot + O_2^- \tag{43}$$

$$RS^- + O_2^- \rightarrow RS\cdot + O_2^{2-} \tag{44}$$

$$2\,RS\cdot \rightarrow RSSR \tag{45}$$

$$O_2^{2-} + H_2O \rightarrow 2\,OH^- + \tfrac{1}{2}\,O_2 \tag{46}$$

in which reaction (43) is rate-determining. This mechanism predicts that the rate of oxidation depends on $[RS^-]$ and this was confirmed by measurements with different solvents. The oxidation rates were first order in thiol, with the strongest base giving the highest oxidation rate. No correction for oxygen solubility in the solvents was attempted.

The experiments were extended to cover a series of thiols, the most acidic thiol being found most resistant to oxidation [110]. The reactivity depended on the organic substituent in the order benzyl > n-butyl > p-aminophenyl > cyclohexyl > phenyl > p-nitrophenyl ≈ 0.

Experiments with simple thiols have been extended by Cullis et al. [113], care being taken to exclude metal contamination from the systems. The oxidation of ethane thiol in sodium hydroxide solutions has been found to be stoichiometric to disulphide, the kinetics of reaction changing at ca. 10—30% of reaction, viz.

$$\text{Initial} - \frac{d[EtSH]}{dt} = k\,[EtSH]\,[O_2] \tag{XIII}$$

$$\text{Final} - \frac{d[EtSH]}{dt} = k_2\,[O_2] \tag{XIV}$$

The energy of activation, E_2, calculated over the temperature range 30—50°C, was equal to 16.5 kcal mole^{-1}. Added diethyl disulphide had no effect on the kinetics.

The order of ease of oxidation of other thiols was found to be n-hexyl > i-butyl > n-butyl > ethyl > benzyl > sec-butyl > phenyl > t-butyl which follows approximately the order of stability of the anions expressed by values of pk_a. n-Hexyl thiol was anomalous, apparently because of possible metal contamination.

Under some circumstances, end products other than disulphides can be identified from the oxidation of thiols alone. Berger [114], for example, has studied the oxidation of n-octane thiol and of thiophenol in t-butanol/potassium t-butoxide mixtures to find that disulphides were produced when thiol was in excess, but that sulphinic and sulphonic acids were produced in excess base. The oxygen uptake rates were zero order in thiol.

Berger suggested a chain mechanism of the type

$$RSO^- + O_2 \rightarrow RSO \cdot OO^- \tag{47}$$

$$RSO \cdot OO^- + RS^- \rightarrow RSO^- + RSO_2^- \tag{48}$$

$$RSO_2^- + RSH \rightarrow RSSR + HO_2^- \tag{49}$$

$$HO_2^- + 2\,RSH \rightarrow RSSR + H_2O + OH^- \tag{50}$$

However, it seems more likely that the acids are produced by hydrolysis of disulphides. Wallace and Schrieshiem [115—117] have shown that thiols may be oxidised to disulphides or sulphonic acids at will, depending on the basicity of solution and on temperature, viz.

$$RSSR \xrightarrow{OH^-} RS^- + RSO^- \tag{51}$$

$$3\,RSO^- \rightarrow RSO_3^- + RSSR \tag{52}$$

$$RSO^- \xrightarrow{O_2} RSO_3^- \tag{53}$$

3.3 METAL CATALYSIS

The oxidation of thiols is accelerated remarkably by traces of catalyst and this reaction forms the basis of petroleum sweetening processes. Although transition metal ions are the most effective catalysts, any additive capable of catalysing electron transfer accelerates the reaction. Nitrobenzene in dimethylformamide/potassium hydroxide [118], 2-nitrothiophene, tetracyanoethylene, and 4-nitropyridine-N-oxide [118] are all good catalysts for the oxidation of 1-butane thiol. The alkaline hydrolysis of disulphides containing aryl, carbonyl, and alpha unsaturated groups also results in catalysis, apparently due to the setting up of a sulphinate—sulphenate redox cycle [119—121].

$$RSSR + H_2O \rightleftharpoons RSH + RSOH \tag{54}$$

$$RSOH + O \cdot \rightarrow RSOOH \tag{55}$$

$$2\,R'SH + RSOOH \rightarrow R'SSR' + RSOH \tag{56}$$

Catalysis obviously requires conditions of alkalinity and temperature that favour disulphide hydrolysis: no such hydrolysis in sodium hydroxide solutions has been observed [113,122].

Early work on the catalytic autoxidation of carboxythiols confirmed the effectiveness of manganese, iron, cobalt, copper, and arsenic, but the first major assault on the mechanism of the reaction was due to Michaelis and Barron [123,124]. The oxidation of cysteine at pH 7—8 was found to be zero order in cysteine and to involve metal—cysteine complexes as active intermediates. Several studies of metal—thiol complexes have been reported [125—131] and the kinetics and rate of oxidation of a given thiol appear to depend on the nature and subsequent reactions of these complexes. Although this work provides valuable pointers, the added complexity resulting from the necessity to buffer the solutions allows more reliance to be placed on the oxidation of simple thiols in unbuffered solutions.

For these simple thiols, there is some difference of opinion as to the importance of thiyl free radicals in the system. The catalytic effect of a number of metal salts has been measured [132] and disulphide has been identified as the major product in aqueous solutions. The original mechanism suggested involves a redox mechanism [133] in which metal—thiol complexes are believed to be important [134], viz.

$$2 M^{2+} + O_2 \rightarrow 2 M^{3+} + O_2^{2-} \tag{57}$$

$$2 M^{3+} + 2 RS^- \rightarrow 2 M^{2+} + 2 RS\cdot \tag{58}$$

$$2 RS\cdot \rightarrow RSSR \tag{59}$$

$$O_2^- + H_2O \rightarrow 2 OH^- + \tfrac{1}{2} O_2 \tag{60}$$

A common feature of all such schemes is the formation of peroxide and this has been confirmed by Holtz and Triem [135] and by Schales [136].

Evidence for the role of thiyl radicals has been obtained from the study of the oxidation of various thiols with ferric octanoate in xylene [137]. The reactions were found to be overall second-order and the presence of thiyl radicals was confirmed by trapping with an olefin [106,137].

On the other hand, Trimm and coworkers [122,138,139] prefer to assign a less important role to thiyl radicals, suggesting that the reaction proceeds primarily through electron transfer reactions involving metal—thiol complexes. Comparisons were made of some kinetic features of the oxidation in alkaline solution of a series of simple aliphatic and aromatic thiols in the presence of a variety of metal catalysts [122]. Although detailed kinetic comparisons were difficult because of differing degrees of ionisation of individual thiols and of differing partition functions between the organic and aqueous layers in the solutions, consideration of the trend of oxidation rates was found to be revealing. For example, it was found possible to explain the order of ease of copper-catalysed oxidation of different thiols in solution in terms of electron-directing and steric effects. For the butane thiols, electron-directing effects would be expected to increase the localisation of an electron on the sulphur atom in the order $Bu^n < Bu^i < Bu^s < Bu^i$. Steric hindrance, on the other hand, would be expected to increase in the order $Bu^n \sim Bu^i < Bu^s < Bu^t$ and the rate of oxidation of thiols would then decrease in this order. The overriding importance of steric effects was confirmed by the experimentally observed order of ease of oxidation, $Bu^i \gg Bu^n > Bu^s \gg Bu^t$, and by the fact that the rate of oxidation of phenylmethane thiol was greater than that of thiophenol.

The isolation of compounds of the empirical formulae $Co(SC_2H_5)_3$, $Ni(SC_2H_5)_3(OH)$, and $Ni(SC_2H_5)_2$, together with the observation that coloured soluble metal complexes existed in solution, led to the suggestion that soluble complexes, at least in these systems, were responsible for the catalytic activity observed [140]. The addition of a range of metal complexes to the solutions, coupled with subsequent filtration of solid material, proved this point and showed that the catalytic activity of a given metal complex was very dependent on the nature of the ligand associated with the metal. Metal ions were shown to accept an electron from a thiol anion with or without the formation of a metal—thiol complex [138,139]. However, outer-sphere electron transfer (with no complex formation) was limited to only a few cases involving very strongly bonded "added" metal complexes. Where this was important, the order of thiol reactivity was found to depend both on the electron directing and the geometric structure of the organic group in the thiol. On the other hand, detailed kinetic measurements showed that the electron transfer reaction was not rate-determining, but that the rate of oxidation of the reduced catalyst could well control the kinetics. This conclusion was in agreement with the overall kinetics of, for example, the oxidation of ethane thiol by ferricyanide, which were found to be zero order in thiol, but to depend on the concentrations of metal and oxygen.

Detailed investigation of the oxidation of ethane thiol in the presence of copper-, cobalt-, and nickel-containing catalysts was also carried out [138]. The reaction was stoichiometric to disulphide, and the dependence of the rates of oxidation on the concentration of individual reactants is summarised in Table 4. It can be seen that the concentrations of "added" metal bear little resemblance to the concentrations of catalytically active metal. The change from initial to final rates usually occurred at about 10—30% of total conversion and was attributed to the formation of disulphides which can compete for coordination sites on the metal ion.

As a result of these experiments, two possible reaction mechanisms were advanced. The first (outer sphere) was suggested to become important when displacement of the original ligand by sulphur-containing species was difficult

$$RS^- + M^{n+}(X)_6 + O_2 \rightarrow RS^- \ldots M^{n+}(X)_6 \ldots O_2$$

$$\rightarrow R\dot{S} \ldots M^{(n-1)+}(X)_6 \ldots O_2 \rightarrow R\dot{S} \ldots M^{n+}(X)_6 \ldots O_2^-$$

$$\rightarrow RS\cdot + M^{n+}(X)_6 + O_2^- \tag{61}$$

Subsequent reactions of thiyl radicals led to the production of some disulphide and of more highly oxidised sulphur-containing species.

Where substitution of a thiol or disulphide group into the coordination sphere of the metal was possible, an "inner sphere" type of reaction mechanism was postulated. For simplicity, this is written as involving only

TABLE 4

The kinetics of the oxidation of ethane thiol catalysed by copper, cobalt, and nickel [138]

System [a]	Soluble [b] (M)	Order in ethane thiol	Order in oxygen	Order in NaOH	k [c] (30°C)	E_a (kcal mole^{-1})	Notes
Uncatalysed		1	1	0	I = 4.9×10^{-2} l mole^{-1} s^{-1} F = 2.0×10^{-1} s^{-1}	16.4	
Cu (10^{-3} M)	n.m.	0	1	0	n.m.	n.m.	Diffusion controlled
Cu (10^{-5} M)	10^{-5}	0	1	0	2.3×10^{-1} s^{-1}	4.3	
Co (10^{-3} M)	6.4×10^{-4}	0	1	0	I = 2.1×10^{-1} s^{-1} F = 1.5×10^{-1} s^{-1}	7.5	
Ni (10^{-3} M)	5.3×10^{-4}	0 (EtSH > 0.5 M) 1	1	n.m. 0	n.m. 3.6×10^{-1} l mole^{-1} s^{-1}	n.m. 8.0	Diffusion controlled

a Metal concentrations as added in brackets.
b Metal concentration in solution, by analysis.
c I = initial rate; F = final rate; n.m. = not measured.

one metal centre, viz.

$$Co(II)\{\overset{*}{X}_4(R\overset{*}{S}\overset{*}{S}R)\}^{2+} + O_2 \rightarrow Co(III)\{\overset{*}{X}_4(R\overset{*}{S}SR)(\overset{*}{O}O)\}^{2+} \tag{62}$$

$$\overset{RS^-}{\longrightarrow} Co(III)\{\overset{*}{X}_4(\overset{*}{S}R)(\overset{*}{O}O)\}^+ + RSSR$$

$$\overset{RS^-}{\longrightarrow} Co(II)\{\overset{*}{X}_4(R\overset{*}{S}\overset{*}{S}R)\}^{2+} + O_2^{2-} \tag{63}$$

where atoms marked with an asterisk are coordinated to the metal ion. The four coordination positions ($\overset{*}{X}_4$) play no part in the chemistry, but would be expected to have a definite influence on the rate of reaction. This would thus be expected to alter as disulphide competes with hydroxy, thiol, or "as added" ligands for these coordination sites, unless the concentration of the original ligand is high enough to preclude such competition. Experimental verification of these predictions has been reported [138,139].

A similar reaction mechanism has been advanced for copper-catalysed systems, with the added driving force that the coordination number decrease on going from Cu(II) to Cu(I) could play a significant part in releasing disulphide from the coordination sphere.

These proposals appear to be internally consistent and to explain many experimental observations. Thus, for example, Kolkoff et al. [141,142] and Gorin and Godwin [143] report that ferricyanide may catalyse thiol oxidation either with or without displacement of one CN^- ligand by a thiol ligand. Overberger et al. [144] also postulate the formation of a ferric—thiol complex during the oxidation of polyvinyl mercaptan by ferric sulphate in dimethyl sulphoxide solution.

Attention has also been focused on the oxidation of thiols in the presence of "solid" catalysts. One of the more comprehensive investigations into systems of this type has been made by Wallace et al. [133,145, 146] with a view to the possible use of phthalocyanine type complexes as commercial sweetening catalysts. Comparisons were drawn with metal pyrophosphates, phosphomolybdates, phosphotungstates, and phosphates. Pyrophosphates were found to be effective catalysts, possible due to the existence of six-membered rings involving the cobalt cation [147], which enhances the ability of the cation to donate an electron to oxygen and stabilises each oxidation state of the cation. For a series of pyrophosphates, the order of activity was Co > Cu > Ni > Fe, an activity pattern which was explained in terms of the stability of the $3d$ electron shells.

The oxidation of thiols by four transition metal oxides in xylene has also been studied in the presence and absence of oxygen [146]. Oxidation resulted in the formation of some organic sulphides, products which were attributed to the reaction of thiyl free radicals with olefins.

Most of this work raises the unresolved question of the relative importance of homogeneous and heterogeneous catalysis. At least part of the catalytic activity of "solid" metal pyrophosphates has been shown to be due to traces of soluble complexes [140] and similar effects may well be important for other "heterogeneous" catalysts. The catalytic activity of traces of soluble metal complexes is so high that spurious "heterogeneous" catalytic effects may well be observed.

The autoxidation of mono- and disulphides has also been studied to some extent, largely because of the role of sulphoxide inhibitors and the possibility of hydrolysis and/or oxidation of disulphides produced in the thiol oxidation reaction. Reaction with conventional chemical oxidants is reviewed by Savige and Maclaren [148] with particular reference to cystine, but it is to the work of Bateman and coworkers that we owe much of our present understanding of the autoxidation of sulphides.

3.4 REACTIONS OF SULPHOXIDE PRODUCTS

As discussed with reference to co-oxidation with hydrocarbons, sulphide oxidation chemistry is complicated by the further reactions of sulphoxide products. The autoxidation of sulphides in the absence of hydrocarbons is a free radical process [149] leading, in the first instance, to hydroperoxides, viz.

$$RCH_2SR' \rightarrow R\overset{\cdot}{C}HSR' \tag{64}$$

$$R\overset{\cdot}{C}HSR' + O_2 \rightarrow \underset{\underset{OO\cdot}{|}}{RCHSR'} \tag{65}$$

$$\underset{\underset{OO\cdot}{|}}{RCHSR'} + RCH_2SR' \rightarrow \underset{\underset{OOH}{|}}{RCHSR'} + R\overset{\cdot}{C}HSR' \tag{66}$$

The hydroperoxides react readily to form sulphoxides

$$\underset{\underset{OOH}{|}}{RCHSR'} + RCH_2SR' \rightarrow RCH(OH)SR' + RCH_2SOR' \tag{67}$$

Subsequent reactions complicate the system, via reactions such as

$$\underset{\underset{OOH}{|}}{RCHSR'} + RCH_2SOR' \rightarrow R_2'S_2 + H_2O + products \tag{68}$$

$$RCH(OH)SR' \rightarrow RCHO + R'SH \tag{69}$$

$$RCH(OH)SR' + R'SH \rightarrow H_2O + complex\ sulphides \tag{70}$$

In addition, reactions of thiols with sulphoxides (see below) can influence the course of reaction.

Kinetic and mechanistic experimental results support this proposed mechanism. Saturated mono-, di-, and tetra-sulphides are much less reactive than unsaturated sulphides, where the relative activity is mono- > di- > tetra-sulphides [88,149] for the initial oxidation. The oxidations are auto-inhibited, inferring sulphoxide intereference with the free radical chain. This is particularly effective for t-butyl sulphoxides [88]: thus, for example, the activity of sulphides CHMe : CH·CHMe·S·R and ⟨◯⟩-SR, where R is an alkyl group, decreases in the order Me > Et > i-Pr > t-Bu (inert).

Steric effects are not of major importance, as shown by the fact that t-butyl-substituted sulphides can be autoxidised in the presence of a catalyst. Rather, it is the inhibitory action of traces of t-butyl sulphoxide which is responsible for the overall lack of oxidation activity. This bears out the observed order of effectiveness of sulphide antioxidants added to oxidising hydrocarbons [97], where t-butyl substituents confer excellent antioxidant properties.

The reaction of thiols with sulphoxides has been the subject of a series of investigations by Wallace et al. [150—154]. The reaction produces disulphides and monosulphides (from the original sulphoxide) and is catalysed by the presence of both acids and bases [152]. The observed ease of thiol oxidation was aryl > aralkyl > alkyl, but the kinetics of reaction were dependent on the acidity of the thiol. Wallace and Mahon [151] were able to postulate a reaction mechanism consistent with kinetic observations in the system, viz.

$$RSH + R_2'SO \rightleftharpoons \left(R_2'\ddot{S}\begin{matrix} {}^{OH} \\ {}_{SR} \end{matrix} \right) \tag{71}$$

$$\left(R_2'\ddot{S}\begin{matrix} {}^{OH} \\ {}_{SR} \end{matrix} \right) + RSH \rightleftharpoons RSSR + R'SR' + H_2O \tag{72}$$

Assuming a steady state concentration of adduct, we may write

$$\frac{d[\text{adduct}]}{dt} = 0 = k_{71}[RSH][R_2'SO] - k_{71}[\text{adduct}]$$

$$- k_{72}[\text{adduct}][RSH] + k_{-72}[RSSR][R'SR'][H_2O] \tag{XV}$$

Substitution into the equation for product formation then gives

$$\frac{d[\text{product}]}{dt} = \frac{k_{72}(k_{-67}[RSSR][R'SR'][H_2O] + k_{71}[RSH][R_2'SO])[RSH]}{k_{-71} + k_{72}[RSH]}$$

$$- k_{72}[RSSR][R'SR'][H_2O] \tag{XVI}$$

Assuming that $k_{-72} \sim 0$ and that $k_{72} \gg k_{-71}$, the equation reduces to

$$\frac{d[\text{product}]}{dt} = k_{71}[\text{RSH}][\text{R}_2'\text{SO}] \tag{XVII}$$

in good agreement with the experimentally observed second-order kinetics.

In terms of the autoxidation of sulphides, these reactions present an additional complication, but it should be pointed out that the reaction does offer an interesting and useful synthesis route in its own right.

Attention has also been focused on various base-catalysed reactions of sulphides. Alkaline decomposition of aliphatic disulphides has been reviewed by Danehy [155,156], but it is the alkali-catalysed oxidation of organic sulphides and disulphides which is of particular interest in the consideration of petroleum sweetening reactions. Disulphides have been discussed above and the base-catalysed autoxidation of α-sulphido carbanions has been investigated by Wallace et al. [157,158] for several benzyl, aralkyl, and dialkyl sulphides. The observed rates of oxygen consumption were found to be dependent on sulphide structure, base strength, and solvent, the results suggesting that the rate-determining step is proton extraction from the sulphide to form an α-carbanion. This ion then reacts readily with oxygen to produce carboxylic and sulphonic acids, apparently via the formation of unstable peroxy ions.

4. Autoxidation of organic compounds containing chlorine

In contrast to the studies of the oxidation of nitrogen- and sulphur-containing compounds, comparatively little work has been completed using chlorinated substrates, particularly with respect to their possible role in the oxidation of other organic materials. Starnes [159] has reported that triphenylmethyl chloride, trimethylmethyl chloride, and vinyl chloride inhibit the oxidation of cumene and of 4-vinylcyclohexene, catalysed by cobalt and manganese salts. In contrast to the autoxidation of 1,1-diphenylhydrazine, which was found to be unaffected by chloroform [160], Starnes noted synergistic effects involving phenyl-substituted methyl chlorides and the phenol $2,4,6\text{-}CH_3[(CH_3)_3C]_2C_6OH$, but no detailed kinetics were reported.

As discussed in the introduction, chlorine substituents may be expected to influence the autoxidation of organic substrates as a result of their electron-directing properties, but these effects may be complex. Thus, for example, Kulicki [161] has reported that halogen and nitrate substituents, particularly in the *ortho* position, have an effect on the autoxidation of cumene by inhibiting the primary oxidation, but they also accelerate the homolysis of any hydroperoxide that is formed: the net result is an overall acceleration of oxidation. Similar effects were noted by Kovalev and

Chervinskii [162] for the oxidation of chlorine-substituted p-xylene, and by Kiiko and Matkovskii [163] in the study of the effect of chlorinated solvents on the oxidation of dimethylnaphthalene.

Some studies of the autoxidation of simple chlorinated compounds have also been reported. Kawai [164] has found, for example, that the oxidation of chloroform, even in the dark, involves formation of a hydroperoxide

$$CHCl_3 \rightarrow Cl_3COOH \begin{cases} \nearrow Cl_2 + CO_2 + HCl \quad\quad (73) \\ \searrow COCl_2 + HCl + [O] \quad (74) \end{cases}$$

The reaction is catalysed by Cl^- and chloroform can be stabilised by rigorous removal of the ion.

Perhaps the most satisfying work using simple chlorinated substrates has been that with trichloroethylene [165,166]. Mayo and Honda [165] have studied the thermally initiated oxidation of trichloroethylene and find that the kinetics are very dependent on the purity of the substrate. Using oxygen at 40 psig at 50°C, ca. 10% of the reacting chloroethylene gave $COCl_2$, HCl, and CO, the remaining yield being equally divided between dichloroacetyl chloride and trichloroethylene oxide. The yield of dichloroacetyl chloride was found to be dependent on the oxygen pressure, decreasing from 56 to 46% as the oxygen pressure decreased from 40 to 10 psig: the yield of epoxide increased over this range from ca. 40% (40 psig) to ca. 50% (10 psig).

The kinetics were typical of a radical chain autoxidation with the rate of oxidation varying from ca. 2.5% h^{-1} at 40 psig oxygen to 0.7% h^{-1} at 10 psig. The induction period observed disappeared in the presence of 2,2'-azobis(2-methylpropionitrile) or of tetramethylsuccinonitrile, but the addition of only a small amount of additive also decreased the rate of oxidation. Thus, for example, the addition of 0.002 M ABN reduced the rate to ca. 40% of that observed in the absence of additive and the addition of 0.08 M ABN was needed before the rate equalled the thermal oxidation rate. No attempt was made to explain these observations in detail, but it was suggested that the additive was interfering with thermal initiation, which itself involved two processes, a slow true thermal initiation and a faster initiation resulting from the build up of a transient intermediate during the induction period.

A mechanism for the oxidation, from which some idea of the magnitude of some kinetic parameters was obtained, was proposed, viz.

$$Cl_2 \rightarrow 2\ Cl\cdot \quad\quad\quad\quad\quad\quad\quad\quad\quad\quad\quad\quad\quad (i)$$

$$Cl\cdot + HClC{=}CCl_2 \rightarrow HCl_2C{-}\overset{\cdot}{C}Cl_2 \quad\quad (75)$$

$$O_2 + HCl_2C{-}\overset{\cdot}{C}Cl_2 \rightleftharpoons HCl_2C\cdot CCl_2O_2\cdot \quad\quad (76)$$

$$HCl_2C-CCl_2O_2\cdot + HCl_2C-\overset{\cdot}{C}Cl_2 \rightarrow (HCl_2C-CCl_2O-)_2 \qquad (77)$$

$$2\,HCl_2C\cdot CCl_2O_2\cdot \overset{\nearrow (HCl_2C-CCl_2O-)_2 + O_2 \qquad (78)}{\searrow 2\,HCl_2C-CCl_2O\cdot + O_2 \qquad (79)}$$

$$HCl_2C\cdot CCl_2O\cdot \overset{\nearrow 0.94\,HCl_2C-COCl + Cl\cdot \qquad (80)}{\searrow}$$

$$0.06\left\{\begin{matrix} & Cl \\ & | \\ H-&C\cdot\ + COCl_2 \\ & | \\ & Cl \end{matrix}\right\} \qquad (81)$$

$$[O] \longrightarrow CO + HCl + Cl\cdot \qquad (82)$$

$$HCl_2C-Cl_2\cdot O_2\cdot + HClC{=}CCl_2$$

$$\overset{\searrow}{\underset{}{}} \quad \begin{matrix} H & Cl & & H & Cl \\ | & | & & | & | \\ Cl-C-&C\cdot O_2\cdot&C-&C\cdot \\ | & | & & | & | \\ Cl & Cl & & Cl & Cl \end{matrix} \qquad (83)$$

$$\downarrow$$

$$HCl_2C-CCl_2O\cdot + HClC\overset{O}{\overset{/\backslash}{-}}C-Cl_2 \qquad (84)$$

This reaction mechanism is very similar to that proposed for the corresponding gas phase reaction, with the last three reactions, accounting for the increased yield of the oxide, being important only in the liquid phase. On the basis of the above, Mayo and Honda derive the rate equation

$$R_o = \left[\frac{(k_i + ek_{75})}{2k_{76}}\right]^{1/2} k_p[C_2HCl_3] \qquad (XVIII)$$

where $e = 0.6$ and k_{76} for ABN is $0.0101\ h^{-1}$ at $50°C$. On this basis, the rate coefficient for thermal initiation is found to be ca. $1.6 \times 10^{-4}\ l\ mole^{-1}\ h^{-1}$ and $(k_{75}/2k_{76})^{1/2}$ is ca. 1.73 in similar units.

Poluektov and Mekhrynshev [166] have also studied the liquid phase oxidation of trichloroethylene, but initiated by γ-ray irradiation from a cobalt-60 source. It is questionable whether they avoided complications from thermal initiation and they did not pay particular attention to purity (as was found necessary by Mayo and Honda [165]), but the oxidations do show similar kinetic features. The rate of oxidation was found to be proportional to the first power of the concentration of the substrate, to the square root of the rate of initial active centre formation, and to the oxygen concentration only at low oxygen pressures. At high pressures, the rate was zero order with respect to oxygen. The overall activation energy was found to be $7.5\ kcal\ mole^{-1}$.

Detailed studies of the kinetics and mechanism of the oxidation of hexachlorobutadiene have been reported by Poluetov and Ageev [167,

168]. The principal reaction products were found to be pentachloro-acetoacetyl chloride, tetrachlorosuccinyl, dichloromaleyl, and dichloro-malonyl chlorides together with trichloroacetyl chloride, carbon tetra-chloride, and phosgene. An induction period was observed, which could be reduced by irradiation with UV light, and the subsequent oxidation rate was independent of the concentration of the diene but dependent on the first power of the dissolved oxygen concentration.

Oxidation of the hexachlorobutadiene is interesting in that the halogen is involved in conjugation and can supply electrons to the pi system of the diene. Chlorine atoms are apparently not involved in the free radical chain and a peroxide-based reaction scheme has been proposed, viz.

$$R\cdot + O_2 \rightarrow RO_2\cdot \tag{2}$$

$$RO_2\cdot + RH \rightarrow R'\cdot \tag{85}$$

$$R'\cdot + O_2 \rightarrow R'O_2\cdot \tag{2'}$$

$$R'O_2\cdot + RH \rightarrow 2\ PO + RO_2\cdot \tag{86}$$

$$ROOR \rightarrow 2\ RO\cdot \tag{87}$$

$$2\ RO_2\cdot \rightarrow ROOR + O_2 \tag{4'}$$

$$RO_2\cdot + R\cdot \rightarrow ROOR \tag{88}$$

$$2\ R\cdot \rightarrow RR \tag{89}$$

where PO is an end product, e.g. oxide or carbonyl compound.

The importance of polymers in the chain is open to some doubt in view of reports that intramolecular peroxides are much more important in this kind of system (see below). However, the authors apply steady-state rela-tionships which give a reasonable approximation to the experimental ob-servations, probably because their kinetic arguments do not distinguish between the reactions of monomer and polymer. The chain length at 180°C was found to be 200 and the overall activation energy to be 20.6 kcal mole^{-1}. Application of steady state arguments leads to the con-clusion that the activation energy for the reaction of substrate radicals with oxygen is ca. 5.6 kcal mole^{-1}: this value seems high in view of the known ease of peroxidation of alkyl radicals in hydrocarbon oxidation systems.

There is considerable interest in the autoxidation of chlorinated polymers and monomers, with particular attention paid to the natures of peroxides formed in the system. As is discussed in more detail below, intramolecular peroxides appear to be formed preferentially whenever the substrate molecule contains a conjugated double bond system, as is the case for hexachlorobutadiene. Where conjugated double bonds are not available, or are sterically protected, then more conventional peroxide formation and peroxy radical chains become of importance. Thus, for example, oxygen-catalysed initiation of vinyl chloride polymerisation [169] apparently proceeds via the formation of peroxy radicals.

A considerable amount of work has also been done on the oxidation of the monomer and polymers of chloroprene. Chloroprene autoxidises rapidly, even at temperatures as low as $0°C$, yielding a polymeric peroxide as the principle product [170,171]. The reaction has been found to be autocatalytic and, up to about 5 mole % oxidation, the mole % oxidation increased as the square of the time [170,172]: above this extent of oxidation, the rate increased even more, apparently due to the subsequent reaction of the peroxide produced. The oxidations were so rapid that conventional initiators and inhibitors had less effect than could have been expected for less labile substrates.

The dependence of the oxygen uptake on the square of the time is frequently observed in autoxidation and is usually accepted to be an approximation to the theoretical equation derived for a long chain length oxidation, initiated by a first-order decomposition (peroxidic products) and terminated by a bimolecular reaction (propagating peroxy radicals), viz.

$$-\frac{d[O_2]}{dt} = \left(\frac{2k_i[R'O_2]}{k_t}\right)^{1/2} k_p[RH] \tag{XIX}$$

where k_i is the decomposition coefficient for the polyperoxide, $R'O_2$, k_p the rate-determining propagation coefficient for the addition of peroxy radicals to chloroprene, RH, and k_t the second-order termination rate coefficient.

Bailey [172] has combined the results obtained for the oxidation of chloroprene in the presence of azobisisobutyronitrile with those in the absence of initiator, to show that k_i varies between $1 \times 10^{-7} \, s^{-1}$ at $0°C$ and $1.65 \times 10^{-6} \, s^{-1}$ at $35°C$, the corresponding chain lengths being 90 and 170. Comparison of the reactivity of several monomers shows that chloroprene is oxidised faster even than styrene and Bailey has shown that the results agree well with predictions based on the reactivity of the double bond and the polarity of monomers involved in co-polymerisation [173]. The system adopted describes the reactivity of monomers in co-polymerisation in terms of the rate coefficient for the addition of monomer 2 (polarity e_2) to the radical of monomer 1

$$k_{12} = P_1 Q_2 \exp(-e_1 e_2) \tag{XX}$$

where P_1 is characteristic of radical 1 and Q_2 is the reactivity of the double bond of monomer 2. Assuming that the rate-determining step is the addition of a peroxy radical to the double bond, the rates of oxidation of seven monomers relative to styrene can be correlated to within a factor of three.

The oxidation of polychloroprene is more complex, largely as a result of the scission and cross-linking of the polymer. The general features of the oxidation have been established by Kuz'minskii and Peschanskaya [174] using a polymer film mounted in a circulating flow apparatus. Oxidation was found to be autocatalytic in the temperature range 60—90°C, 0.3 mole of hydrogen chloride being evolved per mole of oxygen adsorbed. In the absence of oxygen, only 1% of the available HCl was liberated, even on heating to 175°C, and they concluded that the facile oxidative generation of HCl implied the loss of the double bond by peroxide formation.

The oxidative ageing of polychloroprene has also attracted some attention [175], the energy of activation for the process (measured in terms of the percentage of polymer ultimately capable of crystallisation) being calculated to be 8.5 kcal mole^{-1}. Kössler and Svob [176] suggested that the crystallisation properties were more affected by dehydrochlorination in the early stages of ageing than by cross-linking, a suggestion supported by the observation that the energy of activation for HCl evolution (9.2 kcal mole^{-1}) was very similar to that observed for ageing.

Detailed studies of the autoxidation of polychloroprene and of trans-4-chloro-4-octene, a model comparison, have been reported by Bailey [177]. The major part of the hydrogen chloride evolved on heating polychloroprene was confirmed to be associated with oxidative degradation and the kinetics of HCl formation from pre-oxidised polymer heated under nitrogen was investigated.

During the early stages of the oxidation of the polymer, the amount of oxygen adsorbed was again found to increase as the square of the time, viz.

$$(\text{Oxygen adsorbed or HCl evolved}) = [k'(\text{time} - \text{I.P.})]^2 \qquad (XXI)$$

where I.P. is the induction period. Apparent activation energies, obtained over the range 90—120°C, were found to be 17.6 kcal mole^{-1} (oxidation) and 15.8 kcal mole^{-1} (HCl evolution): these values are somewhat larger than the values previously quoted for polychloroprene (9.2 [175] and 13 kcal mole^{-1} [178]). The rate of oxidation passed through a maximum at 5—10% oxidation, apparently as a result of the onset of diffusion limitations caused by changes in the structure of the polymer: this could well account for the difference in activation energy reported.

The effect of added azo-bis(cyclohexane nitrile) was also investigated. The dependence of the rate of oxidation of polymer on the half power of the initiator concentration was in agreement with the rate equation commonly observed for oxidation with "high" pressures of oxygen, viz.

$$-\frac{d[O_2]}{dt} = (k_i I)^{1/2} k_p k_t^{-1/2} [\text{RH}] \qquad (XXII)$$

where I is the concentration of initiator and k_p and k_E are the rate coefficients for peroxy radical propagation and bimolecular termination, respectively. Reasonable approximations on this basis lead to a value of 8.7 kcal mole^{-1} for the energy of activation of propagation, close to that observed [179] for propagation in the oxidation of styrene to polyperoxide (8.4 kcal mole^{-1}).

The model compound, *trans*-4-chloro-4-octene, was chosen because it possessed the —ClC=CH— group of polychloroprene, but without the repeating 1,5-diene structure of the polymer. The autoxidation was similar to that observed for polychloroprene, although no induction period was observed (cf. hexachlorobutadiene oxidation above). The evolution of HCl was proportional to the square of the time, but the oxidation kinetics were approximated more closely by the equation

$$\text{Oxygen adsorbed} = at + bt^2 \qquad\qquad\qquad\qquad \text{(XXIII)}$$

Bailey also found that more hydroperoxides were produced from the octene than form polychloroprene and suggested that chloroprene oxidation proceeded via peroxide formation involving carbon atoms located on non-adjacent double bonds

$$CH_2\overset{Cl}{\underset{\text{O}\text{---}\text{O}}{C}}\text{---}CHCH_2CH_2\text{---}\overset{Cl}{\underset{}{C}}\text{---}CH\cdot CH_2\cdot CH_2\overset{Cl}{\underset{\text{O}\text{---}\text{O}}{C}}\text{---}CH=CH \qquad (90)$$

Considerable evidence exists to support the suggestion that intramolecular peroxides are more important than intermolecular or cross-linked peroxides. In addition to Bailey's work, intramolecular peroxides have been identified during the oxidation of vinyl chloride [180] and of polyvinylene chloride [181]. In the octene, where such peroxides cannot be formed, hydroperoxides are produced by the more conventional radical reaction involving attack on the α-methylene hydrogen atom

$$RO_2\cdot + \text{—}C=CH\text{—}CH_2\text{—} \rightarrow ROOH + \text{—}C=C\text{—}\overset{\cdot}{C}H\text{—} \qquad (91)$$

Acknowledgement

Thanks are due to Dr. H.C. Bailey for valuable comments on the oxidation of chlorine-containing compounds.

References

1 R. Hardy, Rep. Prog. Appl. Chem., 54 (1969) 459.
2 K.U. Ingold, Chem. Rev., 61 (1961) 563.
3 L.R. Kahoney, J. Am. Chem. Soc., 89 (1967) 1896.
4 W.O. Lundberg, Autoxidation and Antioxidants, Interscience, New York, 1962.
5 L. Reich and S.S. Stivala, Autoxidation of Hydrocarbons and Polyolefins, Dekker, New York, 1969.

6   M.G. Evans and J. de Heer, Q. Rev., 4 (1950) 94.
7   C.F. Boozer and G.S. Hammond, J. Am. Chem. Soc., 76 (1954) 3861.
8   G.S. Hammond, C.F. Boozer, C.E. Hamilton and J.N. Sen, J. Am. Chem. Soc., 77 (1955) 3238.
9   A.F. Bickel and E.C. Kooyman, J. Chem. Soc., (1956) 2215, 2217.
10 L.R. Mahoney and F.C. Ferris, J. Am. Chem. Soc., 85 (1963) 2345.
11 W.G. Lloyd and C.E. Lange, J. Am. Chem. Soc., 86 (1964) 1491.
12 C.J. Pedersen, Ind. Eng. Chem., 48 (1956) 1881.
13 K.H. Hausser, Naturwissenschaften, 46 (1959) 597.
14 J.R. Thomas and C.A. Tolman, J. Am. Chem. Soc., 84 (1962) 2930.
15 J.R. Shelton and D.N. Vincent, J. Am. Chem. Soc., 85 (1963) 2433.
16 J.A. Howard and K.U. Ingold, Can. J. Chem., 42 (1964) 2324.
17 S.F. Strause and E. Dyer, J. Am. Chem. Soc., 78 (1956) 136.
18 L.R. Mahoney and M.A. Da Rooge, J. Am. Chem. Soc., 92 (1970) 890.
19 L.R. Mahoney, F.C. Ferris and M.A. Da Rooge, J. Am. Chem. Soc., 91 (1969) 3883.
20 J.R. Thomas, J. Am. Chem. Soc., 82 (1960) 5955.
21 J.Q. Adams, S.W. Nicksic and J.R. Thomas, J. Chem. Phys., 45 (1966) 654.
22 I.T. Brownlie and K.U. Ingold, Can. J. Chem., 45 (1967) 2419.
23 I.T. Brownlie and K.U. Ingold, Can. J. Chem., 45 (1967) 2427.
24 K. Adamic, M. Dunn and K.U. Ingold, Can. J. Chem., 47 (1969) 287, 295.
25 J.A. Howard and K.U. Ingold, Can. J. Chem., 42 (1964) 1044.
26 J.A. Howard and K.U. Ingold, Can. J. Chem., 40 (1962) 1851.
27 J.A. Howard and K.U. Ingold, Can. J. Chem., 41 (1963) 1744, 2800.
28 C.W. Capp and E.G.E. Hawkins, J. Chem. Soc., (1953) 4106.
29 H.E. De la Mere, J. Org. Chem., 25 (1960) 2114.
30 R. Konaka, K. Kuruma and S. Terabe, J. Am. Chem. Soc., 90 (1968) 1801.
31 E.T. Denisov, Kinet. Catal. (USSR), 11 (1970) 262.
32 N.M. Emanuel, Liquid Phase Oxidation of Hydrocarbons, Plenum Press, London, 1967.
33 J.I. Wasson and W.M. Smith, Ind. Eng. Chem., 45 (1953) 197.
34 R.H. Rosenwald, J.R. Hoatson and J.A. Chenicek, Ind. Eng. Chem., 42 (1950) 169.
35 C.D. Lowry, J.C. Morrell and C.G. Dryer, Ind. Eng. Chem., 25 (1933) 804.
36 E.M. Bickoff, J. Am. Oil Chem. Soc., 28 (1951) 61.
37 D.B. Denney and D.Z. Denny, J. Am. Chem. Soc., 82 (1960) 1389.
38 C. Walling and N. Indictor, J. Am. Chem. Soc., 80 (1958) 5814.
39 G.E. Zaikov, Zh. Anal. Khim., 15 (1960) 104.
40 J.L. Bolland and P. Ten Haave, Discuss. Faraday Soc., 2 (1947) 252.
41 F.R. Mayo, A.A. Miller and G.A. Russell, J. Am. Chem. Soc., 80 (1958) 2465, 2480, 2493, 2497, 2500.
42 C. Walling and R.B. Hodgdon, J. Am. Chem. Soc., 80 (1958) 228.
43 H.C. Brown and Y. Okamoto, J. Am. Chem. Soc., 80 (1958) 4979.
44 S. Patai, The Chemistry of the Amino Group, Interscience, New York, 1958.
45 C. Walling, Adv. Chem. Ser., 75 (1968) 166.
46 K.U. Ingold, Can. J. Chem., 41 (1963) 2816.
47 D.V. Gardner, J.A. Howard and K.U. Ingold, Can. J. Chem., 42 (1964) 2847.
48 E. Dyer, O.A. Pickett, S.F. Strause and H.E. Worrell, J. Am. Chem. Soc., 78 (1956) 3384.
49 S.R. Thomas and O.L. Harle, J. Phys. Chem., 63 (1959) 1027.
50 A.L. Buchachenko, O.P. Sykhanova, L.A. Kalashnikova and M.B. Neiman, Kinet. Catal. (USSR), 6 (1965) 538.
51 E. Höft and H. Schultze, Z. Chem., (1967) 137.
52 G.A. Kovtun and A.L. Alexandrov, Izv. Akad. Nauk SSSR, Ser. Khim., (1974) 1274.

246

53 G.A. Kovtun and A.L. Alexandrov, Izv. Akad. Nauk SSSR, Ser. Khim., (1973) 2208.
54 G.A. Kovtun, A.V. Kazantsev and A.L. Alexandrov, Izv. Akad. Nauk SSSR, Ser. Khim., (1974) 2635.
55 G.A. Kovtun and A.L. Alexandrov, private communication.
56 N.N. Pozdeeva, T.I. Sapachova and A.L. Alexandrov, Izv. Akad. Nauk SSSR, Ser. Khim., (1976) 1738.
57 G.A. Kovtun, V.A. Golubev and A.L. Alexandrov, Izv. Akad. Nauk SSSR, Ser. Khim., 793 (1974).
58 G.A. Kovtun, A.L. Alexandrov and E.T. Denisov, Izv. Akad. Nauk SSSR, Ser. Khim., 2611 (1973).
59 T.I. Sapachova, G.G. Agliullina and A.L. Alexandrov, Neftekhimiya, 14 (1974) 450.
60 T.I. Sapachova, G.G. Agiullina, A.L. Alexandrov and E.T. Denisov, Neftekhimiya, 14 (1974) 623.
61 R.G.R. Bacon, Chem. Ind. (London), (1962) 19.
62 O. Meth-Cohn and H. Suschitzky, Chem. Ind. (London), (1969) 443.
63 L.N. Denisova and E.T. Denisov, Izv. Akad. Nauk SSSR, Ser. Khim., (1966) 2220.
64 D.F. Minor, W.A. Waters and J.V. Rambsbottom, J. Chem. Soc. B, (1967) 180.
65 H. Aebi, B. Dewald and H. Suter, Helv. Chim. Acta., 48 (1965) 656.
66 V.N. Vetchinkina, I.P. Skibida and Z.K. Maizus, Izv. Akad. Nauk SSSR, Ser. Khim., (1971) 711.
67 G.A. Kovtun, A.V. Kazantsev and A.L. Alexandrov, Izv. Akad. Nauk SSSR, Ser. Khim., (1974) 2635.
68 B.F. Sagar, J. Chem. Soc. B, (1967) 428, 1047.
69 J.R. Bolland, Q. Rev., 3 (1949) 1.
70 G.A. Russell, A.J. Moye, E.G. Janzen, S. Mak and E.R. Talaty, J. Org. Chem., 32 (1967) 137.
71 J. Bakke, Acta. Chem. Scand., 25 (1971) 2509.
72 D.H. Johnson, Chem. Ind. (London), (1953) 1032.
73 M.A.T. Rogers, Chem. Ind. (London), (1953) 1033.
74 M.N. Hughes, H.G. Nicklin and K. Shrimanker, J. Chem. Soc. A, (1971) 3485.
75 D.J. Cowley and W.A. Waters, J. Chem. Soc. B, (1970) 96.
76 H.H. Stroh and L. Ebert, Chem. Ber., 97 (1964) 2335.
77 W.F. Taylor, H.A. Weiss and T.J. Wallace, Chem. Ind. (London), (1968) 1226.
78 K.H. Pansacker, J. Chem. Soc., (1950) 3478.
79 L. Horner and K.H. Knapp, Makromol. Chem., 93 (1966) 69.
80 K. Berneis, M. Kofter and W. Bollag, Helv. Chim. Acta, 46 (1963) 2157.
81 A.P. Terentjew and Ja.D. Mogiljanski, Zh. Obshch. Khim., 31 (1961) 326; Chem. Abstr., 55 (1961) 22192.
82 L. Bateman and K.R. Hargrave, Proc. R. Soc. (London), Ser. A, 224 (1954) 389.
83 L. Bateman and K.R. Hargrave, Proc. R. Soc. (London), Ser. A, 224 (1954) 399.
84 K.R. Hargrave, Proc. R. Soc. (London), Ser. A, 235 (1956) 55.
85 D. Barnard, L. Bateman, E.R. Cole and J.I. Cunneen, Chem. Ind. (London), (1958) 918.
86 L. Bateman, M.E. Cain, T. Colclough and J.I. Cunneen, J. Chem. Soc., (1962) 3570.
87 G.W. Kennerly and W.L. Patterson, Ind. Eng. Chem., 48 (1956) 1917.
88 L. Bateman and J.I. Cunneen, J. Chem. Soc., (1955) 1596.
89 L. Bateman, Q. Rev., 8 (1954) 147.
90 L. Bateman and F.W. Shipley, J. Chem. Soc., (1955) 1996.
91 Y. Ogata and S. Suyama, Chem. Ind. (London), (1971) 707.
92 M. Sexton, private communication, 1976.

93 M.S. Kharasch, A.T. Read and F.R. Mayo, Chem. Ind. (London), 57 (1938) 752.
94 J.F. Ford, R.C. Pitkethly and V.O. Young, Tetrahedron, 4 (1958) 325.
95 A.A. Oswald, J. Org. Chem., 24 (1959) 443; 26 (1961) 842.
96 A.A. Oswald, B.F. Hudson, G. Rodgers and F. Noel, J. Org. Chem., 27 (1962) 2439.
97 A.A. Oswald and T.J. Wallace, in N. Kharasch (Ed.) Organic Sulphur Compounds, Pergamon Press, Oxford, 1961, Vol. 1, p. 205.
98 M. Dixon and H.E. Tunnicliffe, Proc. R. Soc. (London), Ser. B, 94 (1923) 266.
99 W.A. Schulze and F.E. Frey, U.S. Pat. 1,964,219, 1934.
100 J. Gevers, Belg. Pat. 628,409, 1963.
101 J.M. Brooke, U.S. Pat. 3,117,077, 1964.
102 M.S. Kharasch, in W.A. Waters (Ed.), Vistas in Free Radical Chemistry, Pergamon, Press, Oxford, 1959, p. 101.
103 R.H. Rosenwald, Pet. Process., 6 (9) (1951) 969.
104 R.H. Rosenwald, Pet. Process., 11 (10) (1956) 91.
105 L.M.Rampino and M.J. Gorham, Pet. Process., 10 (8) (1955) 1146.
106 M.S. Karasch, W. Nudenberg and G.J. Mantell, J. Org. Chem., 16 (1951) 524.
107 A.A. Oswald, F. Noel and A.J. Stephenson, J. Org. Chem., 26 (1961) 3969.
108 T.J. Wallace, N. Jacobson and A. Schriesheim, Nature (London), 201 (1964) 609.
109 T.J. Wallace and A. Schriesheim, J. Org. Chem., 27 (1962) 1514.
110 T.J. Wallace, A. Schriesheim and W. Bartok, J. Org. Chem., 28 (1963) 1311.
111 J. Xan, E.A. Wilson, L.D. Roberts and N.H. Horton, J. Am. Chem. Soc., 63 (1941) 1139.
112 R. Benesch and R. Benesch, J. Am. Chem. Soc., 77 (1955) 5877.
113 C.F. Cullis, J.D. Hopton and D.L. Trimm, J. Appl. Chem., 18 (1968) 330.
114 H. Berger, Rec. Trav. Chim. Pays-Bas, 82 (1963) 773.
115 T.J. Wallace and A. Schriesheim, Tetrahedron Lett., (1963) 1131.
116 T.J. Wallace and A. Schriesheim, Tetrahedron, 21 (1965) 2271.
117 T.J. Wallace and A. Schriesheim, J. Appl. Chem., 17 (1967) 48.
118 T.J. Wallace, J.M. Miller, H. Pobiner and A. Schriesheim, Proc. Chem. Soc., (1962) 384.
119 D.C. Harrison, Biochem. J., 21 (1927) 1404.
120 S. Smiles and J. Stewart, J. Chem. Soc., (1921) 1792.
121 A. Schoeberl and E. Ludwig, Ber. Dtsch. Chem. Ges. B, 70 (1937) 1422.
122 C.F. Cullis, J.D. Hopton, C.J. Swan and D.L. Trimm, J. Appl. Chem., 18 (1968) 335.
123 L. Michaelis and E.S.G. Barron, J. Biol. Chem., 81 (1929) 29.
124 L. Michaelis and E.S.G. Barron, J. Biol. Chem., 83 (1929) 191.
125 L. Michaelis and E.S.G. Barron, J. Biol. Chem., 84 (1929) 777.
126 L. Michaelis and M.P. Schubert, J. Am. Chem. Soc., 52 (1930) 4418.
127 N. Tanaka, I.M. Kolthoff and W. Stricks, J. Am. Chem. Soc., 77 (1955) 1996.
128 N. Tanaka, I.M. Kolthoff and W. Stricks, J. Am. Chem. Soc., 77 (1955) 2004.
129 M.S. Kharasch, R.R. Legault, H.B. Wilder and R.W. Gerard, J. Biol. Chem., 113 (1936) 537.
130 D.L. Leussing and I.M. Kolthoff, J. Am. Chem. Soc., 75 (1953) 3904.
131 H. Lamfron and S.O. Nielsen, J. Am. Chem. Soc., 79 (1957) 1966.
132 F. Bernheim and M.L.C. Bernheim, Q. Rev. Biol., 7 (1939) 174.
133 T.J. Wallace, A. Schriesheim, H. Hurwitz and M.B. Glaser, Ind. Eng. Chem. Process Des. Dev., 3 (1964) 237.
134 I. Pascal, Ph.D. Thesis, University of Rochester, 1956.
135 P. Holtz and G. Triem, Z. Physiol. Chem., 248 (1937) 1.
136 O. Schales, Ber. Dtsch. Chem. Ges. B, 71 (1938) 447.
137 T.J. Wallace, J. Org. Chem., 31 (1966) 1217.
138 C.J. Swan and D.L. Trimm, J. Appl. Chem., 18 (1968) 340.
139 C.F. Cullis and D.L. Trimm, Discuss. Faraday Soc., 46 (1968) 144.

248

140 C.J. Swan and D.L. Trimm, Adv. Chem. Ser., 76 Pt. II (1968) 182.
141 I.M. Kolthoff, E.J. Meehan and M.S. Tsao, J. Polym. Sci. Part A, 3 (1965) 3957.
142 I.M. Kolthoff, E.J. Meehan, M.S. Tsao and Q.W. Choi, J. Phys. Chem., 66 (1962) 1233.
143 G. Gorin and W. Godwin, J. Catal., 5 (1966) 279.
144 C.G. Overberger, K.H. Burg and W.H. Daly, J. Am. Chem. Soc., 87 (1965) 4125.
145 T.J. Wallace, A. Schriesheim and H.B. Jonassen, Chem. Ind. (London), (1963) 734.
146 T.J. Wallace, J. Org. Chem., 31 (1966) 3071.
147 J.R. van Wazer, Phosphorus and its Compounds, Interscience, New York, 1958, Vol. 1.
148 W.E. Savige and J.A. Maclaren, in N. Karasch and C.Y. Meyers (Eds.), The Chemistry of Organic Sulphur Compounds, Vol. II, Pergamon Press, Oxford, 1966, p. 367.
149 L. Bateman, J.I. Cunneen and J. Ford, J. Chem. Soc., (1956) 3056.
150 T.J. Wallace, J. Am. Chem. Soc., 86 (1964) 2018.
151 T.J. Wallace and J.J. Mahon, J. Am. Chem. Soc., 86 (1964) 4099.
152 T.J. Wallace and J.J. Mahon, J. Org. Chem., 30 (1965) 1502.
153 T.J. Wallace, Chem. Ind. (London), (1964) 501.
154 T.J. Wallace and H.A. Weiss, Chem. Ind. (London), (1966) 1558.
155 J.P. Danehy, in N. Karasch and C.Y. Meyers (Eds.), The Chemistry of Organic Sulphur Compounds, Vol. II, Pergamon Press, Oxford, 1966, p. 337.
156 J.P. Danehy and W.E. Hunter, J. Org. Chem., 32 (1967) 2047.
157 T.J. Wallace, H. Pobiner, F.A. Baron and A. Schriesheim, J. Org. Chem., 30 (1965) 3147.
158 T.J. Wallace, H. Pobiner, F.A. Baron and A. Schriesheim, Chem. Ind. (London), (1964) 945.
159 W.H. Starnes, U.S. Pat., 3,557,232, 1971.
160 F.W. Wassmundt and H. Goldstein, Tetrahedron Lett., 52 (1970) 4565.
161 Z. Kulicki, Zesz. Nauk Politech. Slask. Chem., 36 (1967) 3; Chem. Abstr., 67 (1967) 116355.
162 V.I. Kovalev and K.A. Chervinskii, Khim. Tekhnol. (Kharkov), 20 (1971) 46; Chem. Abstr., 76 (1972) 24401.
163 I.I. Kiiko and K.I. Matkovskii, Khim. Tekhnol (Kharkov), 20 (1971) 15; Chem. Abstr., 75 (1972) 24402.
164 S. Kawai, Yakugaku Zasshi, 86 (1966) 1125; Chem. Abstr., 71 (1967) 10455j.
165 F.R. Mayo and M. Honda, Am. Chem. Soc., Div. Pet. Chem. Prepr., 13 (1968) C5.
166 V.A. Poluektov and Yu.Ya. Mekhrynshev, Kinet. Katal., 12 (1971) 833.
167 V.A. Poluektov and N.G. Ageev, Kinet. Catal. (USSR), 11 (1970) 480.
168 V.A. Poluektov and N.G. Ageev, Kinet. Catal. (USSR), 12 (1971) 19.
169 S.K. Minsker, A.S. Shevlyakov and G.A. Razuvaev, J. Gen. Chem. (USSR), 26 (1965) 1227.
170 W. Kern, H. Jockush and A. Wolfram, Makromol. Chem., 3 (1949) 223.
171 A.L. Klebanskii and R.M. Sorokina, Zh. Prikl. Khim., 35 (1962) 2735.
172 H.C. Bailey, Adv. Chem. Ser., 75 (1968) 138.
173 D.E. Van Sickle, F.R. Mayo, R.M. Arluck and M.G. Syz, J. Am. Chem. Soc., 89 (1967) 967.
174 A.S. Kuzminskii and R.Ya. Peschanskaya, Dokl. Akad. Nauk SSSR, 85 (1952) 1317.
175 I. Kössler, B. Matyska and J. Polacek, J. Polym. Sci., 53 (1961) 107.
176 I. Kössler and L. Svob, J. Polym. Sci., 54 (1961) 17.
177 H.C. Bailey, Rev. Gen. Caoutch. Plast., 44 (1967) 1495.
178 J. Dvorak and B. Matska, Collect. Czech. Chem. Commun., 28 (1963) 2387.
179 J.A. Howard and K.U. Ingold, Can. J. Chem., 43 (1965) 2729.
180 G.A. Razuvaev and K.S. Minsker, J. Gen. Chem. (USSR), 28 (1958) 957.
181 A.A. Berlin and R.H. Aseeva, Chem. Abstr., 72 (1970) 13356g.

Index

254